BEING NUCLEAR

BEING NUCLEAR

AFRICANS AND THE GLOBAL URANIUM TRADE

GABRIELLE HECHT

THE MIT PRESS
CAMBRIDGE, MASSACHUSETTS
LONDON, ENGLAND

First MIT Press paperback edition, 2014

© 2012 Massachusetts Institute of Technology

Distributed by Wits University Press in the Southern African region (South Africa, Namibia, Botswana, Swaziland, Lesotho, and Zimbabwe) by arrangement with the MIT Press.

For information about quantity discounts, email special_sales@mitpress.mit.edu.

Set in Engravers Gothic and Bembo by Toppan Best-set Premedia Limited. Printed and bound in the United States of America.

Library of Congress Cataloging-in-Publication Data

Hecht, Gabrielle.
Being nuclear : Africans and the global uranium trade / Gabrielle Hecht.
   p.   cm.
Includes bibliographical references and index.
ISBN 978-0-262-01726-8 (hardcover : alk. paper)—ISBN 978-0-262-52686-9 (pbk.)
1. Uranium industry—Africa. 2. Uranium industry—Political aspects—Africa.
3. Uranium mines and mining—Africa. 4. Nuclear energy—Africa. I. Title.
HD9539.U72A43   2012
382'.4249096—dc23

2011030531

10 9 8 7 6 5 4 3

for Janet

CONTENTS

When I began this project in 1998, I sought to challenge conventional narratives of "the nuclear age" as a technological and a geopolitical rupture. According to those narratives, splitting the atom promulgated a new world order that replaced imperialism with "the bomb." But it was clear that colonialism remained central to the nuclear order's technological and geopolitical success. Even a short list of atomic test sites makes the point: Bikini Atoll, Semipalatinsk, Australian Aboriginal lands, the Sahara, French Polynesia. Yet these spectacular displays told just part of that story.

The history of uranium production, rarely addressed in any depth, suggested that imperialism was much more deeply and lastingly woven into the fabric of the nuclear age. Congolese uranium powered the Hiroshima bomb. Uranium in the Commonwealth helped Britain maintain nuclear relations with the United States after the war. France mined uranium in its African colonies. Uranium also played a key role in more recently established colonial relationships, such as that between South Africa and Namibia. Examples abounded: East Germany and the Soviet Union, Adivasi lands in India, First Nation lands in Canada, and on and on.

What, I wondered, had global nuclear development meant for local communities in the so-called third and fourth worlds? How did it reflect and shape relationships between "developing nations" and "nuclear powers"? To keep the project manageable as I set out to explore this history, I limited myself to Western imperial relationships. Because I wanted to learn from the people who worked in or lived near the mines, I chose places where conducting oral interviews and field research was feasible. In addition to Gabon, Madagascar, Namibia, and South Africa, my initial sites included the Navajo Nation and Australia's Northern Territory.

As my research continued, I concluded that a single book covering all these places would be too unwieldy. The evidence I'd collected in Gabon, Madagascar, Namibia, and South Africa had begun to coalesce around an argument that addressed not only the power of nuclear things, but also the positions of African nations in transnational technological systems and the complex challenges faced by African workers who participated in those systems. Still, my geographic coverage was limited. Violence around uranium sites and other political tensions meant that neither Niger nor the Democratic Republic of Congo was conducive to archival research or fieldwork on this topic. I could therefore address them only in the limited way made possible by secondary literature and documentation available in US and European archives.

My work was strongly shaped by my collaboration with Bruce Struminger, a medical doctor with an interest in anthropology who worked at the Indian Health Service in Shiprock, New Mexico. Together we filmed a series of interviews with people who'd worked at uranium mines in Namibia, South Africa, and the Navajo Nation. I owe him a very special acknowledgment for teaching me about occupational health research and for being part of this project, off and on, for four years.

I am deeply grateful to the many people who shared their histories in Gabon, France, Madagascar, Namibia, South Africa, and the UK. Particularly generous with their help, their memories, and their insights were those mentioned in the appendix (not always by name, by their choice), especially Georges Heurtebize. I also thank Jacques Blanc, Robert Bodu, Bernard Keiffer, Jozua Ellis, Paul Fitzsimons, Rob Heard, Alison O'Gorman, Juste Mambangui, Charles Scorer, David Salisbury, Annie Sugier, and Mike Travis for facilitating my access to uranium mines, plants, and other nuclear sites. Staff members at the CRIIRAD, the Institute for Science and International Security, the Laka Foundation, the OECD, Sherpa, the South African History Archive, TradeTech, and the World Nuclear Association kindly provided data, illustrations, and other information. I also thank the many librarians, archivists, and others who smoothed my way.

None of this research would have been possible without generous funding. I was the fortunate recipient of two fellowships and a collaborative research grant from the National Endowment for the Humanities, two grants from the National Science Foundation (awards SES-0848568 and SES-0237661), a Frederick Burkhardt Fellowship from the American

Council for Learned Societies, fellowships from the University of Michigan's Eisenberg Institute for Historical Studies and the Stanford Humanities Center, and a grant from Stanford University's Office of Technology Licensing. Any opinions, findings, and conclusions or recommendations expressed in this material are mine and do not necessarily reflect the views of the National Science Foundation (or any other funding agency). I especially appreciate the good guidance of NEH and NSF program officers through the application and grant management process. Margy Avery of the MIT Press patiently nurtured the book for many years, cheerfully advising me on structure, style, and other matters.

Research assistants—many of them excellent scholars in their own right—helped me cope with the scope of this project. I extend thanks to David Backer, Millington Bergeson-Lockwood, Letitia Calitz, Michelle Escobar, Anne Fellinger, Andy Ivaska, Adriana Kale Johnston, Matt Raw, Emmerentia van Rensburg, Tasha Rijke-Epstein, and Nafisa Essop Sheik. Clapperton Mavhunga and Sezin Topçu proved to be especially gifted researchers. So did Dan Hirschman, who also commented on drafts and offered essential help in the final weeks.

This project found an extraordinary home at the University of Michigan, and especially in the Department of History, the Program in Science, Technology, and Society, and the African studies community. From the moment I arrived, Kali Israel offered me her friendship and her vision of Ann Arbor. Susan Douglas and TR Durham made life tasty and never wavered in their encouragement. Doug Northrop helped me keep word counts and fight writing demons; he and Michelle McClellan broadened Michigan's horizons for me. Geoff Eley, Joel Howell, and Mary Kelley generously offered all manner of counsel at critical moments. The manuscript was much improved by comments from John Carson, Robyn d'Avignon, Joshua Grace, Nancy Hunt, Steve Jackson, Elise Lipkowitz, Amanda Logan, Davide Orsini, Emma Park, Derek Peterson, Liz Roberts, Stephen Sparks, Alex Stern, and other participants in the STS reading group, the African History and Anthropology Workshop, and my graduate seminar on Bodies, Technologies, and Nature in African History. I'd have been lost without Connie Hamlin, Karen Higgs-Payne, and the other wonderful staff members who keep things humming. Others who enriched my life and work in Ann Arbor include Kelly Askew, Kathryn Babayan, Charlie Bright, Kathleen Canning, Pär Cassel, Sueann Caulfield, David

Cohen, Joshua Cole, Juan Cole, Fred Cooper, Susan Crowell, Christian de Pee, Angela Dillard, Mamadou Diouf, Tirtza Even, Dario Gaggio, Dena Goodman, Steve Gutterman, Rima Hassouneh, Sue Juster, Valerie Kivelson, Terry McDonald, Gina Morantz-Sanchez, Rachel Neis, Marty Pernick, Phil Pochoda, Helmut Puff, Sonya Rose, Rebecca Scott, Scott Spector, Howard Stein, Tomi Tonomura, Penny Von Eschen, Butch Ware, and Mary Ellen Wood. Cynthia Esseichick and Sue Watts helped me keep body and soul together.

Keith Breckenridge and Catherine Burns shared their home and their community in South Africa and offered advice and critique at many key points; this project would have turned out differently without their unparalleled generosity. Nolizwi Mpandana Stofela lived with this project for more than a year, and traveled over 30,000 kilometers in the process; I will always be grateful to her, and to Cyprian Stofela and Kokwana Mpandana. Others who offered hospitality or help in (or about) Gabon, Namibia, Madagascar, and South Africa include Gretchen Bauer, Florence Bernault, Renfrew Christie, Barry Ferguson, David Fig, Jeff Guy, Verne Harris, Patricia Hayes, Werner Hillebrecht, Jonathan Hyslop, Mike Kantey, Sonya Keyser, Pier Larson, Karen Middleton, Dunbar Moodie, Julie Parle, Vololona Rabeharisoa, Jeremy Silvester, and Sandy Thompson.

The ideas in this project were also nurtured by exchanges with many other colleagues, including Itty Abraham, Warwick Anderson, Yannick Barthe, Joel Beinin, Alain Beltran, Michel Callon, Lynn Eden, Mats Fridlund, Pascal Griset, Hugh Gusterson, Toby Jones, Kairn Klieman, Arn Keeling, Cathy Kudlick, Susan Lindee, Morris Low, Joe Masco, Donna Mehos, José Manuel Mendes, Stephan Miescher, Laura Mitchell, Suzanne Moon, Ruth Oldenziel, Sara Pritchard, Peter Redfield, Richard Roberts, David Rosner, Sonja Schmid, Johan Schot, Helen Tilley, Hans Weinberger, and Luise White. Those who read portions of the manuscript at various stages and helped to improve its arguments include Soraya Boudia, Kai-Henrik Barth, Tim Burke, Rebecca Herzig, Arne Kaijser, Paul Landau, Grégoire Mallard, Chandra Mukerji, Ben Némery, Chris Sellers, Keith Shear, Bill Storey, Lynn Thomas, and Brad Weiss. I owe a special debt to colleagues who took significant time away from their own work to critique the entire draft in exacting and remarkably constructive detail: John Krige (who has offered encouragement since the earliest days of the project), Julie Livingston, and Michelle Murphy. Michael Adas, Ken Alder, and Bill Leslie have long been generous and steadfast supporters.

My friendship with Nina Lerman has shaped my thinking and scholarship in fundamental ways. Certain conceptual aspects of this project bear the traces of a paper we wrote together over two decades ago; she has read drafts of much of my work since then, and still always finds something insightful to say. This book would be a different (and weaker) beast without the firm editorial hand of Jay Slagle. His dedication during the final months added new depth to a friendship that began with us counting cosmic rays in a college lab. Book writing requires breaks, and I thank the friends and family outside my academic orbit who helped me enjoy life outside this project: Ivan, Ginny, and Karin; Lauren, Maya, Elaine, and Don; Dori, Eli, Isaac, and Rick; Todd, Ilona, and CS; and the amazing Edwards clan, especially the children: Annie, Jesse, Maddie, and Spencer.

Janet Edwards is in a class of her own. She has offered insight in domains ranging from proposal writing to parenting, and much in between. She has flown to the rescue more times than I can count. Her wisdom and optimism will always inspire me. I dedicate this book to Janet; I could not have written it without her.

Nor could I have written it without Paul Edwards, who accompanied me up, down, and around the world for the entire duration of the project, often putting his own work on hold. He took notes, made copies, set up tents, kept things running, and maintained his faith in the project. He couldn't possibly have wanted to read so many words about uranium, but he read them anyway and made them better. By example and in his critiques, he held me to the highest standards. Most of all, he gave me courage. So did our son Luka, bringer of light and joy and hope. A special copy of this book awaits you, Luka, whenever you're ready. Just like you asked.

ACRONYMS AND ABBREVIATIONS

ADL
Arthur D. Little, Inc.

AEB
Atomic Energy Board (South Africa)

AEC
Atomic Energy Corporation (South Africa)

AEC
Atomic Energy Commission (United States)

AEOI
Atomic Energy Organization of Iran

AGR
Archives Générales du Royaume (Belgium)

AMWU
African Mineworkers Union

ANC
African National Congress

ANF
Archives Nationales de France

AUA
African Uranium Alliance

BCLAS
Bodleian Library of Commonwealth and African Studies, Oxford University

CAAA
Comprehensive Anti-Apartheid Act (United States)

CALPRODS
Calcined Products (South Africa)

CATRAM
Collectif des anciens travailleurs miniers de COMUF (Gabon)

CDA
Combined Development Agency

CDT
Combined Development Trust

CEA
Commissariat à l'Énergie Atomique (France)

COGÉMA
Compagnie Générale des Matières Nucléaires (now Areva)

CNS
Council for Nuclear Safety (South Africa)

COM
Chamber of Mines—official archives (South Africa)

COM-ES
Chamber of Mines—Environmental Section archives (South Africa)

COMUF
Compagnie des Mines d'Uranium de Franceville (Gabon)

CRIIRAD
Commission de Recherche et d'Information Indépendantes sur la Radioactivité (France)

DNSA
Digital National Security Archive

DOE
Department of Energy (United States)

DRC
Democratic Republic of Congo

DVL
Dust and Ventilation Laboratory (South Africa)

EDF
Electricité de France

EURATOM
European Atomic Energy Community

FCO
Foreign and Commonwealth Office (South Africa)

FNRBA
Forum of Nuclear Regulatory Bodies in Africa

GME
Government Mining Engineer (South Africa)

GML
Government Metallurgical Laboratory (South Africa)

GOLDFIELDS
Collection of the Goldfields Mining Company, Cory Library for Historical Research, Rhodes University (South Africa)

IAEA
International Atomic Energy Agency

ICRP
International Commission for Radiological Protection

IDC
Industrial Development Corporation (South Africa)

ILO
International Labour Organisation

IRB
Institutional Review Board

IRSN
Institut de radioprotection et de sûreté nucléaire (France)

LARRI
Labour Resource and Research Institute (Namibia)

MAE
Ministère des Affaires Etrangères (France)

MDM
Médecins du Monde

MINTECH
Ministry of Technology (United Kingdom)

MNJ
Mouvement des Nigériens pour la Justice

MPL
maximum permissible level

MUN
Mineworkers Union of Namibia

MWU
Mine Workers Union (South Africa)

NARA
National Archives Records Administration (United States)

NASA
National Archives of South Africa

NECSA
Nuclear Energy Corporation South Africa

NIEO
New International Economic Order

NNR
National Nuclear Regulator (South Africa)

NUFCOR
Nuclear Fuel Corporation (South Africa)

NUM
National Union of Mineworkers (South Africa)

RAMS
Rössing Archives—mine site

RAS
Rössing Archives—Swakopmund headquarters

SG
Shaun Guy private papers

SWAPO
South West Africa People's Organization

TNA
The National Archives (United Kingdom)

UCOR
Uranium Enrichment Corporation (South Africa)

UKAEA
United Kingdom Atomic Energy Authority

UN
United Nations

UNCN
United Nations Council for Namibia

UNSCEAR
United Nations Scientific Committee on the Effects of Atomic Radiation

UNSCR
United Nations Security Council Resolution

WHO
World Health Organization

WLNA
Witwatersrand Native Labor Association (South Africa)

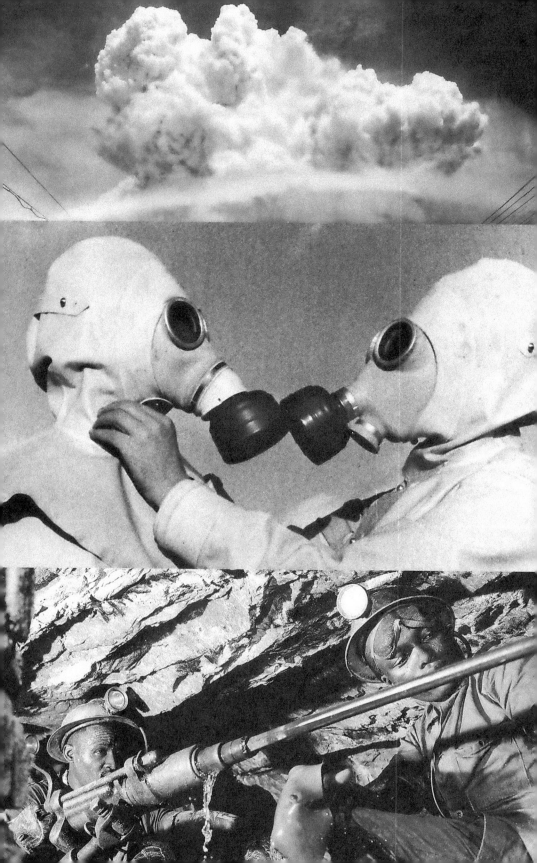

1 INTRODUCTION: THE POWER OF NUCLEAR THINGS

In late 2002, US President George W. Bush announced that Iraqi dictator Saddam Hussein had "recently sought significant quantities of uranium from Africa." The implication? Iraq planned to build nuclear weapons, and the world must act.

The scenario seemed plausible. Weapons inspectors had uncovered a clandestine program after the 1991 Persian Gulf War. Surely Saddam would try again. Bush and his advisors had been implying this for months, releasing assorted evidence to make the case. US national security advisor Condoleezza Rice insisted, for example, that Iraq had imported aluminum tubes whose only conceivable use was in a nuclear weapons program.

The administration's display of evidence gained only modest traction in the media. Behind the scenes, many intelligence officials disputed the validity of the claims.[1] With the case for a military intervention foundering, the notion of "uranium from Africa" appeared promising. It could be fleshed out to 500 tons of "yellowcake from Niger," which certainly sounded scarier than "aluminum tubes." Sidelining disputes among intelligence agencies, administration officials declared that the British government had provided corroborating evidence. No need to wait months for UN inspectors to comb the country in search of a smoking gun. After all, as Rice had warned, "we don't want the smoking gun to be a mushroom cloud."

on facing page: (top) The Nagasaki mushroom cloud, 15 minutes after the explosion and 9.4 kilometers from the hypocenter (photograph by Hiromichi Matsuda, courtesy of Nagasaki Atomic Bomb Museum). (middle) Atomic air raid wardens, Bonn, Germany, 1954 (Bettmann/Corbis). (bottom) Miners drilling into rock at Sub Nigel East Gold Mine, 1961 (Ron Stone/Fox Photos/Getty Images).

In early March 2003, experts from the International Atomic Energy Agency obtained the thin folder substantiating the yellowcake claim. It took them only a few hours to determine that the documents were forgeries—and not even good ones. But by then it was too late. In public discourse, "uranium from Africa" had topped the list of evidence pointing to Iraqi weapons of mass destruction. War plans were in motion. On March 19, 2003, a "coalition of the willing" launched an assault on Iraq. A thorough search after the invasion found no evidence that Iraq had restarted its nuclear weapons production or entered into a uranium deal.[2]

Tangled tales of intrigue emerged during the next few years. The forgeries had come to the US through unknown channels from Italian intelligence services, which had obtained them from a former agent named Rocco Martino, who said he got them from a woman employed by the Nigérien embassy in Rome. Martino had been working for French intelligence (perhaps as a double agent) and supposedly tried to sell the documents first. His French handlers had immediately spotted the fakes, doubtless because one signature was purportedly from a Nigérien foreign minister who had left office over a decade before the date of the agreement. These fakes notwithstanding, British Prime Minister Tony Blair insisted that separate evidence revealed Saddam's intentions to get uranium *elsewhere* in Africa, a claim that British intelligence has yet to substantiate.

Had disgruntled CIA agents seeking to trap the Bush administration forged the documents? How about corrupt operatives with business ties to the Iraqi opposition? Conspiracy theories abounded. In the US, the media focused on Joseph Wilson and his wife, Valerie Plame. A former Foreign Service officer who'd begun his diplomatic career in Niger, Wilson had been sent to the capital city of Niamey by the CIA in February 2002 to investigate whether Niger had sold uranium to Iraq. He found no trace of the sale. When Bush claimed the contrary a few months later, Wilson assumed that the president meant some *other* uranium-producing African nation. Upon realizing that Bush meant Niger, Wilson went public. To discredit him, the administration outed Plame as a CIA operative. A truly impressive quantity of ink, pixels, and bytes was devoted to the drama, pitting Bush stalwarts against defenders of Wilson. Bickering over personal credibility drowned out the pivotal issue of whether the president had misled the nation into war.

It also obscured the significance of the Niger episode for global nuclear relations.

Consider the political and technical parameters of the Bush administration's claims. Officials repeatedly stated that Iraq had sought uranium "from Africa." Had Saddam been suspected of approaching Kazakhstan, would they have asserted that he'd sought uranium "from Asia"? In the Western public imagination, Africa remains the "dark continent," mysterious and politically corrupt—the perfect source for black-market nuclear goods. Consider also the assumption that acquisition of uranium constituted prima facie evidence of bomb building. Before uranium attains weapons grade, it must be mined as ore, processed into yellowcake, converted into uranium hexafluoride, enriched, and pressed into bomb fuel. "Uranium" is as underspecified technologically as "Africa" is politically.

The Niger episode reflects the ambiguities of the nuclear state, and the state of being nuclear. What exactly is a "nuclear state"? Does a uranium enrichment program suffice to make one of Iran, as its president Mahmoud Ahmadinejad claimed in early 2010? Or are atomic bomb tests the deciding factor, thereby justifying Israel's insistence that it will not be the first nuclear state in the Middle East? Such ambiguities cannot be dismissed as doublespeak or grandiose rantings. They matter too much to be discounted so easily.

The nuclear status of uranium is an important aspect of these ambiguities. When does uranium count as a nuclear thing? When does it lose that status? And what does Africa have to do with it? This book argues that such issues lie at the heart of today's global nuclear order. Or disorder, as the case may be.

The questions themselves sound deceptively simple. Understanding their significance and scope requires knowing their history. Yellowcake from Niger may not have entered Iraq in 2002, but uranium from Africa was (and remains) a major source of fuel for atomic weapons and power plants throughout the world. Uranium for the Hiroshima bomb came from the Belgian Congo. In any given year of the Cold War, between a fifth and a half of the Western world's uranium came from African places: Congo, Niger, South Africa, Gabon, Madagascar, and Namibia.

This book argues that views from Africa matter not only on their own terms, but also because they transform our perspective on the power of nuclear things. They help us see that *nuclearity*—a term I introduce to

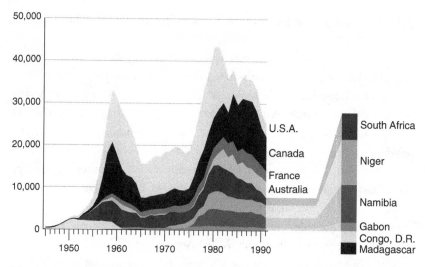

Uranium production (tons per year). (figure prepared by Evan Hensleigh with Dan Hirschman)

signal how places, objects, or hazards get designated as "nuclear"—
has often been contentious. Designating something as nuclear—whether
in technoscientific, political, or medical terms—carries high stakes.
Fully understanding those stakes requires layering stories that are usually
kept distinct: atomic narratives and African ones, histories of markets
and histories of health. Part I of this book follows the path of uranium
out of Africa, tracing some of its historical trajectories through the
nuclear world and examining the invention of the global uranium
market.

Part II enters these nuclear worlds and excavates their histories, focusing
on occupational health among African mine workers. What did nuclear
things mean to Fanahia, a Malagasy worker who extracted uranium ore
(*vatovy*) from the desert of southern Madagascar in the 1950s and 1960s?
Fanahia and his co-workers were taught that if they didn't wear dosimeters
they might get fired, but they didn't realize that *vatovy* would end up in
French bombs and power plants. How did Marcel Lekonaguia, who mined
uranium in eastern Gabon for over three decades, experience nuclear
things before he and hundreds of other Gabonese workers learned that
radioactive elements had penetrated their bodies, their houses, their water,
and their land?

In a very different setting, thousands of migrant workers who toiled in mixed gold–uranium shafts of the South African Witwatersrand never knew about their exposures. Most, like Kokwana Mpandana, believed that gold was the only treasure at their fingertips. Their white supervisors also remained in the dark. Under apartheid, studies documenting high radon levels remained restricted to a handful of scientists and industry officials. After apartheid ended, the mining industry lobbied for exemption from nuclear regulation, insisting that South African mines were *not* nuclear places, radiation notwithstanding. That was a far cry from the experience of workers at Namibia's Rössing uranium mine. Defying their employer and seeking independent expertise on matters radioactive, Namibian labor leaders commissioned a study on the effects of low-level radiation.

I have tried to make this work of scholarship engaging for a non-academic audience. Prologues to the chapters lay out the multiple contexts of this unruly history. This introductory chapter sets the historical stage, explains concepts, and summarizes the book's arguments. Readers interested in how markets get invented, or in matters of political economy more generally, may wish to focus on part I. Those interested in occupational health or labor history will probably prefer to skip to part II. Chapters 2, 6, and 8 cover the production of scientific knowledge and technological infrastructures. Chapters 4 and 7 cover Francophone Africa; chapters 3, 5, 8, and 9 cover Southern Africa. Chapter 10 concludes the book and explores the implications of my analysis for contemporary conundrums. The appendix offers an overview of the sources upon which I based the book.

There are yet more limits to what I could do in a single volume. My discussion of uranium conversion and enrichment is limited to South African efforts. My analysis of labor processes is mostly restricted to worker experiences with radiation and nuclearity. I do not delve deeply into the broader social and economic histories of uranium mines. My examination of radiation exposure is confined to workers; with the exception of one section in chapter 7, I do not discuss environmental contamination produced by mine and mill tailings. There is much to say about all these topics, but I must leave those challenges for later—or to others.

Still, much remains to be done. We require a multitude of starting points. Let's begin with an atomic one.

NUCLEAR EXCEPTIONALISM

The atom bomb has become the ultimate fetish for our times.[3] World order has been created and challenged in its name and for its sake. Salvation and apocalypse, sex and death: the bomb's got it all. In the two decades following World War II, "the bomb" became the ultimate political trump card, first for the superpowers (the US in 1945, the Soviet Union in 1949) then for waning colonial powers (the UK in 1952, France in 1960). Other nations soon followed (China in 1964, Israel in the mid 1960s). Geopolitical status seemed directly proportional to the number of nukes a nation possessed.

Although more than 28,000 nuclear warheads now populate the planet, they somehow retain their singularity. We still hear about "the" bomb, as in "When could Iran get the Bomb?"[4] The implication is that nuclear things are unique, different in essence from ordinary things. I call such insistence on an essential nuclear difference—manifested in political claims, technological systems, cultural forms, institutional infrastructures, and scientific knowledge—*nuclear exceptionalism*.

As a recurring theme in public discourse since 1945, nuclear exceptionalism often transcended political divisions, allowing both Cold Warriors and their activist opponents to portray atomic weapons as fundamentally different from any other human creation. The rupture in nature's very building blocks, wrought during fission, propelled claims of a corresponding rupture in historical space and time. "Nuclear" scientists and engineers enjoyed far more prestige, power, and funding than their "conventional" colleagues. Morality-speak inevitably accompanied debates, rendering nuclear things either sacred or profane. Yet whatever the political leaning, exceptionalism expressed the sense that an immutable ontology distinguished the nuclear from the non-nuclear. The difference, or so it seemed, came down to fission and radioactivity.

The technopolitical qualities of being "nuclear" made this form of exceptionalism remarkably robust. Yet nuclear exceptionalism could be made, unmade, and remade. In the early decades, exceptionalism emanated mainly from atomic energy experts and the journalists whose imaginations they captured. The utopian dreams that had accompanied the advent of railways and airplanes found their apotheosis in atomic fantasies. "Our children will enjoy in their homes electrical energy too cheap to meter,"

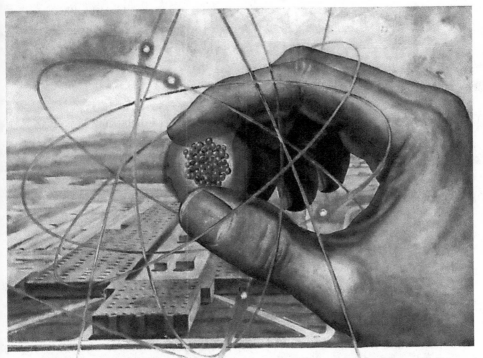

What does Atomic Energy really mean to you?

Dramatic new developments in medicine, agriculture,
and industry promise long-time benefits for us all

A 1952 Union Carbide magazine ad promised a beautiful atomic future.

the chairman of the US Atomic Energy Commission proclaimed in 1954. "It is not too much to expect that our children will know of great periodic regional famines in the world only as matters of history, will travel effortlessly over the seas and under them and through the air with a minimum of danger and at great speeds, and will experience a lifespan far longer than ours, as disease yields and man comes to understand what causes him to age."[5] Shattering the atom had apparently put humanity's ageless dreams within grasp. These were the many promises of nuclear things, and the promise of many nuclear things: limitless electricity, atomic-powered transportation, huge increases in crop yields, cures for disease, and if not eternal life at least one much longer and far more comfortable.

Utopias can be infectious. Atomic fantasies spread quickly on both sides of the Iron Curtain. Nuclear nationalism comforted state leaders anxious about their country status. The French compared reactors to the Arc de Triomphe and the cathedral of Notre Dame. The Russians likened them to samovars. In Communist China, leaders spoke of "the people's bomb"; in India, of the "Smiling Buddha."

Utopian dreams breed dystopian nightmares, though, and few were more terrifying than nuclear war. Photos of Hiroshima and Nagasaki, censored for two decades, trickled out to haunt the public imagination with spectacles of horrifying burns, peeling skin, and ashy landscapes. Shortly after the atomic arms race began, the superpowers upped the ante on public anxiety by testing vastly more destructive thermonuclear weapons in the waters around the Marshall Islands and on the plains of the Kazakh Soviet Socialist Republic. As geneticists studied chromosomal aberrations caused by radiation, gigantic ants and towering lizards began to wreak havoc, at least in the reels of B movies.

Apocalypse, no longer the preserve of religion, now lay within humanity's technological grasp. Authors and directors spun out scenarios, grim and comic, for reaching the tipping point at which someone, somewhere, pushed the button to end it all. Books and movies imagined the few remaining humans taking refuge in a world sizzling with fallout. Sometimes the two apocalyptic modes merged, famously so in Walter Miller's 1959 novel *A Canticle for Liebowitz*. Set centuries after a devastating nuclear war, the novel opens by depicting a monastic order whose mission is to preserve and illuminate the remnants of scientific texts, including a blueprint signed by a soon-to-be-sainted engineer named Liebowitz. By the end of the book, humanity has reinvented the bomb and again stands poised on the brink of self-destruction.

Exuberant or ghastly, nuclear exceptionalism was full of contradictions. For all the efforts at making nuclear things exceptional, there were opposing attempts to render them banal. Government propagandists assured citizens that simple gestures offered protection if the bombs did fall. American schoolchildren could take refuge under their desks, sang Bert the Turtle in the famous "Duck and Cover" ditty. Fallout shelters promised the perpetuation of suburban lifestyles in the event of nuclear war. The hyper-organized Swiss went so far as to pass building codes *requiring* fallout shelters. In the late 1970s, as a teenager, I lived in the suburbs of Zürich.

Far-fetched depictions of how radiation exposure might change the human race expressed the inherent ambiguities in 1950s atomic fantasies. (*Mechanix Illustrated*, December 1953)

My parents ignored the basement shelter, with its massive lead-lined door, leaving it devoid of the canned goods and blankets prescribed for nuclear survival. Secretly I feared the place. How and what would we breathe if the bombs fell?

The spread of commercial nuclear power brought new expressions of exceptionalism and banality, especially in the 1970s. Environmental activists seized on nuclear energy as the symbol of ruthless capitalism and its pollution. They countered the promises of cheap, abundant electricity with the prospects of meltdowns and radioactive leaks. The industry insisted that radioactivity was part of nature, nuclear power just a form of energy like all others. It published reassuring charts that compared the radiation received from the sun, airplane flights, bananas, medical procedures, and reactor proximity. When accidents at Three Mile Island (1979) and Chernobyl (1986) challenged claims to banality, nuclear experts reasserted exceptionalism in the guise of extraordinary safeguards. The nuclear industry spent *more* money than any other on accident prevention and risk mitigation, at least in the West. Chernobyl, they insisted, could be chalked up to Soviet sloppiness.[6]

With the end of the Cold War, nuclear exceptionalism shifted terrain. The "clash of civilizations" replaced the "superpower struggle," and climate change replaced nuclear war as the greatest global fear.[7] In 1989, French public intellectual Régis Debray opined that "broadly speaking, green [meaning Islam] has replaced red as the rising force." This was especially frightening because "the nuclear and rational North deters the nuclear and rational North, not the conventional and mystical South."[8] Anthropologist Hugh Gusterson calls this sort of discourse "nuclear orientalism," arguing that it has crossed left-right political divides to become part of "common sense" in the West.[9] Sure enough, at the dawn of the twenty-first century, George W. Bush's "axis of evil" formulation escalated fears that nuclearity might escape the control of the "rational North."

Discourse surrounding the "nuclear renaissance" of the early twenty-first century has hewed to the standard industry script by playing down the terrifying longevity of radiation. The prospect of the imminent apocalypse of global warming has allowed nuclear power to reemerge as a commonsense and desperately needed energy source. Predictably, within hours of the 2011 Fukushima reactor disasters, the industry scrambled to maintain a sense of banality. Exceptionalism, nuclear power advocates

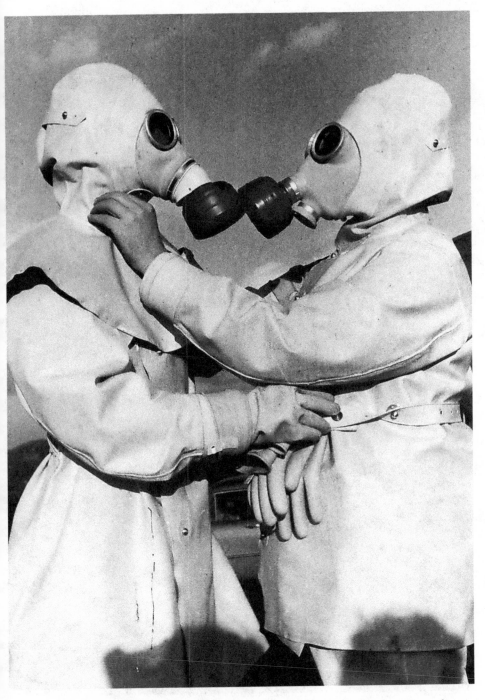

Atomic air raid wardens, Bonn, Germany, 1954. (Bettmann/Corbis Images, used with permission)

Anti-nuclear poster for Verenigde Aktiegroep Stop Kernenergie, Belgium, ca. 1978–1982. (collection of Laka Foundation; used with permission)

insisted, lay in the earthquake's magnitude and the tsunami that followed—
not in the technology.

So much for public discourse. But historians and other scholars have
also fetishized "the bomb" and its builders. Witness the obsession with the
historical minutiae of "*the* decision to drop *the* bomb," the endless stream
of biographies of Manhattan Project scientists, and the insistence on the
uniqueness of moral dilemmas posed by atomic activities. Scholars who've
managed to move beyond the 1950s remain caught in the trappings of
nuclear exceptionalism, concentrating on electricity production and the
high-tech systems surrounding weapons. Their work remains geographically centered on the Cold War superpowers and Europe, only occasionally
extending to South Asia and Japan. Most treat the "nuclear" as exceptional
and self-evident. I include myself among the culprits.

Here's the problem. This unreflective reflex, this certainty about which
things do or don't fall into the domain of the "nuclear," simply doesn't
correspond to historical realities. That can be difficult to see from the
vantage point of a European reactor or a North American weapons lab.
Standing in an African uranium mine makes the contingent character of
nuclearity much more visible.

Consider: Yellowcake from Niger made Iraq nuclear in 2003. But in
1995 yellowcake didn't make Niger itself nuclear. According to a major
US government report on proliferation that year, neither Niger nor Gabon
nor Namibia had *any* "nuclear activities." Yet together these nations
accounted for more than one-fifth of the uranium that fueled power plants
in Europe, the US, and Japan that year.[10] Experts noted decades ago that
workers in uranium mines were "exposed to higher amounts of *internal*
radiation than . . . workers in any other segment of the nuclear energy
industry."[11] But neither workers' radiation exposures nor their role in the
global nuclear power industry was enough to render uranium mining in
these countries a "nuclear activity."

So what things make a state "nuclear," what makes things "nuclear," and
how do we know? Are the criteria for nuclearity scientific? Technical?
Political? Systemic?

These questions are matters of ontology, questions about the things and
categories of things that exist. Historical actors often deployed an ontology
that appeared fixed, incontrovertible, and transparently empirical, in which
essential qualities rigidly separated the nuclear from the non-nuclear.

Scholars have generally left this assumption unchallenged. Yet close examination shows that the boundary between the nuclear and the non-nuclear has been frequently contested. The qualities that make a nation, a program, a technology, a material, or a workplace count as "nuclear" remain unstable, even today. There isn't one nuclear ontology; there are many.[12] My term for this contested terrain of being, this unsettled classificatory scheme, is *nuclearity*.

Nuclearity, this book argues, is a contested technopolitical category. It shifts in time and space. Its parameters depend on history and geography, science and technology, bodies and politics, radiation and race, states and capitalism. Nuclearity is not so much an essential property *of* things as it is a property *distributed among* things.[13] Radiation matters, but its presence does not suffice to turn mines into nuclear workplaces. After all, as the nuclear industry is quick to point out, people absorb radiation all the time by eating bananas, or sunbathing, or flying over the North Pole. For a workplace to fall under the purview of agencies that monitor and limit exposure, the radiation must be man-made rather than "natural." But is radiation emitted by underground rocks natural (as mine operators sometimes argued), or man-made (as occupational health advocates maintained)?

For mines to be treated as "nuclear" workplaces in any meaningful scientific, political, or cultural sense, their radiation levels must be detected and recorded using instruments, laboratories, and comparison data. If these devices and institutions don't exist, if they break down, if the connections between them are weak, then the mines devolve into ordinarily dangerous workplaces rather than specifically nuclear ones. This is one reason why I argue that history and geography have shaped nuclearity. Mining in Madagascar began under French colonial rule; uranium in South Africa came from the gold mines whose labor systems formed the template for apartheid; Namibian uranium became tied up with the struggle for independence from South African occupation. These circumstances all shaped the institutions and technologies of uranium production. They thus shaped how a given mine did—or did not—become identified as a nuclear workplace.

Inherently fractured, nuclearity was achieved by laborious degrees. Treating mines in France as nuclear didn't automatically confer nuclearity on French-run mines in Madagascar. Malagasy ore may have achieved a *geological* nuclearity by way of Geiger counters and geologists. But this didn't translate into *medical* nuclearity that Malagasy workers could invoke to make political or economic claims. Colonial rule (and its legacies), grounded in presumptions of racial difference, made that translation particularly difficult to achieve. Making medical nuclearity politically useful would have required that Malagasy radiation exposures become visible through a denser network of instruments, labs, and the like. It would have required state agencies and courts through which claims could be filed. And it would have required that broader manifestations of nuclearity—such as the countless images and scenarios that made "the nuclear age" an "age" in some parts of the world—acquire cultural and political relevance in Madagascar. By shaping the things onto which nuclearity was distributed, time and place shaped nuclearity itself.

Put differently: *Radiation* is a physical phenomenon that exists independently of how it is detected or politicized. *Nuclearity* is a *technopolitical* phenomenon that emerges from political and cultural configurations of technical and scientific things, from the social relations where knowledge is produced. *Nuclearity is not the same everywhere*: it is different in the US and France, in Namibia and Madagascar, in South Africa and Gabon. *Nuclearity is not the same for everyone*: it has different meanings for geologists and physicists, geneticists and epidemiologists, managers and workers, Nigériens and Canadians. *Nuclearity is not the same at all moments in time*: its materialization and distribution in the 1940s and 1990s differed markedly.

To understand nuclearity, we must explore its spatial and temporal variations. Nuclearity took different shapes and had different heft in Gabon, Madagascar, Namibia, Niger, and South Africa. By excavating the historical contingencies, however, I am *not* claiming that bombs and radiation have no specific physical properties. Radiation exposure can cause diseases; atomic bombs could destroy the planet. Such properties matter to the formation of nuclearity, of course, but they do not *by themselves* determine the nature or power of "nuclear" things.

Equally important, my critique of nuclear exceptionalism is *not* an accusation of "atomic alarmism."[14] I do not discount the historical and material significance of nuclear things. Rather, I aim to show the consequences of rendering such things exceptional or dismissing them as banal.

Designating something as "nuclear" is not a straightforward act of classification. Ambivalence and ambiguity, as political scientist Itty Abraham argues, are structural features of nuclear technologies.[15] Agreements and disagreements about *degrees* of nuclearity have significant consequences. They structure global control over the flow of radioactive materials. They constitute the conceptual bedrock of anti-nuclear movements and nuclear power industries. They affect regulatory frameworks for occupational health and compensation for work-related illnesses. And sometimes they send nations off to war.

The ambiguities underlying recent struggles over the nuclear state of the world are too important to be dismissed as mere political wrangling. They are part of the "nuclear age," the claim that nuclear technologies define a phase of human history. Largely because of our mooring in time and space, we haven't known how to view these ambiguities. Our fetishes keep us close to bombs and reactors and far from other places where nuclearity gets made and unmade. We have become complacent and complicit in the equation between nuclearity and "development."

Nuclearity, like many categories, can be deployed as a tool of empowerment or disempowerment. Its significance depends on its technopolitical distribution. Its contingencies are particularly visible in African places . . . provided we don't lump all African places into a single undifferentiated geography. The temptation to do so offers another starting point for our history.

AFRICA AND TECHNOLOGY

"Africa" has also been a fetish in Western imaginations, and for far longer than the atom bomb. Savage and starving, inferior and infantile, superstitious and corrupt—the list of pejoratives goes on and on. The image of Africans as irrational took root in the Enlightenment and took off during the imperialism that followed. Europeans built political philosophies premised on the radical Otherness of Africans.[16] Armed with Maxim guns and industrial goods, they saw artisanally produced African technologies as proof of a primitive existence.[17] "Africa" became seen as a place without "technology." Colonialism, the conquerors were convinced, would transform the continent through European science, technology, and medicine.[18] During the decades of decolonization and Cold War, modernization theorists followed suit, updating the language and tools of the colonial "civilizing mission" but sticking to its core vision: humanity perched along a ladder of development, with well-meaning Westerners at the top and Africans at the bottom.[19]

Such perceptions infused Cold War pop culture, which sometimes placed its atomic fixations and "savage Africa" in the same narrative frame. Uranium mines provided the most legitimate reason for setting atomic stories in Africa. In the 1953 film *Beat the Devil*, Humphrey Bogart and Gina Lollobrigida set off with a band of rogues to stake a uranium claim in British East Africa. An episode of the campy 1950s television series *Sheena: Queen of the Jungle*, set in Kenya, has the buxom heroine protecting "her natives" and a white-owned uranium mine from a nefarious prospector and his African sidekick, Leopard Man.

African jungles and feuding superpowers pervaded comic books too, merging again in stories about uranium mines found amid ignorant "natives" in loincloths. My favorite example comes from a 1954 *Jungle Action* comic featuring Lo-Zar, a blond, muscle-packed Tarzan clone. The lord of a remote African jungle inhabited by pygmies, Lo-Zar learns that "human beings from a red power" have invaded his "sanctuary." "Behold, little men of the Matubi tribe," he says after capturing a map from a red agent, "plans for the location of a new material for which rats like these invade our jungle and kill, scheme, and rob . . . Uranium!" Lo-Zar immediately knows what "uranium" means, even though the Matubi find the word "strange." "In the world," he intones, "there are two types of men . . . those on the side of democracies who would use it to protect their rights . . . and creatures called reds who seek destruction and terror with it!" Upon which he grabs a vine and swings off to defeat the Reds, along the way battling dinosaurs, "sentries from prehistoric ages" that signal the primitiveness of the place.[20]

Black Africans had no agency in these narratives. Their homes were sites of Cold War struggle; white heroes protected them and their resources from falling into the wrong hands.[21] Black superheroes didn't achieve distinction until the *Black Panther* series in the 1970s, over a decade into decolonization. This time uranium was rendered as "vibranium," which could "change the body structure of humans and transform them into living horrors." The African kingdom of Wakanda guarded the mysterious metal. "Wakanda history is the history of vibranium," explained T'Challa,

Lo-Zar in *Jungle Action*, 1954 (copyright Marvel Entertainment, LLC; used with permission)

the Black Panther's alter ego. Wakandans "survive and prosper because they've never been abused." Absent the depredations of colonial rule, they became a technologically advanced society dedicated to protecting the human race from vibranium's harmful effects. Their goal was financed by the sale of the metal "to research laboratories for astronomical prices."[22] In this fantasy, Africans profited technologically and financially from their resource.

Americans might have been ready to imagine Africans as technological agents in the 1970s, but apartheid South Africa marched to a different historical rhythm. "Bantu education" sought to exclude black Africans from scientific and technological knowledge. Apartheid elites viewed their nation as the product of a dialectic: nature and geography made it African, industrialization made it part of Western civilization.[23] Purple prose from the 1979 official history of the South African Atomic Energy Board, *Chain Reaction*, bore an eerie resemblance to comic book text—though the South African author was deadly serious:

In terms of human social advancement, much of the vast African continent is poor; the civilisation of today has not even reached the more remote areas and a subsistence existence is still the lot of millions of its inhabitants. But beneath the dripping jungles and the searing desert sands, in the hills and mountains and the far-reaching grassland and scrubland lie rich mineral deposits which are the envy of many nations—oil, coal, gold, uranium, diamonds, copper, chrome, cobalt and a myriad of other base, precious, and exotic minerals. . . . [T]he Republic of South Africa, with its advanced technology, is far ahead of the rest of the Continent in cataloguing and exploiting its mineral resources. . . . Although coal is believed to have been used by the Zulus several centuries ago when they exploited outcrops of it to replace charcoal for smelting iron ore, the mining of minerals really dates only from the last century; small-scale coal recovery started in the early 1800s, copper in the mid-century, diamonds nearly twenty years later and then, in 1886 came the opening up of the Witwatersrand gold fields.[24]

After gold came uranium. In this rendition, precolonial Africa was a place without technology. Even Zulu coal use seemed a matter of conjecture, not "really" a part of the continent's history. Only Europeans could fully appreciate the vast potential of African minerals. Mining, with its ability to generate wealth, thus figured as Africa's historical destiny.

Scholars have fought vigorously against the fetishizing singularity of "Africa." As historian Lynn Thomas observes, the academic field of African history "partly came into being [at the height of the Cold War] by challenging racist, teleological, and condescending presumptions embedded in . . . conceptions of the modern."[25] Countering stereotypes of Africa as static and tradition-bound, historians demonstrated the dynamism and diversity of precolonial polities.[26] In the 1970s, scholars inspired by dependency theory argued that Europeans and Americans had achieved their supposedly exemplary industrialization thanks to slavery and imperialism. This exploitation, rather than any innate inferiority, explained the "lack" of technological development in Africa.[27]

Other writers challenged conceptions of African manufacturing and agriculture as inefficient. Both before and after the arrival of Europeans, Africans made technological choices well adapted to their social and environmental contexts. West African textile industries may not have been mechanized, for example, but thanks to their materials and skills they matched or exceeded European cloth in quality.[28] Evidence concerning early smelting and metalworking techniques demonstrates the sophistication of precolonial African innovations.[29] Social scientists have recently begun to examine technological creativity in colonial and postcolonial times, portraying Africans as skilled in designing and re-purposing a full range of technological objects and systems, from guns to electricity meters.[30]

Many of these insights have particular salience for the history of mining in Africa. Mineral extraction and metallurgy predated the arrival of Europeans by centuries, archeologists have shown, with gold, copper, and iron integral to political dynamics in many parts of precolonial Africa.[31] Beginning in the late nineteenth century, however, Europeans dramatically increased the scale of mining, fundamentally transforming many African societies and landscapes. By the early twentieth century, some 200,000 men migrated annually to work in South African gold mines. Colonial states obligingly imposed taxes to facilitate recruitment, pushing millions of Africans into wage labor.[32] Roads, railways, and ports served as symbols of the breadth of colonialism's "civilizing mission," but these sociotechnical systems were often designed to meet the narrow needs of mining and other colonial industries. The political, economic, and technological legacies of these infrastructures outlasted colonial rule.

The exploitation and violence accompanying these transformations are an inescapable part of Africa's past and, all too often, its present. But

Africans weren't passive victims of mining capital. However constrained by colonial or postcolonial conditions, miners brought their own notions of collectivity and identity to their workplaces, making choices and fashioning their own lifestyles. The economic and cultural effervescence of life in the compounds enabled miners to escape total control by management, even under the repressive conditions of apartheid South Africa. Mineworkers generated new forms of gender and ethnic expression, new modes (and expectations) of modernity. In some places, the universalizing promise of "modernization" and "development" that accompanied the start of decolonization gave labor unions means of claiming political rights.[33] All these forms of ferment varied by time and place, group and circumstance.

Demonstrating African historical, political, and technological dynamism is important for combating stereotypes about "Africa." Yet, as anthropologist James Ferguson argues, the *idea* of "Africa" as a singular place persists, replete with pessimism about its technological future. Most writers on globalization omit or dismiss African places in constructing their theories of global connectivity, describing the continent as the "black hole" of the information age.[34] Journalists follow suit, as do policy makers, financial investors, and others.

Responding to this relentless marginalization, Africanists have demonstrated how diverse places on the continent have long been connected to other parts of the world. Making such connections visible disrupts the illusion of smooth, flowing networks invoked by contemporary usage of the word "global."[35] Political scientist Jean-François Bayart uses the term "extraversion" to describe how Africans strategically seek international connections and resources in waging battles for sovereignty and survival. Historian Frederick Cooper cautions that appeals to "universal" values and supranational authority, though often powerful, also expose "the limits of the connecting mechanisms" and the "lumpiness" of power.[36]

Fruitful as this scholarship has been, it has largely left unexplored the technological systems that are so often invoked by globalization theorists as the material channels for global power in the contemporary world. So while Africanists have examined technological creativity, mining's complex history, the power of universals, and the continent's uneven connections with the rest of the world, they have yet to put all these elements together.

Technology's absence from analyses of African political agency, though doubtless not deliberate, makes it appear exogenous—a global force that buffets ordinary Africans and turns them into victims. Such a view makes it difficult to grasp how technological entanglements permeate industrial labor in postcolonial Africa, how these entanglements both open and close political possibilities, and how their contradictions sometimes serve as sources of hope. By exploring the political, technological, and medical life of nuclearity in Africa, this book offers purchase on such questions.

Along the way, we must also remember that discourse portraying "Africa" as a place without "technology"—a trope that says as much about perceptions of what counts as "technology"[37] as it does about perceptions of "Africa"—has real political and economic effects. Although the continent contains more than fifty countries, "Africa" (like "the bomb") retains its rhetorical singularity. However misleading it may be, this perception of singularity has concrete consequences for foreign investment, for diplomatic decisions, for how many Africans see themselves, and for a wide range of other things.[38] Including some nuclear ones.

IN A (POST)COLONIAL REGISTER

In one form or another, empire has long been central to nuclear geographies. Congolese ore exploding over Hiroshima was only the beginning. Britain's weapons program exploited imperial ties to uranium-rich regions in Africa.[39] Uranium reserves gave apartheid South Africa a material role in the "defense of the West." France's nuclear program depended on ore from its African colonies. Australia, Canada, and the US found uranium in Aboriginal, First Nation, and Navajo lands. Soviet bombs used uranium produced by prison labor in East Germany or dug from mountains in Uzbekistan and Kyrgyzstan. The list goes on.

As empires crumbled, the rhythm and rhetoric of decolonization affected the power of nuclear things. Less than three months after the US bombed Hiroshima, the United Nations charter proclaimed "the principle of equal rights and self-determination of peoples." In principle if not in practice, a new world order would be built upon a foundation of equality. Independence would free Africans and Asians from the shackles of white rule. Formerly colonized people could choose their leaders, pursue economic prosperity, educate their children, and join the global community as peers. New nation-states would serve the interests of their people, who for the first time would be citizens rather than subjects. The 1948 Universal Declaration of Human Rights was hailed as a leap forward for humankind, a moral-historical rupture, just like atomic power.

Political leaders blended nuclear and postcolonial discourses about rupture and morality in various ways. Postwar French and British leaders not only hoped that the atom bomb would substitute for colonialism as an instrument of global power, but also saw in it a means of preventing their own colonization by the superpowers. In 1951, Winston Churchill's chief scientific advisor, Lord Cherwell, said: "If we have to rely entirely on the United States army for this vital weapon, we shall sink to the rank of a second-class nation, only permitted to supply auxiliary troops, like the native levies who were allowed small arms but no artillery." Across the Channel that same year, French parliamentary deputy Félix Gaillard chimed in: "Those nations which [do] not follow a clear path of atomic development [will] be, 25 years hence, as backward relative to the nuclear nations of that time as the primitive peoples of Africa [are] to the industrialized nations of today."[40] In claims such as these, nuclearity signified power, its absence signified colonial subjugation, and the undifferentiated mass of Africa remained the metonym for backwardness.

For Europeans, such acts of technopolitical mapping had deep roots, extending the assumptions and practices of the "new imperialism" to the nuclear state and the state of being nuclear. Colonial warfare rested on the assumption that different moral structures underlay the rules for conflict with "civilized" nations and with "savages." Aerial bombing followed the machine gun as a tool of extermination, claiming its first victims in oases outside Tripoli (1911) and villages in Morocco (1913). Even as ecstatic prophets in Europe and America proclaimed the airplane's ability to ensure world peace, the RAF experimented with strategic bombing in Baghdad (1923) and the French bombarded Damascus (1925).[41]

For prescient science fiction writers, it was only a matter of time before atomic weaponry followed suit. In a Pacific war with virulent racial overtones, several hundred thousand Japanese became the first victims of the "white race's superweapon."[42] As the Atomic Bomb Casualty Commission industriously erected colonial scientific structures to study the explosions' aftermath,[43] the US and Britain had already begun to scour African colonies in a desperate bid to monopolize the magic new stuff of geopolitical power: uranium.

The equation of nuclearity, nationhood, and geopolitical power also drove atomic ambitions in new countries eager to recover from empire—especially India, as Itty Abraham has forcefully argued. India's first Atomic Energy Act was passed on 15 August 1948, a year to the day after independence. It received eloquent support from Prime Minister Jawaharlal Nehru, who declared humanity "on the verge . . . of a tremendous development." He continued:

Consider the past few hundred years of human history: the world developed a new source of power, that is steam—the steam engine and the like—and the industrial age came in. India with all her many virtues did not develop that source of power. It became a backward country because of that. The steam age and the industrial age were followed by the electrical age which gradually crept in, and most of us were hardly aware of the change. But enormous new power came in. Now we are facing the atomic age; we are on the verge of it. And this is something infinitely more powerful than either steam or electricity.[44]

In Nehru's rendition, Abraham points out, "India became colonized because of its lack of technological sophistication."[45] Indian scientists subsequently saturated their atomic energy program with postcolonial significance, proclaiming it (and themselves) engines of modern statehood. The desirability of Indian nuclear things wasn't in doubt. For the next two decades, debates focused instead on whether India should build a bomb or pursue a distinctively Gandhian—that is, peaceful—nuclearity.

Meanwhile, US President Dwight Eisenhower's 1953 "Atoms for Peace" speech to the United Nations acclaimed the emergence of atomic power plants and medical radioisotopes that served "the peaceful pursuits of mankind." The centerpiece of the initiative would be an agency that would run a fuel bank supplied principally by the existing stockpiles of "normal uranium and fissionable materials" held by governments in both East and West. "Experts would be mobilized to apply atomic energy to the needs of agriculture, medicine, and other peaceful activities," Eisenhower proclaimed, adding that "a special purpose would be to provide abundant electrical energy in the power-starved areas of the world."[46] "First World" nuclearity would thus solve "Third World" problems.

Although the fuel bank proved unfeasible, other elements of Eisenhower's proposal morphed into the International Atomic Energy Agency. Discussions about the agency's membership structure began in 1954. For Western leaders, maintaining political credibility amid rising Cold War

Eisenhower's 1953 Atoms for Peace speech to the United Nations was carefully staged as a major world event. (UN photo by MB)

tensions required the IAEA's Board of Governors to have adequate representation from both East and West. Achieving technical credibility meant that atomic expertise had to play an important role in the selection of board members. Postcolonial nations, however, challenged this nuclear geography.

Indian delegates warned that if atomic governance relied solely on technical achievement and a Cold War, East–West balance, the agency would reproduce the global imbalances perpetrated by colonialism and industrialization.[47] The warning had become a leitmotif for Indian interventions in international forums. A year before Eisenhower's famous speech, India had spearheaded the creation of the UN's Disarmament Sub-Committee, calling for "the prohibition of atomic weapons and the use of atomic energy for peaceful purposes only."[48]

Nehru had brought a similar message to the 1955 Afro-Asian Conference in Bandung, Indonesia, an event credited as the birthplace of the Non-Aligned Movement's search for a third way in the bipolar Cold War. Bandung's Final Communiqué specifically "urged the speedy establishment of the International Atomic Energy Agency, which should provide for adequate representation of the Asian-African countries on the executive authority of the Agency."[49] During the long process of IAEA statute negotiations, India followed through on its message with a proposal that qualification for a seat on the Board of Governors should combine nuclear "advancement" with regional distribution.

The sentiments from Bandung reverberated well beyond the IAEA negotiations. The Final Communiqué had also called for peaceful uses of atomic energy and made numerous references to nuclear weapons, asserting that "the nations of Asia and Africa . . . have a duty towards humanity and civilization to proclaim their support for disarmament and for prohibition of these weapons." The conference condemned racial discrimination, especially in apartheid South Africa, and called for a worldwide study of "the way radioactivity from tests of nuclear and thermonuclear weapons spreads through the atmosphere and in the waters of the ocean."[50] Facing the threat of fallout from French atomic tests in Algeria a few years later, Ghanaian Prime Minister Kwame Nkrumah, internationalist activist Bayard Rustin, and other Pan-Africanists built on the Bandung declaration, denouncing the "desecrat[ion of] the soil of Africa in the interests of a new "nuclear imperialism."[51]

Finalized in 1957, the IAEA's founding statute reflected India's influence by allocating five permanent board seats to member states deemed the "most advanced in the technology of atomic energy including the production of source materials." Five seats were distributed according to geographical region.[52] Uranium producers in Eastern-bloc and Western-bloc nations would rotate through another two seats, and "suppliers of technical assistance" would rotate through one seat. The ten final spots would be distributed among the eight IAEA regions by election.

Protest in Accra against French nuclear testing in Algeria, September 1960. (Bettmann/Corbis Images; used with permission)

The emphasis on "advancement" transformed the Cold War obsession with technological rankings into a structural feature of the IAEA. Yet geography and national history also mattered. The regional framework accommodated—even encouraged—postcolonial fantasies of nuclear nationalism. So what would make a nation count as "most advanced in the technology of atomic energy including the production of source materials"? What were "source materials," and how significant a manifestation of nuclearity were they? In the 50 years since these phrases laid the foundation for the global nuclear order, their meanings have been negotiated and renegotiated in treaties, contracts, and practices. A few examples suffice to illustrate the high stakes of nuclear exceptionalism.

Consider the role of apartheid South Africa, whose delegate was responsible for including "source materials" in the IAEA statute as an indicator of technological "advancement." By 1956, contracts with the US and Britain had made uranium production vital to South Africa's economy.[53] Anticipating that the IAEA would play a central role in shaping the future uranium market, South Africans were determined to obtain a statutory seat on its board. But the apartheid state represented the antithesis of the postcolonial settlement pursued by India. Only by presenting a depoliticized, technical vision of nuclearity could South Africa hope to secure its seat.

When IAEA statute discussions took place in 1954-56, South African nuclearity was limited to uranium production underwritten by a small research program. This was an increasingly tenuous basis for a claim to superior "advancement," especially since uranium's nuclearity was in flux in the mid 1950s. Before that period, the uranium narrative went something like this:

• Uranium was the only naturally occurring radioactive material that could fuel atomic bombs. These, in turn, were weapons of a fundamentally new kind, capable of rupturing not only global order but the globe itself.
• Uranium ore was rare. If the West could monopolize its supply, it could keep the Communists at bay and make the world safe for democracy. The West therefore had to secure all sources of uranium around the world. Nothing mattered more.
• Uranium's significance made it imperative to proceed as secretly as feasible. Geological surveys, actual and potential reserves, means of production, and terms of sales contracts were state secrets one and all.

If uranium's nuclearity imposed secrecy, that secrecy in turn reinforced the ore's nuclearity. Uranium thus became the only ore subject to legislation specifically targeted at ensuring the secrecy of its conditions of production. By the mid 1950s, however, it had become clear that, although high-grade pitchblende was rare, lower grades of ore were everywhere. The Soviets had found their own sources, making a Western monopoly on "source material" impossible. The real challenge lay not in *finding* ore but in *processing* it to weapons-grade quality.

In IAEA statute discussions, one sign that the nuclearity of uranium ore had eroded was that nations whose primary claim to nuclearity lay in uranium production would have to rotate seats on the IAEA board. Indeed,

India had tried to relegate South Africa and Australia to mere "producers" rather than "most advanced" in their regions. Prevailing on their powerful American and British customers, South African delegates succeeded in having "source materials" count as an indicator of "advancement," even though South Africa was no more technologically "advanced" in 1957 than, say, Portugal, which also mined uranium.[54] Their difference lay in technopolitical geography. Portugal was in Western Europe, a region at the pinnacle of nuclear "advancement." South Africa was in the IAEA's Africa/Middle East region, where its competitors for nuclearity—Israel and Egypt—carried political baggage even heavier than its own.

In a time and place where the Cold War trumped racial injustice, South African "source materials" made the country nuclear enough to drown out the increasingly vocal opposition of postcolonial nations to the apartheid state. For the purposes of IAEA board membership, South Africa's uranium production served as the pinnacle of African nuclearity. That status did not falter in 1958 when the Belgians, under the auspices of the US Atoms for Peace program, built a research reactor on what is now the campus of the University of Kinshasa. Congo was then still under colonial control, and only nation-states could achieve representation on the IAEA board. Yet South African prominence did not diminish in 1960 after the Republic of the Congo achieved independence. Apartheid South Africa would not get ejected from the IAEA's Board of Governors until 1977, when pressure from the international anti-apartheid movement proved too strong to resist.

TRADING NUCLEAR THINGS

Today's media see the IAEA primarily as the UN's "nuclear watchdog," conducting inspections to certify that civilian installations haven't been diverted to military ends. But this function emerged over time. The IAEA was originally formed to *facilitate* the circulation of certain nuclear things.[55]

The South Africans craved a seat on the IAEA board because they wanted to sell uranium and shape its market. Their seat secured, they lost no time exploring these commercial possibilities. Donald Sole, the South African delegate for over a decade, used his IAEA contacts to deepen relationships with potential uranium customers. In 1959, Sole escorted two representatives of the South African Atomic Energy Board (AEB) all over

Western Europe. This "sales survey team" sought to forecast supply and demand for the upcoming decade, guess at the probable price structure of commercial contracts, and assess how safeguards might constrain the sale of uranium.[56] The tour proved so fruitful that the AEB's sales committee repeated it regularly, building on Sole's expanding network of contacts.[57]

South Africans were by no means alone in using the IAEA in this way. From its inception, the agency served as a forum (in the Roman sense of marketplace) for its members to learn about competing technologies and materials, make commercial contacts, and offer or apply for technical assistance. It was *as part of all this*, I argue, that IAEA members began discussing international rules for regulating the flow of atomic knowledge and things. I'll get to those discussions shortly, but first let's take a quick look at some of nuclear trade deals concluded inside and outside the halls of the IAEA.

In 1955, even before the IAEA got underway, the gigantic Atoms for Peace conference in Geneva was part international scientific conference, part trade show, and part intelligence-gathering operation.[58] The US had long been selling radioisotopes for medical use.[59] After Eisenhower's 1953 speech, it began exporting research reactors too, selling 25 by 1965.[60] Westinghouse and General Electric, as vendors of commercial reactors, competed fiercely for domestic and foreign customers beginning in the 1960s. The Soviets worried that Atoms for Peace deals would lead to weapons proliferation, but they worried even more about American hegemony. They began their own reactor sales to Eastern Europe in the mid 1950s, using a fuel take-back system to prevent the production of weapons-grade material. These exports, ideologically construed as instruments of modernization, served as flagships for a Soviet-style "civilizing mission" in Eastern Europe.[61]

Other countries also used nuclear exports to expand their technopolitical reach. Some of those transactions justified the Soviet fears about the proliferation risks of exporting dual-use technologies. The French began to export nuclear technology in the mid 1950s, starting with the sale of a reactor to Israel. Emulating French design, the reactor was ostensibly geared toward electricity generation but was in reality optimized for the production of weapons-grade plutonium.[62] Israeli scientists, like their French colleagues before them, promptly began churning out plutonium for a secret bomb program.[63] In the next two decades, customers for French reactors and other nuclear systems included Spain, South Africa, Iraq, and Iran.

Canada had also developed a reactor design, and in the mid 1950s it sold a research reactor to India. Expecting to supply the fuel, the Canadians were surprised when the Indians manufactured their own fuel rods, which entitled them to keep the spent fuel. From this, the Indian scientists secretly extracted weapons-grade plutonium for use in their "peaceful nuclear explosion" of May 1974. The test was widely interpreted as part of a weapons program, rather than as an emulation of the superpowers' programs to use nuclear explosives in large-scale construction projects.[64] It outraged and embarrassed Canada, which promptly implemented a strict safeguards policy. But by then it was too late.

SAFEGUARDING THE NUCLEAR ORDER

The problem with the trade in nuclear things was the exceptionalism of things nuclear. How to buy and sell technologies that carried such heavy moral baggage and destructive potential? States not only had to agree on *how* to regulate trade, they also had to agree on *whom* and *what* to regulate. Who could be trusted with which systems? Which materials, knowledges, and systems were unique to atomic weapons? Which served civilian systems? Which were dual-use? How did technologies that served both nuclear and non-nuclear systems fit in? It seemed understood that strongly nuclear materials should be subject to stricter controls than weakly nuclear ones. Banalizing certain things promised to ease some of the tensions between promotion and proliferation. But what would banalization mean in practice?

It was clear that "fissionable materials" should be controlled, but where in its multiple stages of processing did the "source material" of uranium ore become "fissionable material"? The distinction mattered enormously because the two categories were subject to different controls. In the words of one South African scientist who participated in the IAEA statute discussions, "the definitions would have to be essentially practical, rather than 'textbook' in nature, . . . legally watertight, and must take account of certain political implications."[65] In 1957, the IAEA abandoned the more ambiguous term "fissionable material" (preferred by the Indian delegates) in favor of three other categories: "source materials," "special fissionable materials," and "uranium enriched in the isotope 235 or 233."[66]

The new definitions alone didn't determine methods of control. The US promoted a pledge system in which purchasers agreed not to pursue military ends and agreed to accept international inspections to verify compliance. While most other nations *selling* nuclear systems paid lip service to such a scheme, buyers rejected the prospect of controls. Arguments on both ends obscured mundane political and commercial issues. The US, the UK, and the Soviet Union simply refused to accept inspections on their soil. Western European designers of nuclear systems, fearing that inspections would open the door to commercial spying, accused the US and the UK of seeking competitive advantage. They argued that Western Europe should also benefit from inspection exemption, and that Euratom, the recently created European nuclear agency, could offer sufficiently strong safeguards.

South Africa wanted to avoid any commercial disadvantage caused by mandatory controls on uranium end-use. They suspected that the Israelis, enticed by a French offer to sell them uranium with no strings attached, had broken off negotiations to buy South African uranium in 1962 because of safeguards meant to placate the US and the UK.[67] Within India, experts disagreed over whether to build a bomb at all, but at the IAEA they tried to keep their options open by arguing that regulations would perpetuate colonial inequalities and undermine national sovereignty. Overall, "Third World" nations deemed such regulatory proposals straightforward moves by the North to dominate the global South by writing the rule book in its own favor.[68]

The 1968 Treaty on the Non-Proliferation of Nuclear Weapons (usually referred to as the Non-Proliferation Treaty, or NPT) expressed all these tensions.[69] Under the NPT, "nuclear weapons states" pledged not to transfer atomic weapons or explosive devices to "non-nuclear weapons states." The latter, in turn, renounced atomic weapons and agreed to accept IAEA safeguards and compliance measures. Strikingly, the NPT invoked human rights language and the rhetoric of development:

1. Nothing in this Treaty shall be interpreted as affecting the *inalienable right* of all the Parties to the Treaty to develop research, production and use of nuclear energy for peaceful purposes. . . .

2. All the Parties to the Treaty undertake to facilitate, and *have the right to participate in*, the fullest possible exchange of equipment, materials and scientific and technological information for the peaceful uses of nuclear energy. Parties to the Treaty in a position to do so shall also cooperate in contributing alone or together with other States or international organizations to the further development of the applications of nuclear energy for peaceful purposes, especially in the territories of non–nuclear-weapon States Party to the Treaty, *with due consideration for the needs of the developing areas of the world.*[70]

In an effort to accommodate postcolonial morality and palliate the ascendancy of the Cold War paradigm, the NPT essentially declared that nuclearity of the "peaceful" persuasion was a fundamental right. As far as I can tell, no other international treaty has ever referred to a scientific or technological activity as an *"inalienable* right" of special importance to "the developing areas of the world."[71]

The NPT codified global nuclearity but left the IAEA to implement the vision. The agency launched a major "technical assistance" program aimed at developing nations. It tried to design a safeguards system but had trouble determining which things to include. South Africa pushed to exclude mines and ore processing plants from official definitions so as to minimize external oversight of its industry. The IAEA's 1968 safeguards document *specifically excluded* uranium mines and mills from the classification of "principal nuclear facility," which were seen as "a reactor, a plant for processing nuclear material irradiated in a reactor, a plant for separating the isotopes of a nuclear material, a plant for processing or fabricating nuclear material (excepting a mine or ore-processing plant)."[72] The 1972 safeguards document further excluded uranium ore from the category of "source material," thereby exempting its production from the ritual of inspections.[73] International authorities thus didn't consider uranium "nuclear" until it underwent the conversion processes that turned it into feed for enrichment plants or fuel for reactors. By the 1970s, in other words, the nuclearity of uranium ore and yellowcake had plummeted.

Inspections and safeguards offered mechanisms to balance the spread and containment of nuclear things. But how would exporters know what they could sell? In 1971, a group of NPT signatory states appointed

representatives to the newly formed Zangger committee, tasked with drafting lists of things nuclear enough to trigger safeguards.[74] The first "trigger
list," which appeared in September 1974, included reactors, fuel fabrication
and reprocessing plants, and enrichment plant equipment. India's "peaceful
nuclear explosion," though, rendered the list obsolete four months before
it was published. In the wake of the Indian test, the Nuclear Suppliers
Group (which included non-NPT states like France) formed to establish
more complex lists and practices.[75] Yet NSG compliance remained voluntary even as its lists grew longer and more detailed.[76]

Much remained unresolved or underspecified. Did uranium ore count
as "source material" or not? It depended on which IAEA document one
looked at. Did yellowcake count as "natural uranium" for export purposes?
Also unclear. These fine-grained and ever-shifting distinctions framed
global trade by separating the things that could be sold from those
that could not. In and of themselves, uranium ore and yellowcake were
deemed sufficiently banal to be bought and sold without exceptional
inspections. Even NPT signatories could export yellowcake without IAEA
intervention. Safeguards on uranium sales, if they existed, consisted only
of contractual promises not to use ore for military ends, an accommodation between the exceptionalism of nuclearity and the banality of
commerce.

This accommodation, in turn, laid down the technopolitical conditions
under which "the uranium market" could exist. In the 1940s and 1950s,
the US and the UK had strongly resisted the notion of a "market value"
for uranium.[77] Invoking the specter of Soviet supremacy, they'd strong-
armed suppliers into cost-plus pricing arrangements and kept contract
terms secret. Cold War ideology had thus placed uranium beyond "the
market."[78] Only after safeguards on uranium ore became defined as end-use
pledges written into sales contracts did the "uranium market" emerge as
an object and a practice of political economy.

That's where part I of this book begins.

PROLIFERATING MARKETS: ARGUMENTS AND THEMES OF PART I

Social scientists and humanists writing about the nuclear age have spent
little time on markets. We might think of this absence as the scholarly
fallout of nuclear exceptionalism. Markets, commerce, commodities, and

prices seem like ordinary topics, banal in comparison to the threat of apocalypse, the thrill of "electricity too cheap to meter," the weirdness of "atomic cocktails," the fearsome threat of radiation.

Part I of this book challenges the absence of political economy from most scholarship on nuclear topics. Specifically, it explores the place of African ores in the global uranium trade. Chapter 2 examines efforts to commodify uranium through the invention of "the uranium market," taking note of how the design of market-making tools excluded black Africans from data production and decision making. Chapters 3–5 delve into the transnational technopolitics of uranium from Africa. Chapter 3 focuses on South Africa, Britain, and Namibia. Chapter 4 looks at Gabon, Niger, and France. Chapter 5 turns to Namibia, Europe, and the US. Together, the four chapters in part I develop the following arguments:

• Uranium was re-invented as a banal commodity. Beginning in the 1960s, mining companies, brokerage firms, geologists, economists, national institutions, and international agencies all sought to facilitate the sale and purchase of yellowcake by turning uranium ore into a commodity governed by economic mechanisms instead of political ones. They created what Michel Callon and other sociologists have called market devices—technologies that generate knowledge and practices which create markets and define their means of commercial exchange. These devices served as tools for de-nuclearizing yellowcake, turning it into a banal commodity subject to the "laws of the market."[79]

• De-nuclearizing uranium offered a way to assert power over the terms of its trade. Despite claims to the contrary, politics and economics remained tightly bound in uranium's market devices. Invoking the "free market" validated a political geography in which imperial powers could continue to dominate former colonies after independence. In Britain and South Africa, the "free market" took on a valence of moral rectitude in assertions that anti-apartheid and anti-colonial sentiments should not affect the uranium trade. France invoked the "free market" to maintain its privileged access to African uranium when postcolonial Gabonese and Nigérien governments tried to assert sovereignty over natural resources.

• For African uranium producers, the shifting boundary between exceptionalism and banality was deeply entangled with the technopolitics of

state sovereignty. Precisely because it was profoundly political, the denuclearization of uranium was heavily contested. The international anti-
apartheid movement and the Namibian liberation struggle invoked the
politics and nuclearity of uranium in seeking sanctions against South
Africa. In the wake of the 1973 oil crisis, both Niger and Gabon protested
French neo-colonial pricing practices that undermined their sovereignty
and undervalued their ores. They diverged in their responses, though.
Niger emphasized the political value of uranium for French atomic programs to ask for price increases. Gabon pursued mechanisms of banalization that would give it more control over uranium sales.
• Transnational entanglements between licit and illicit transactions suffused
the technopolitics of African uranium. In the mid 1970s, Niger sold yellowcake to Libya and Pakistan, and Gabon tried to sell some to Iran. These
were legitimate transactions by some lights, but not by others. Around the
same time, the UN declared Namibian uranium illicit because of South
Africa's continued colonial occupation. To get around accusations of illegitimate commerce, uranium hexafluoride conversion plants in Europe and
the US used "certificates of origin" as market devices. These certificates
erased the Namibian origin of Rössing yellowcake. The converted product
thereby traversed the unstable boundary from illicit to licit thanks to a
new nationality that enabled its commercial circulation.

That's only half of my story, though. The history of uranium is not just
about the political economy of yellowcake. It's also about the people who
dug out the rocks. It's about their labor, their bodies, and the radon they
didn't know existed. To see all this, we need to shift scales and find a new
entry point.

THE (NON)NUCLEAR LIVES OF MINES

Uranium ore may be high or low grade. Its host rock can be hard or soft,
can contain other valuable elements (thorium, radium, copper, gold) or
nothing else worth extracting. Shallow deposits are mined in open pits,
whereas deep deposits require tunnels and underground shafts. Some mines
extract uranium exclusively; others produce it as a by-product. Many
operations include mills that crush the ore and perform the first phase of
extraction using various technologies and solvents, depending on the rock,

the materials available locally, and the metallurgical expertise of plant designers. The mines then either ship the ore out for further processing or take the next step themselves and produce yellowcake.

The bottom line: uranium mines, in all their diversity, looked a lot like other mines. Underground operations were dangerous, dark, and dusty. French supervisors at the Mounana mine may have scoffed at Gabonese workers reluctant to work underground for fear of evil spirits. But regardless of whether or not spirits lurked, the work was terrifying, with its constant danger of shafts collapsing or methane exploding. A 1970 flood at Mounana trapped five miners in a cul-de-sac. The spectacle of their bloated bodies, retrieved after six excruciating days of searching, remained seared in Gabonese memories for decades.[80] And other insidious hazards lurked: if the host rock had a high silica content, for example, miners could contract silicosis.

Open pits also carried dangers, including rock falls, heat exhaustion, and ever-present dust. For miners in Madagascar (and elsewhere), the most memorable moments were the narrow escapes, like this one recounted by Jeremy Fano:

I was in the quarry, it was eight o'clock in the evening. . . . They put the dynamite in place and lit it, and it exploded. A bit of rock flew off and fell on my foot. . . . A few toes were left, and I tore them off. . . . There was blood everywhere. . . . Rebem went to go see the boss to say that I badly hurt. He arrived and said don't touch it, we'll take you to the hospital and the doctor will do it. . . . But I said no, it's too far, these toes are just here, better rip them off now. . . . Then they could bandage me in the truck on the way to the hospital. . . . They gave me ten shots in the hospital, that's when I knew I was alive again. . . . I had fainted.[81]

Three decades later, when Fano heard that a foreigner was asking questions about the *vatovy* years, he walked 10 kilometers from his home to the town where I was staying. His one concession to the old injury: unlike most other inhabitants of the region, he wore shoes.

Scary. Arduous. Painful. Dangerous. Debilitating. But nuclear? Even in places where uranium epitomized nuclearity, the association did not necessarily extend to the mines themselves.

Some experts thought it should, even as early as the 1940s. In a 1942 medical textbook, Dr. Wilhelm Hueper of the US National Cancer Institute wrote that "the oldest known source of occupational pulmonary

Workers in a uranothorianite mine in the Androy, Madagascar, 1950s. CEA veterans liked the exotic look of this image. (courtesy of Cogéma)

malignancy resulting from the exposure to radioactive ores is represented by the cobalt mines . . . in Saxony, Germany." The Renaissance physician Paracelsus had noticed the ailment in the sixteenth century, but "it was not until 1879 that this occupational illness was diagnosed properly as a malignant tumor."[82] Duncan Holaday, an epidemiologist with the US Public Health Service who struggled to introduce radon monitoring into American uranium mines (starting in the 1950s), liked to say that "the effects of exposure to excessive amounts of radon and its daughters were observed and studied long before the fission of uranium was discovered." Holaday often reminded his colleagues that uranium miners were "exposed to higher amounts of internal radiation than . . . workers in any other segment of the nuclear energy industry."[83] After a visit to uranothorianite mines in Madagascar in 1960, a metallurgist for the French atomic energy commission ruefully described an example of such exposures: "The concentrates are spread out in the sun on large sheets of metal and are turned over periodically by a worker armed with a trowel or a rake. This procedure is clearly archaic, long, and above all dangerous because the worker is exposed to dust and radiation."[84]

In 1976, the first *Manual on Radiological Safety in Uranium and Thorium Mines and Mills*, a joint effort of the IAEA and the International Labour Organization, went even further than Holaday, declaring that "uranium mining is unique in the nuclear industry in that it is the only component of the nuclear production cycle that has associated with it a significant incidence of occupational illness"—an assertion that workers who'd been sickened at Rocky Flats and other weapons production sites could easily dispute. Nonetheless, the manual went on to state that although uranium mills were much safer, they shared at least one thing in common with mines: "both are more nearly descendants of earlier technologies than part of the modern 'nuclear industry' and its recognized innovations."[85]

The implications of this otherwise irrelevant observation were clear. The dangers of uranium mines were linked to their historical roots rather than their modern nuclearity. Two decades later, the South African Chamber of Mines pursued another line of reasoning in its official comments on a post-apartheid proposal to give nuclear regulators jurisdiction over uranium mining. The Chamber insisted that "radiation protection is essentially a health issue and not a nuclear energy issue, and therefore does not belong within the Nuclear Energy Act."[86] This meant, of course, that mines should not fall under the jurisdiction of nuclear regulators.

Even the strongest advocate for nuclear regulation of South African uranium mines didn't think that high radon levels necessarily mandated mine closures. "It's no use regulating and then saying well, sorry chaps, but we're going to make ten thousand of you redundant because the levels are too high," health physicist Shaun Guy told me in 2004. Guy had spent years uncovering radon levels that the South African mining industry had tried to hide, fighting hard to obtain regulation. But, he said, "I don't agree that you close down mines because the levels are too high—that's my personal opinion. . . . A lot of people lose their jobs. . . . [With a] First World industry in a Third World country like South Africa the impact can be enormous, whereas in America and the UK the government just says close it down and that's it, there's no argument about it because there are various safety nets. . . . But there's no safety net here."[87] South African economic conditions, Guy felt, required a different approach to risk management.

Placing a "First World" industry in a "Third World" environment required accepting higher exposure, sometimes by invoking fantasies about the efficacy of "First World" safety nets. Sometimes it meant denying the danger outright. In chapter 7, I tell the story of Dominique Oyingha and his brother, who extracted uranium for three decades at the French-run mine in Gabon. When Oyingha confronted the mine doctor regarding his sick brother in the 1960s, the doctor responded, "Are you crazy? . . . Who told you that uranium made people sick?"[88]

So knowledge about the health effects of working in a uranium mine existed "long before" the discovery of fission, at least as early as 1879. Yet throughout the twentieth century, experts had to assert and reassert the existence of radiation hazards, as though they'd been newly discovered. A cognitive dissonance emerged from inversions of historical time, from the presumed rupture between the modern and the traditional, from the presumably related gap between the First and Third Worlds. How could the *least modern* part of the nuclear industry carry the greatest danger of internal exposure to the *most modern* of hazards?

As part of the *nuclear* industry, uranium mining seemed oddly out of place in Africa. But as part of the *mining* industry, it seemed right at home. The ongoing need to reassert the unique dangers of uranium mining also reflected constant contests over those dangers. Perhaps managing radiation exposures in mines wasn't "a nuclear energy issue" at all? Perhaps those lung cancers and other ailments had as much to do with arsenic, or smoking, or dust? Was the mere *potential* of contracting lung cancer worth an economic sacrifice for everyone?

THE NUCLEAR LIFE OF RADON

Much of the wrangling concerned a seemingly straightforward question: Does radon exposure cause cancer? Uranium atoms decay into radon, which then decays into other elements that, when inhaled, lodge in the lungs and bombard soft tissue with alpha particles. Determining if a particular lung cancer was triggered by radon exposure gets complicated, though. Let's break the question down into its three constituent parts: radon exposure, cause, and cancer.

First, radon exposure. How much radiation *from radon* did mineworkers absorb? Before the 1980s, personal dosimetry captured only external exposure from gamma rays, the type of radiation that penetrates clothes and skin, causing burns and internal cellular damage at high doses. Unlike gamma levels, which are easy to predict from the ore grade and can be detected using a simple piece of film, alpha levels are inherently unpredictable and vary throughout the mine with the type of rock, the mining activity, and the ventilation system. At first, only heavier instruments stationed throughout mining areas could capture radon readings. Most mines chose to average these results, which undervalued the dangers of "hot spots" far from the air vents, where reduced ventilation meant elevated radon daughter levels and higher temperatures—the kind of place where (for example) white foremen stationed black workers in South African mines. As we'll see in greater detail throughout part II, measuring radon exposure presented significant technopolitical complications.

Next, cause. According to standard scientific practice, determining causality requires isolating the effects of a contaminant. Did illness in uranium miners come only from radon exposure, or did other contaminants (such as tobacco) contribute or even serve as the primary trigger? Researchers proceeded in two ways. The epidemiological approach favored by the US Public Health Service involved calculating exposures, tracking individual histories, recording physiological characteristics, and statistically analyzing data for meaningful correlations. Collecting data presented endless challenges. Even if scientists reached consensus on their choices, doubts about the conclusiveness of the analyses persisted among mine operators, atomic energy agencies, and experts from other fields. Statistics offered correlation, not certainty about cause.

By contrast, animal experimentation, such as that favored by French atomic energy scientists, *was* equipped for causal determination. It seemed simple: Expose rats to radon and see if they got cancer. They did. Did tobacco smoke make it worse? Yes. Case closed? Not quite. People aren't rats. Just because rats got cancer didn't mean that people would. Lung size, air intake rate, and bronchial mucosa all mattered for *how* and *how much* radon caused cancer. Therein lay the vulnerability of animal experimentation. More research, please! For both the epidemiological and experimental approaches, conclusive causality proved endlessly elusive.

We'll explore the problems of establishing causality in more detail in chapter 6. It's worth noting that similar issues have long plagued studies of other occupational and environmental contaminants. The effects of asbestos, lead, silica, vinyl chloride, and pesticides have all been thrown into doubt by corporations pitting epidemiology against experimentation and invoking the specter of uncertainty. Tobacco industry officials cynically boasted that "doubt is our product."[89] Doubt produces delays in setting standards, in creating regulations, in testing, and in enforcement. It lowers operating costs and raises profits. Contaminants may be recognized and regulated in one place but not another; many industries distribute their hazards across international borders for precisely that reason. So radon and radiation were by no means unique, although they were paradigmatic—at least in the US, where cultural anxiety about radiation during the Cold War alerted labor leaders and the public to the possibility that invisible dangers lurked in industrial activities.[90]

At last we come to cancer itself, which is difficult to see and treat, especially in Africa. First there's the problem of time. Radon exposure takes 10 to 30 years to instigate disease. That's a long time to track people in a scientific study. It's also enough of a lag to generate doubt about the link between exposure and illness. In France, Gabon, and Namibia, miners often worked for the same company for decades. In the US, Madagascar, and Australia, mine operations typically lasted less than 10 or 15 years, making it difficult to find out whether workers subsequently contracted cancer.

Next comes the problem of detection. Even a mining site that has dedicated doctors and clinics can't detect cancer without the right diagnostic tools. Mine clinics are rarely outfitted with the proper tools, often because officials aren't interested in finding cancer. So argues Jacqueline Gaudet, from whom we'll hear more in chapter 7. Both of Gaudet's parents contracted cancer while working at a mine in Gabon. Misdiagnosis by mine clinicians meant that effective treatment was too late by the time the family returned to France. The lack of diagnostic tools at the mine, Gaudet insists, enabled operators to claim that there were never any cancers at all. When even French citizens lacked access to life-saving diagnosis and treatment, what hope was there for African workers?

Finally, as historian Julie Livingston argues, the global health community has until recently treated cancer like a "First World" disease. Many researchers have assumed that Africans simply don't live long enough to contract most types of cancer. Oncology got a promising start in post-independence Uganda, but by the 1970s it had withered under the effects of political violence and structural adjustment. Since then, oncological research and development has been geared toward patients in rich countries with extensive medical infrastructures, an approach that has favored expensive treatment plans.

In contrast, public health in most African countries has been influenced by colonialism, missionary work, mineral extraction, and other external interests. Its focus on infectious disease, malnutrition, and fertility has shaped statistical collection and policy planning in ways that make it difficult to introduce new dimensions.[91] Among other repercussions, this has led to a near-total absence of national cancer and tumor registries. As we'll see in chapter 9, when Namibian labor leaders began to worry about cancer and called for research into the health effects of local uranium mining, the lack of a Namibian registry posed an insurmountable stumbling block. How could anyone know whether uranium had caused *excess* cancer without a baseline against which to measure the surplus?

The question of causality—"Does radon cause cancer?"— has always been a historical and geographical question. It has no single, abstract answer above and beyond the politics of expert controversy, labor organization, capitalist production, or colonial difference and history. That answers depend on the friction between these, however, is most clearly visible at the margins of nuclearity.

NUCLEAR WORK: ARGUMENTS AND THEMES OF PART II

Uranium mines—especially in Africa—were at the margins of an industry driven by claims to exceptionalism. Compared to reactors and bombs, they appeared banal and peripheral, more closely allied (technologically, politically, and geographically) to other forms of mining than to other nuclear things. Indeed, many aspects of the stories I tell about African uranium miners resemble the histories of labor and occupational disease in asbestos or gold mining.[92] That's part of my point. The nuclearity of uranium mines was not self-evident. It was not handed down from on high. Nor was it their only significant characteristic. The real, material similarities with other mining workplaces often made nuclearity *more* difficult and laborious to produce.

Part II of this book examines the considerable work required to make
African uranium mines nuclear. Chapter 6 presents a history of "global"
data and standards for radon and radiation exposure, paying attention to
the invisibility of African uranium miners in this process for over five
decades. Chapters 7–9 explore how and why miners in Madagascar, Gabon,
South Africa, and Namibia did or did not construe their labor in nuclear
terms. Together, the four chapters in part II develop the following related
arguments:

• Standards for radon exposure were fundamentally technopolitical. Radon
exposure standards reflected the tensions in reconciling scientific research
results, technological systems for measuring and containing radiation,
national imperatives, corporate profit, international organizations' quests for
global authority, and shifting power relations between experts, corporations, and labor. Since the 1970s, the International Commission for Radiological Protection (ICRP) has promoted the exposure philosophy of
ALARA: As Low As Reasonably Achievable. "As low as" reflects the rough
consensus that all radiation exposure has some health effect; "reasonably
achievable" represents a concession to economic and political imperatives
(and power). Buried deep in the ICRP's philosophy is the assumption that
human lives have different values in different places. As we'll see, this
philosophy has been interpreted as legitimation for spending less to protect
workers in poor nations who have remained invisible to experts.
• Invisibility was systemic but not always deliberate. Invisibility resulted
from what historian Michelle Murphy calls regimes of perceptibility—that
is, assemblages of social and technical things that make some hazards and
health effects visible but leave others invisible.[93] Such regimes had local,
national, and global dimensions that included dosimeters and protective
equipment, laboratories for analyzing exposure results, mechanisms for
communicating those results, national regulatory systems, manuals, guidelines, and conferences. We'll see, for example, that radiation experts in
apartheid South Africa deliberately avoided studying radon exposures of
black miners. In Europe and North America, experts accepted the South
African rationales for excluding black workers because these rationales
matched standard epidemiological criteria for selecting study
populations.
• The stakes of inclusion or exclusion were scientific, political, and corporeal. For varying reasons, radon exposures endured by miners in South

Africa, Madagascar, and Gabon never became scientific data. This absence shaped biomedical knowledge, allowed for greater exposure, and permitted the absence of occupational health regulation. For example, radiation levels in South African mines remained unregulated for decades, with untold results for miners. Where regulatory principles did exist, actual practices diverged significantly. The standards and rules at Mounana, for instance, weren't necessarily tied to state supervision and weren't always implemented.

• Some African miners eventually developed politically usable forms of nuclearity; others didn't. In none of the countries I examine did uranium miners achieve "biological citizenship," a term that anthropologist Adriana Petryna uses to describe how Chernobyl victims used their radiation exposures to fashion new identities and lay claim to health care, welfare, and other resources.[94] Some miners, though, came closer than others to making their exposures politically, socially, and medically meaningful. Because uranium production in southern Madagascar ended long before workers could file claims in transnational arenas, it never achieved a nuclearity that allowed Malagasy exposures to serve as a resource for postcolonial claims making. Although Gabonese workers remained largely unaware of their specific exposures, they eventually developed their own sources and contexts of knowledge about radiation, which enabled them to seek compensation and remediation after the mines closed. For their part, Namibian uranium workers used political alliances formed during the liberation struggle to develop a sophisticated sense of nuclear exceptionalism and its political possibilities.

AFRICA AND THE NUCLEAR WORLD

This book argues that nuclearity has never been defined by purely technical parameters. Like other master categories that claim global or universal purview, the "nuclear" both inscribes and enacts politics of inclusion and exclusion. Neither technical function nor radiation sufficed to make African nations and their mines nuclear. Part I argues that the nuclearity of African uranium—along with the banalization of uranium ore more generally—was closely tied to the political economy of the nuclear industry. This had consequences for the legal and illegal circulation of uranium and for the global institutions and treaties governing nuclear systems. Part II argues that the historical and geographical contingencies shaping the "nuclear" as a category also had profound consequences for the lives and health of mineworkers.

In no way can we point to neat divisions between "global" processes and "local" examples. Institutions and agreements that claimed "global" purview, like the IAEA, the ICRP, and the NPT, were themselves "local" by virtue of their inclusions and exclusions, by the ways they circumscribed knowledge, defined expert communities, and conducted debate. The concepts, standards, and practices they produced changed *every time* they were implemented, either on the ground or underground.

Anthropologist Anna Tsing uses the notion of *friction* as a metaphor for the creative and destructive power generated by "universal" concepts and practices when they travel.[95] This friction calls attention to the unevenness with which knowledge travels, the inequalities that shape its motion, the always-local circumstances that change its content along the way, and the material consequences of its motion. The production and dissolution of nuclear things in African places, I argue, occurred in the friction between transnational politics and (post)colonial power, between abstract prescriptions and embodied, instrumentalized practices.

Along the way, I demonstrate not just the uneven spatial distribution of nuclearity, but also its uneven temporalities. There was no moment in global time when the nuclearity of uranium mines became forever settled everywhere. Variations depended in part on clashes between different historical rhythms: decolonization and Cold War; knowledge production and capital flows; mine openings and closures; apartheid, transnational activism, and postcolonial politics.

The stakes of Africa's absences from the nuclear world continue to accumulate. In the uranium boom currently in progress across the African continent, mine operators and state officials—invoking the need for "social judgments" advocated by the ICRP and other international sources of authority—pit the immediate urgency of "development" against the long-term uncertainties of exposure. This book documents the historical and ongoing struggle to see Africa in the nuclear world, and the nuclear world in Africa.

I PROLIFERATING MARKETS

MARKET AVERSIONS

In the 1930s, the Homer Laughlin China Company of West Virginia introduced a popular line of ceramics glazed with a yellow powder that turned orange-red upon firing. Although permanently discontinued in 1973, uranium-glazed Fiesta dishes are still available for purchase on eBay. Were you to drink your daily coffee from a vintage "Fiesta red" cup, your lips would get an annual dose of 400 millirems (mrem) of beta-gamma radiation, your fingers a dose of 1,200 mrem, and your guts another dose from the uranium leachate you'd ingested.[1]

It feels odd to think of dinnerware as nuclear. These objects seem more like radioactive curiosities, collectables left over from the days of atomic enthusiasm. But they demonstrate that uranium had a modest market life long before a fat lump of it exploded over Hiroshima.

The Homer Laughlin China Company experienced the nuclearity of this abundant, innocuous-seeming pigment when the US government confiscated its entire supply in 1943, effectively keeping "Fiesta red" plates off the market until 1959, when manufacturing resumed using depleted uranium.

For US officials, the health risks of radioactive table settings were the least of their concerns. Their goal was to secure supply chains for American atomic weapons and prevent other nations from accessing bomb-building materials. They dreamed of controlling the world's uranium supply, of setting the economic and political terms of its circulation. Put another way, the last thing the Americans wanted was for uranium to have a market value.

The dream began with the illusion of scarcity.[2] In 1940 the Belgian industrialist Edgar Sengier shipped over 1,250 tons of high-grade uranium ore from the Belgian Congo to a warehouse in Staten Island, New York.

Manhattan Project planners got wind of this and bought it all in one stroke. In the next few years, they devoted considerable energy and resources to finding and securing additional supplies.

To this end, the Americans and the British formed a procurement partnership known as the Combined Development Trust (CDT). Its largest supply contracts were with South Africa, Australia, and Sengier's mine in the Belgian Congo. The United States signed separate contracts with Canadian mines.

Hardly novices in the mining business, these early suppliers requested "market price" contracts. The CDT, however, rejected the notion that uranium could have a market value.[3] A rare mineral critical to powerful weapons, uranium required special pricing arrangements. Such was the CDT's argument, one that allowed it to call the shots.

Invoking the specter of Soviet supremacy, the CDT strong-armed suppliers into cost-plus pricing arrangements that granted each of them the same profit margin. They kept contracts secret lest the Soviets discern the quantity of uranium purchased and divine the extent of American and British weapons development. But contract confidentiality prevented suppliers from discovering variations in the prices paid by the CDT.

Suspecting discrepancies, the Belgians repeatedly requested price parity with South Africa and Canada. This would have yielded them considerably higher profits thanks to low labor costs in colonial Congo, due in no small part to the atrocious working conditions at the Katanga mines.

In this and other instances, though, the CDT successfully cleaved to its policy of profit parity. To avoid invoking the unsavory anti-market connotations of the word "Trust," the partnership changed its name to Combined Development *Agency* (CDA).[4]

Through the 1940s and 1950s, suppliers took consolation in the prospect of long-term benefits that ranged from expectations about sharing nuclear technologies to fantasies about colonial benevolence. Sengier, for example, wrote to David Lilienthal, chairman of the US Atomic Energy Commission: "My whole life has been devoted to the industrial and social development of Katanga, the mining district of the Belgian Colony, with a policy of putting the general interest of the country and the welfare of white and native populations above the financial interest I was responsible for. The pleasure derived from such a policy turned out to be the most valuable dividend I received for my work."[5] For Sengier, nuclear exceptionalism and the pleasures of empire offered consolation for limitations on profit.

Then the CDA switched course. After the war, the US fostered uranium exploration on American soil, most notably by offering discovery premiums and price guarantees. These measures effectively ensured state control over uranium production and sales within the US. They also meant that uranium was no longer scarce.

The ensuing exploration boom made it clear that uranium wasn't confined to particular geological formations or geographical locations. The stuff was everywhere. The US had access to more uranium than it could use, not just for its military program but also for civilian reactors. The United Kingdom, meanwhile, hadn't developed civilian nuclear power as quickly as planned. It didn't need as much uranium as it had contracted for, nor did it need the material quite so urgently.[6]

In 1959 the CDA initiated discussions with South Africa and other suppliers to defer delivery of the ore and "stretch out" their uranium contracts. Once the renegotiated contracts expired, the US halted foreign purchases altogether.

The sudden cessation of US purchases stunned suppliers who had built up mine capacity in the expectation of ongoing contract renewals. After they recovered from the shock, they began to wonder: Was a free market in uranium now possible? What would it take? It appeared as if the Americans and the British were poised to gain full control over uranium supplies in the Western world. That wasn't something the French could abide.

From its founding in 1945, France's Commissariat à l'Énergie Atomique (CEA) made uranium prospecting a top priority. The search began in the metropole, where uranium ores had been discovered in 1799. After a rocky start, prospectors found pitchblende (high-grade ore) in 1948. In short order France was producing modest quantities of uranium oxide.

The colonies were next. Anticipating uranic revelations in Madagascar, which was reputed to be a geological paradise, French atomic scientists had persuaded Charles de Gaulle to hang on to the colony regardless of political cost.[7] In 1956 CEA geologists began to use aerial survey techniques to scour Africa for the stuff that promised national radiance in the twilight of empire. By 1960—the year most French African colonies gained independence—the CEA was operating uranium mines and mills in metropolitan France, Madagascar, and Gabon. It had also begun exploration in Niger and elsewhere in central and northern Africa.

The CEA's first priority was to ensure adequate supplies for its nascent nuclear programs, both military and civilian. Although it was one of the best-funded institutions in France, the agency's resources weren't infinite. Jacques Mabile, the young engineer in charge of the CEA's mining division, tried to limit costs. With the nation's reputation at stake, though, how much was too much to pay for uranium?

Once it became clear that France had enough material to fuel its own bomb and reactor programs, Mabile began to think about the rest of the world. Could uranium serve the "radiance of France" on a global scale, turning the nation into a major mineral exporter? The idea was plausible. CEA prospecting teams fanned out from the former African colonies to Afghanistan, Turkey, Iran, even Canada. One thing was clear: to become a leading uranium supplier, France had to match or beat the production costs of its competitors. But costs were often state secrets covered by special legislation.

The CEA approached both Canada and South Africa about purchasing uranium. Although Mabile wanted to get as much uranium as his budget would allow, he also wanted to find out how much other suppliers were charging. How better than to express interest in buying? Mabile was thinking in the long term. He knew South Africa's production costs were low because almost all uranium came from gold mines where capital investments were amortized by gold sales. If the CEA could negotiate a good price, France could buy South African uranium for immediate use while stockpiling its own higher-cost product until worldwide demand increased. At the very least, a contract proposal would tell Mabile just how low the South Africans could go. Crunching the data, Mabile began to set yearly cost targets for French mines.[8]

Meanwhile, the CEA began to get serious about predicting the world's nuclear future. The stretch-outs negotiated by the US and the UK suggested low demand through at least the 1960s. Mabile believed that the rise of civilian nuclear power throughout the world would make this downturn temporary. Gut feeling, however, couldn't justify high sums for prospecting and mine development. Adopting a South African method, CEA geologists began using statistical techniques that allowed for faster, more sweeping projections of deposit sizes. Data on European and

American reactor planning supported Mabile's optimism about demand. These forecasts persuaded CEA directors to continue sinking substantial sums into prospecting programs—even at the expense of domestic production.[9] In the stretch-out years of the mid 1960s, they were the only ones doing so.

Jacques Mabile blended commerce with matters of state. Setting cost targets, estimating reserves, and forecasting demand ensured CEA financial viability and demonstrated to national and international constituencies that France could engage in commercial ventures on par with the "Anglo-Saxons" (a rubric for the white, industrialized, English-speaking world). In the long term, Mabile hoped to expand French uranium production beyond its postcolonial orbit. In the 1960s, however, only Africa offered the sufficiently rich and reliable sources that made it possible for France to imagine a uranium market.

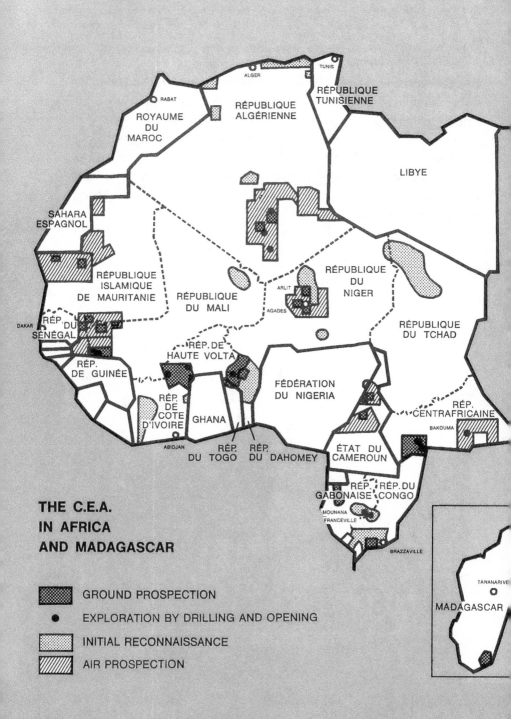

TUNIS

ALGER

RABAT

RÉPUBLIQUE
TUNISIENNE

RÉPUBLIQUE
ALGÉRIENNE

ROYAUME
DU
MAROC

LIBYE

SAHARA
ESPAGNOL

RÉPUBLIQUE
ISLAMIQUE
DE MAURITANIE

RÉPUBLIQUE
DU MALI

ARLIT

AGADES

RÉPUBLIQUE
DU
NIGER

RÉPUBLIQUE
DU TCHAD

DAKAR

RÉP. DU
SÉNÉGAL

RÉP. DE
HAUTE VOLTA

RÉP.
DE GUINÉE

RÉP.
DE
COTE
D'IVOIRE

GHANA

FÉDÉRATION
DU NIGERIA

RÉP.
CENTRAFRICAINE

BAKOUMA

ABIDJAN

RÉP.
DU TOGO

RÉP.
DU DAHOMEY

ÉTAT DU
CAMEROUN

RÉP.
GABONAISE

RÉP. DU
CONGO

MOUNANA
FRANCEVILLE

BRAZZAVILLE

THE C.E.A.
IN AFRICA
AND MADAGASCAR

TANANARIVE

MADAGASCAR

GROUND PROSPECTION

EXPLORATION BY DRILLING AND OPENING

INITIAL RECONNAISSANCE

AIR PROSPECTION

Before the mid 1960s, the idea of "the uranium market" was aspirational, little more than an expression of desire: the desire to sell to any buyer, the desire to buy from any seller, the desire to be free of restrictions while conducting transactions. The desire, in short, to treat uranium like an ordinary, profit-generating commodity.

The desire needed expressing, because so much blocked its fulfillment.

The desire for commodification—of uranium, and of other nuclear things—had shaped the formation and development of the IAEA. Defining "source materials," debating safeguards, dispensing technical assistance. . . . These activities all involved making certain nuclear things *ordinary* enough for commercial exchange. The debates, as we saw in the previous chapter, were endless. They shaped the creation of the IAEA, the drafting of the NPT, the trigger lists devised by the Nuclear Suppliers Group, and much more.

These global institutions and arrangements offered corporations and nation-states instruments for navigating between the imperatives of national security and the allure of economic expansion. In this sense, they framed the technopolitical and moral *conditions of possibility* for trade. They made it possible to imagine buying and selling a material that could, in the right form, pulverize the planet.

But conditions of possibility do not create markets by themselves. Markets require knowledge, data, instruments, standards, and other components. Beginning in the mid 1960s, private companies, state agencies,

on facing page: CEA activities in Africa, 1970. (courtesy of Commissariat à l'Énergie Atomique) and international organizations began seeking ways of systematizing uranium supply and demand. They devised forms of knowledge: forecasts, reserve estimates, contracts, prices. They built plants and set standards for uranium in all its forms: ore, yellowcake, tetrafluoride, hexafluoride, enriched. They created vehicles for exchange: brokerages, cartels, lobbying associations.

In this chapter, I treat these various technologies and practices as *market devices*, instruments through which the uranium market was imagined, structured, contested, and—however haltingly—brought into existence.[10] Market devices included *reserve estimates* of how much uranium a particular country or region possessed, *forecasts* for how much uranium would be required in the future, and *prices* paid for a pound of yellowcake. All three were claimed to be *descriptions* of an existing state of affairs. Experts and industry leaders portrayed the uranium market as an autonomous entity "out there." They treated reserves, forecasts, and prices as ways of generating knowledge about that independent entity.

"The uranium market," however, did not exist prior to this knowledge-making. Reserves, forecasts, and prices served as technologies for *creating* the market. I argue that the workings of this market were shaped—not "discovered"—by these devices.

This distinction is crucial, because market devices mapped moral and technopolitical geographies of trade in which "global" meant "capitalist," "the West" included Japan and South Africa, and colonial relations persisted in the technologies of production and the infrastructures of exchange. Corporate and state officials from apartheid South Africa played important roles in devising market devices with global purview. Officials from newly decolonized states governed by black Africans did not. For them, as we'll see in chapter 4, market devices functioned as instruments of exclusion.

The status of uranium as a commodity—and indeed the very question of whether "the uranium market" truly existed—remained contested for decades. Should something with the power to destroy the planet or sicken those who engaged in its manufacture really be traded like cotton or grain? Was it realistic to expect that governments might refrain from intensely regulating its production or trade? Could uranium even count as a commodity without being listed on any exchanges? As a commodity, would it be mineral or fuel? What did it mean to speak of "the uranium market" or "the market price of uranium" in the absence of any futures trading?

These ongoing debates demonstrated that "the uranium market" was not a stable entity. The concept meant different things to industrialists, brokers, mine operators, policy makers, and state officials. At the risk of frustrating some readers, I make no effort to pinpoint moments in which uranium behaved like a market object according to abstract criteria. Instead, I am primarily concerned with how the process of making a market generated politics that were embedded and enacted in knowledge and devices. How did invoking "the uranium market" serve to arrange and rearrange power relationships? What was included—and excluded—in producing data that claimed merely to describe "the market"?

The exclusions mattered as much as the inclusions. For market makers, commodification promised to remove uranium from the constraints of nuclear exceptionalism and the demands of decolonization. Uranium could fuel reactors or power weapons. For the light-water reactors that came to dominate nuclear power production, the branching point came at the moment of enrichment. Before that point, uranium passed through many stages and places. Rendering it a banal commodity—one that didn't require special nuclear accounting mechanisms, and whose geographic origin was irrelevant to its value or its trade—would make that passage less cumbersome and more profitable.

Making uranium banal involved distinguishing between "economic" and "political" forces. Placing "political" factors outside "the market" turned nuclear proliferation, government regulation, anti-nuclear activism, postcolonial natural resource sovereignty, the struggle against apartheid, and occupational health legislation into exogenous forces whose irksome presence inappropriately impeded the flow of ore and profits. Such political displacements were hardly unique to the uranium industry, of course. Industry leaders were skilled in the strategic displacement of politics— and nowhere more so than in apartheid South Africa, where apolitical invocations of "the market" justified trade in the face of escalating sanctions.

Then there was the problem of morality. As principles for global order, nuclear exceptionalism and decolonization carried tremendous moral force readily marshaled by those whom market makers sought to exclude. In fending off the "intrusions" and "distortions" of politics, market makers

invoked an alternate moral order. Made virtuous by Cold War discourse
that pitted capitalist freedom against state-run shackles, market morality
was about honoring contracts, producing instruments of reliability, and
creating trust among producers and buyers.

As we'll see in this and the next three chapters, market devices required
perpetual maintenance. The commodification of uranium constantly challenged the border between exceptionalism and banality.

FROM "MARKETING URANIUM" TO "THE URANIUM MARKET"

How would Western European nations handle the challenges posed by US
technological imperialism as their empires crumbled?[11] The emergence of
civilian nuclear power gave European nations good grounds for fearing
that the US would translate military might into commercial nuclear
supremacy. In the late 1950s, the US Department of State discouraged
independent French and German nuclear programs in favor of reliance
on US reactor technology.[12] Although the US professed support for
IAEA-based multilateral safeguards over civilian nuclear technology, it
persisted in signing *bilateral* safeguards agreements as conditions for reactor
sales.

At the end of the 1950s, the US also seemed poised to control the
uranium trade. The US Atomic Energy Commission's supply contract
"stretch-out" presaged the influence of the US as a uranium consumer. If
the US dominated global commercial reactor sales, that influence would
extend into the civilian arena. Unlike the gas-cooled or heavy-water reactors built by Canada, France, and Britain, which used non-enriched,
"natural" uranium, the light-water reactors that American companies sold
(both domestically and internationally) used enriched uranium. The AEC's
monopoly on uranium enrichment services meant that any country buying
an American light-water reactor had to route uranium through US facilities. This, in turn, increased their likelihood of buying American ore.

In 1964, the US further tightened its grip on uranium flows by amend-
ing atomic energy legislation to prohibit the use of foreign uranium in
American reactors.[13] Furthermore, the AEC could continue enriching
foreign ore for reactors abroad only under long-term contract. The legislation thus protected American uranium producers and locked foreign customers into continued dependence.[14]

Mining

Choices about open-pit versus underground mining in Africa depended on the grade of the ore, the geology and extent of the deposit, and the costs of production in a given area.

Waste products

Waste rock was often left in large piles or used in building and road construction.

CRUSHED ORE

Milling and Refining

After crushing, a series of chemical processes turned raw ore into a form of uranium oxide, known as yellow-cake because of its color. Mills were located at the mine site; often yellowcake plants were too.

YELLOW CAKE IN 55 GAL DRUMS

Known as tailings, waste chemicals and materials were often processed into sludge. Disposal ranged from lined, sealed tailings ponds (best practice, but quite rare) to open-air piles or direct dumping into rivers.

Conversion

Plants in Europe and North America converted yellowcake to uranium tetrafluoride (UF_4) and hexafluoride (UF_6) gas in preparation for enrichment.

UF_6 IN STEEL CYLINDERS

Enrichment

Plants in Europe and North America enriched UF_6 to increase the proportion of fissionable uranium (^{235}U) in the fuel. Commercial reactors require 3-5% ^{235}U while bombs require 90% ^{235}U.

Weapons

Highly enriched uranium was used for the manufacture of atomic bombs.

Energy

Low-enriched uranium was pressed into fuel pellets or rods for use in reactors.

The uranium fuel chain. (figure prepared by Evan Hensleigh)

Outside the US, uranium producers were livid. Power plant construction was proceeding more slowly than atomic apostles had prophesied.
Fuel sales were slow and not especially profitable. Mines that had developed massive production potential in response to US military needs struggled to stay afloat. Canadian mines were hit hard because they lacked the
South Africans' cushion of gold extraction.

In September 1965, South African, Canadian, and British mining executives met to discuss ways to free the uranium market from American control.
Rio Tinto Zinc (RTZ), which operated uranium mines in Canada, South
Africa, and Australia from its headquarters in Britain, took the lead by proposing the creation of a Uranium Producers' Clearing Agency (UPCA) to
"assist in *the orderly development of the uranium market* and the nuclear power
industry." The UPCA would take the form of "a centralised trading agency
in Switzerland acting, under suitable control, for all the producers."[15] The ore
producers would negotiate contracts and the UPCA would handle the for-
malities. To drive prices up, UPCA members would develop an escalating
price schedule for short- and medium-term contracts.

Canadian companies worried that contracts written in the UPCA's
name would dilute their corporate brand and deny them the fruits of their
"pioneering work" in market surveys, promotion, and development. The
South Africans liked RTZ's proposal, however, and insisted that *all* producers had done this work and no company could lay claim to "its own special
sphere of influence." The parties also disagreed about price. The South
Africans, who could extract uranium from gold tailings cheaply, wanted
to start the price schedule below $5 per pound to undercut American
producers. Canadians, however, couldn't afford to start that low. Discussion
resolved nothing and the idea was shelved.[16]

UPCA proposals eventually evolved into a broad-based cartel in the
early 1970s. Soon after that, RTZ would establish its own Swiss agency
to handle politically controversial contracts. Initially, though, the unsuc-
cessful attempt to create a centralized trading agency marked a subtle shift
in how producers approached uranium sales. In their first reaction to the
American "stretch-out," they had scrambled to "market uranium" wherever
they could. The UPCA discussions suggested a more concerted approach
that grappled with treating "the uranium market" as an entity that could
be conceptualized and described.

State and corporate officials involved in civilian nuclear power shared the mining industry's eagerness to commoditize uranium. As we saw in chapter 1, South African officials excluded mines and ore-processing plants from official IAEA definitions of "principal nuclear facilities," exempting yellowcake production from direct safeguards and inspections. But de-nuclearizing uranium was only one step in facilitating the transnational trade. Nuclear power planners needed *calculability*: Manufacturers and utilities could justify massive capital investment only if they could predict uranium supplies and costs without interference from military imperatives or over-eager policy makers.

Calculability became complex as light-water reactor designs gained traction in Europe. By the early 1970s, even France had purchased a license from Westinghouse for a light-water design following a huge battle between rival experts and institutions.[17] The new reactors required low-enriched uranium, which for the time being was available only from the AEC. Loosening the American grip would require a huge capital investment in enrichment plants. To build partnerships that spread the risk, European nations and corporations needed credible predictions about uranium supply to assure potential investors about future profitability.

Such calculations required more data. Clearly, uranium was more abundant than US military planners had initially imagined. How much was out there? Where was it? In what form? How hard was it to extract? How much would it cost to obtain? Atomic energy agencies and state geological services had developed secret estimates for the nations and colonies in their political orbit. Commodification, though, required reliable, accessible data.

Enter our first market device: international resource estimates.

MAPPING "SUPPLY AND DEMAND"

In 1965, the nuclear energy arm of the Organisation for Economic Co-operation and Development (OECD) pooled data to develop resource estimates worldwide (that is, among capitalist nations). The IAEA soon joined the effort. The resulting assessments were published roughly every two years and became informally known, thanks to their crimson covers, as the Red Books.[18]

Composed primarily by geologists, the Red Books offered an accounting of uranium deposits, providing baseline data for calculating the future of nuclear power and serving as market-making devices for the uranium trade.[19] While boldly publishing former state secrets, they simultaneously hid the diverse imperial relationships that structured uranium exploration and mining. Presented as mere descriptions of ore deposits, the Red Books became technopolitical instruments for imagining and structuring the transnational trade in uranium.

Geology, economics, and technology were visibly intertwined in Red Book accounts. The first few volumes specified three target price categories and offered estimates of "Reasonably Assured Resources" and "Possible Additional Resources" within each category.[20] Reasonably Assured Resources—also dubbed "reserves"—were defined as "material which occurs in known ore deposits of such grade, quantity and configuration that it can, within the given price range, be *profitably* removed from the earth and processed with *currently proven* mining and processing technology."[21] For example, in 1965 geologists predicted with reasonable confidence that South African mines could profitably extract and process 140,000 tons of uranium oxide at less than $10 per pound. Possible Additional Resources were more speculative. They represented guesses about uranium deposits that were "surmised to occur" and "expected to be economically discoverable and exploitable."

Discussions of data rarely made politics visible. The text on South African reserves, for example, didn't call attention to the exceptionally low labor costs that made this uranium so profitable. And while titles and graphs referred to "world uranium production" and "world uranium demand," the "world" was highly circumscribed. Contrary to the implication of their colorful covers, the Red Books saw the "world" as "all countries with the exception of the USSR, Eastern Europe and China"—that is, only countries conducive to the market the books sought to devise.[22]

Organizing data by country, the Red Books hid from view the imperial conditions of data production, such as the fact that France provided data

on facing page: Estimating reserves began with the difficult work of prospecting. In the 1950s and 1960s, the CEA drew on considerable local knowledge when prospecting for uranium in Madagascar, Niger, and Gabon. (photos courtesy of Cogéma)

for Gabon and Madagascar in the 1960s, or the fact that South African data included uranium from Namibia in the 1970s. Country categories made it difficult to see the non-national ways in which jurisdiction over uranium was structured, such as the fact that RTZ controlled mines in southern Africa, Canada, and Australia.

The Red Books' performative agenda, however, was crystal clear. The reports regularly raised the specter of insufficient uranium, urgently calling for increases in exploration. The more data the books produced, the more persuasive their forecasts of looming shortages. Far from being merely descriptive, the Red Books explicitly promoted uranium exploration and market formation.

Successive reports displayed an increasingly sophisticated vision of "the uranium market." At first, the phrase itself barely appeared, serving mostly as a stand-in for (low) demand. By 1970, though, geologists had become more proficient in market-talk. They had learned that concepts like "market price" and "market stability" held sway over buyers and sellers, and they began framing their arguments in such terms:

Consumers could improve the prospect of market stability either by advance purchases or by negotiating long-term supply contracts; this would encourage suppliers to proceed with plans to prepare new production capacity and to search for the additional reserves to sustain future needs, and thereby to avoid the mismatches of supply and demand which, in the long term, seldom benefit either producers or consumers.[23]

Continuing geological exploration thus became a matter of market efficiency, which raised the question of how to measure whether the market was functioning correctly. As the OECD gained traction, it began producing demand forecasts along with reserve estimates. These were intended to stimulate progress in prospecting techniques, encouraging exploration through cost effectiveness. This, in turn, would produce better information and better prices.[24]

After all, lack of demand for uranium—due to stretch-outs and slow reactor sales—translated into lack of demand for data. Absent plausible demand figures, the first two Red Books struggled to make cases for increased exploration. The 1969 Red Book managed to quantify demand as well as production. It projected nuclear electricity production through

1980, giving high and low estimates broken down by reactor type, thereby translating gigawatts into a forecast of uranium demand. The accompanying graphs, trending steeply upward, radiated optimism. "For the first time in its history," the 1969 Red Book proclaimed, "the uranium mining industry may look forward to a stable and promising commercial market, on which reasonably firm plans for exploration and production can be based."[25] Subsequent reports multiplied demand data and meshed them with figures on reserves and production. Generating ever more detailed, number-laden forecasts, successive Red Books (1970, 1973, 1975) intensified their arguments for stepping up exploration to meet *projected* demand. A serious shortfall was brewing, they insisted.

The Red Books also served as market devices by standardizing practices that were critical to fashioning commodities.[26] They made disparate geological data commensurable, translating information about the many types of ore-bearing rock formations into a single quantity: tons of uranium oxide. They were on shakier ground, however, when it came to translating those tons into prices.

The problem stemmed in part from the very notion of a uranium price. The 1973 Red Book, for example, expressed doubt about lumping uranium reserves into a single "less than $10/lb" category when most transactions appeared to take place under $6 per pound. By 1975, Red Book authors decided that categorizing resources by price was impossible. Geologists preferred to "maintain a certain stability in the resource categories employed" and to produce valid estimates regardless of "significant price changes" that reflected "the market situation" rather than "actual recovery costs." The authors therefore "decided to abandon the notion of 'price categories' in favour of 'cost categories.'"[27]

So while discussing "price trends" more than its predecessors, the 1975 Red Book did not integrate those numbers into its resource estimates. So where *did* the figures for those price trends originate? Who *could* claim to have a handle on uranium price data? Had a "market price" for uranium actually emerged? No one could say for sure. But by 1975 one entity was doing its best to claim ownership of the concept of market price: the Nuclear Exchange Corporation, a uranium brokerage firm headquartered in Menlo Park, California.

Enter the next market device: price.

A PRICE FOR URANIUM

The Nuclear Exchange Corporation was formed in 1968 by two veterans of the US uranium mining industry.[28] Known as Nuexco, the company aimed to broker uranium oxide sales and to serve as a clearinghouse for "information on market activity."[29] At first such information was scarce. Attributing this to the *absence* of a market,[30] Nuexco tried to persuade its members—American utilities and fuel-cycle companies that paid a fee to participate in Nuexco-brokered transactions—that it had the ability to manufacture market stability:

> Most of the representatives of the nuclear industry are looking to the Exchange to establish a stable, orderly market upon which nuclear fuels may be traded with confidence and at predictable prices. To date, a true market level has been difficult to establish because the purchasing side has hesitated to make concrete bids which can be placed against existing asking prices. It would appear that an open market might result in better values for both buyers and sellers than those obtained by complex, private negotiations.[31]

To achieve this ideal state of affairs, members should treat Nuexco as an obligatory passage point: "Consummate your nuclear fuel transactions through your Exchange to the maximum practical extent. The cost to you is very low, and by so doing you help establish a true competitive market."[32] Nuexco would then collect the data in "the strictest confidence" and use it to produce price information for all to see. The more transactions that went through Nuexco, the better its information and the more reliable its "exchange value."

This "exchange value" gradually became Nuexco's signature product, distributed to members by monthly newsletter. But in the absence of actual transactions, it was a puzzling product. What did the value reflect? Nuexco explained that the figure represented a projection of the price for a *spot* transaction: the purchase of a single lot of uranium for future delivery. The value did *not* reflect prices in long-term contracts that covered the vast majority of transactions.[33] Nevertheless, American mines and utilities seemed to find the value tangible. In mid 1970, Nuexco claimed that its "price information includes most of the transactions which are consummated domestically."[34]

Nuexco first used price quotes from suppliers to derive its exchange value. Given the paucity of actual transactions in the late 1960s, it had

little alternative. These quotes were unreliable, not the least because "suppliers often quote somewhat different prices in answer to different inquiries within a short span of time."[35] By late 1970, business had picked up,
and Nuexco began to use prices from actual transactions. Similar constraints still applied, though. The number didn't allow for "special terms
and conditions," nor did it reflect "very large . . . or long term contracts."[36]
Within those limitations, Nuexco insisted, its exchange value offered high
reliability. Newsletters noted which transactions were "consistent" with
the value, explained those that weren't, and adjusted the value almost
monthly.

In 1973, Nuexco modified the concept of the exchange value and
reified the notion of the market itself. "The market," it announced, had
experienced a "turnaround." The oil crisis had spurred renewed enthusiasm for nuclear energy. After "five uninterrupted years of a buyer's market
in uranium," at long last "a solid seller's market" had emerged. Because
offers to sell were now "practically non-existent," they could no longer
provide a basis for calculating the exchange value. Henceforth, this number
would "represent prices which are being bid"—that is, prices that sellers
could hope to get, rather than those that buyers could hope to pay.[37] For
the next few years, the value continued to rise.

Nuexco's members were all US firms. In October 1970, though, it
concluded an agreement with the German firm NUKEM to establish "a
Nuclear Exchange to be operated in Europe and Africa." The two companies would provide Nuexco subscribers with "current market information from European and African sources, opportunities to sell or lend
uranium and plutonium in the European market through an organized
marketplace, and opportunities to borrow or buy overseas if, as, and when
AEC regulations permit." In late 1972 Nuexco concluded a similar agreement with the Marubeni Corporation of Japan.[38] It would henceforth be
able to "tie the world nuclear fuel markets together through a group of
cooperating organizations which operate with the same general practices
and principles."[39]

Nuexco and its partners, claiming the right to speak for African uranium,
would thus do much more than *report* on market conditions. They would
become "a commodity market for nuclear raw materials."[40] Their global
reach, furthermore, seemed to leave little room for other actors to maneuver. Could members conclude that the Exchange wasn't merely "a" market

68   PART I

but "the" market? Did that, in turn, make the exchange value "the" market price?

A lot depended on one's point of view. Outside the US, the largest uranium producers had little reason to relinquish control over information about uranium prices, let alone over the prices themselves. To them a "free" market meant, above all, freedom from US dominance. And by the early 1970s they had concluded that the best way to achieve freedom was by making their own markets.

PROLIFERATING MARKET MAKERS

Brokerage firms such as Nuexco and its associates appealed to small-scale producers, including many US mines. Needing long-term contracts to stay afloat, large-scale producers seeking global purview preferred to negotiate directly with buyers. To do so effectively, they had to become—or generate—their own market-making entities. Rio Tinto Zinc did this in house, collating data from its mining affiliates, generating supply and demand forecasts, and re-interpreting information from the Red Books and Nuexco. Large Canadian producers acted similarly.

For their part, the South Africans and French merged commerce with matters of state. In both countries, new institutions aimed to create a national uranium oxide brand with a distinctive technopolitical identity that would attract those reluctant to buy from North American suppliers. (I will discuss those technopolitics in depth in subsequent chapters. Here I'm just laying the groundwork.)

When uranium production began in South Africa, in 1952, the state arrogated ownership over all uranium to itself. Partnering with the state to build uranium processing plants, the gold industry reaped handsome rewards but couldn't sell the stuff itself. In 1967, a revised Atomic Energy Act transferred rights to the industry, giving uranium the same status as other minerals. A consortium of mines, the Nuclear Fuels Corporation of South Africa (NUFCOR), was formed to coordinate the uranium output, operate the centralized plant that processed the South African brand of uranium oxide, and market the product overseas. Although contracts still required government approval, NUFCOR enjoyed great autonomy as the commercial face of South African uranium, generating forecasts, managing production, and negotiating with clients to become the nation's uranium market maker.

The director of the French CEA had also decided to treat uranium as a "banal civilian fuel" subject to commercial considerations.[41] This perspective led the CEA in 1976 to slice off its fuel-cycle activities into a separate company called the Compagnie Générale des Matières Nucléaires (Cogéma). In France, however, institutionalizing the banality of nuclear fuel was contentious. Many felt strongly that the CEA should retain control of the stuff that had reframed France's global "radiance" in nuclear terms.[42] Supporters had to proceed slowly. The first, cautious step in 1969 was to create Uranex, a semiprivate brokerage firm tasked with selling uranium oxide produced in the metropole and the former colonies.

In staking France's claim to "the market," Uranex exemplified the changing texture of France's international relations and the place of former colonies therein. The firm aimed to sell a staggering 24,000 tons of oxide between 1970 and 1974.[43] To achieve this goal, it had to negotiate long-term contracts on behalf of actual uranium mines in France, Gabon, and Niger, as well as potential mines in Francophone Africa (such as the unexploited Bakouma deposit in the Central African Republic). By constituting itself as an obligatory passage point for Gabonese (and, soon, Nigérien) uranium, Uranex ensured that Francophone Africa remained the private nuclear hunting ground for its former colonial master. Although published statistics differentiated between uranium from Africa and uranium from France, computers models of supply and demand churned it all as French data.

These market-making efforts notwithstanding, sales remained slow up to the 1970s. NUFCOR, Uranex, and RTZ signed a few small contracts, hardly what they'd hoped for. They agreed that whatever it was, no matter what it was called, the market was a mess.

In October 1970, frustrated executives from all three firms met to revive the idea of "orderly marketing."[44] Joined by representatives from Canadian and Australian firms, the initial discussions mirrored the UPCA's failure to produce a concrete plan. Then in the fall of 1971, the US AEC confirmed its embargo of foreign uranium throughout the 1970s. Worse, it planned to use thousands of tons of stockpiled uranium through a complex manipulation of its enrichment operations.[45] The net effect, for non-American producers, was that selling uranium would become harder than ever.

That was the last straw.

ORDER, COLLUSION, CARTEL

Banding together in 1971, Rio Tinto Zinc and representatives of mining interests in Canada, Australia, South Africa, and France decided to take "the market" into their own hands. Discreetly headquartered in the CEA's Paris offices, their cartel was supervised by André Petit of the CEA's international relations division. Beginning in 1972, an annual quota system restricted each member to a total sales tonnage. The cartel set a minimum bidding price that was expected to increase in step with demand and was renegotiated during periodic meetings in Paris, Johannesburg, or London. The scheme, expected to run through December 1980, received approval from the South African, French, Canadian, and Australian governments.

Participants saw their cartel as a countermeasure to US protectionism, a market free from American dominance. As one French report remarked in late 1973,

The uranium market is rigged by American legal measures that forbid importing uranium for domestic consumption. This embargo in no way stops the export of American uranium, notably in the case of American-made, "turnkey" reactors. Because American demand represents half of all uranium demand, the truly free market is reduced by almost half, while still remaining open to American surplus. During periods of over-production this aggravates the imbalance between supply and demand outside of the USA.[46]

Their convictions of moral rectitude notwithstanding, cartel members remained anxious to keep their operational details secret. They called themselves a "club" rather than a "cartel," mindful of American attitudes toward OPEC and perhaps recalling the anti-trust case that the US had filed against DeBeers after World War II. Even before the OPEC-triggered oil crisis, Louis Mazel, RTZ's lead representative, cautioned a colleague in Melbourne not to use the word "cartel":

In your letter you mention a word which we would not even like to mention as some members of the club are rather worried about informal price agreements. I would like to stress very strongly that under all circumstances there can only be an unofficial agreement and whatever agreement is struck it should be on a strictly confidential basis. For the outside world all Paris and subsequent meetings will be in connection with the exchange of marketing information.[47]

Mazel often exhorted club members to refer to themselves as merely an "informal association."[48] If details about the club's quota and bidding system came to light, companies with American affiliates might be vulnerable to US anti-trust legislation. The fact of the club's existence did filter out into the trade press (including Nuexco's monthly newsletters), which periodically reported that a meeting had occurred, a floor price been set.[49] The details and extent of the arrangements, however, remained secret for good reason.

The heart of the cartel-as-market-device was its system of rigged offers. Each prospective client was assigned a lead bidder that offered yellowcake at the minimum bidding price and a runner-up that came in a bit higher. Middlemen and Asian (especially Japanese) utilities received higher quotes than others. Contracts for deliveries before 1977 had a fixed price that escalated predictably. In setting quotas, the club drew on the Red Books as the only nonproprietary prediction for supply and demand.[50] Club members also colluded on the *force majeure* clause in contracts: buyers suffered much higher penalties than sellers for reneging on contracts.

The system masked collusion and put steady upward pressure on prices. Long-term optimism was built in. For deliveries from 1978 onward, contract offers could be pegged to a "world market price" negotiated between buyers and sellers.[51] Producers anticipated that their system would lead to the emergence of a "free" market. This would go hand in hand with the emergence of mechanisms for price discovery (as clients saw it) or determination (as cartel members handled it).

Although cartel members worked to turn uranium oxide into a commodity with a fixed price and a controlled market, they were also competitors who might undercut floor prices or exceed quotas. With internal policing going only so far, the benefits of nuclear exceptionalism helped them to keep an eye on each other. Co-opting legislation that subjected uranium to security oversight, the club convinced the governments of France, South Africa, Canada, and Australia to reject contracts that failed to meet their price and quota criteria.[52] Even better, their price fixing remained secret as "Acts of State," making them immune to anti-trust laws in Australia and Canada. Neither France nor South Africa, with their profusion of parastatal companies, felt a need to police market morality. Club members with American affiliates hoped this immunity would extend to the US as well.

In 1973 orders began picking up: whether because of "orderly market-ing" or global events, the club members didn't know.[53] In Australia, a new Labor government suspended mine development pending the resolution of conflicts over Aboriginal land rights in the uranium-rich Northern Territory. In the US, enrichment plants began requiring utilities to commit to services for periods up to 10 years.[54] October brought a new Arab-Israeli war and OPEC's dramatic price hike, giving large-scale nuclear power development new urgency in Europe and Japan.[55] The cumulative effect of these events made Red Book warnings about a looming uranium shortage increasingly credible.

With Australian sales on hold, other club producers rapidly filled their quotas. By the end of 1973, NUFCOR had fielded so many inquiries that it pressed South African mines to increase their capacity.[56] Uranex managed to sell a whopping 13,000 tons from its inventory. After the French government announced plans in 1974 to build reactors on a massive scale, Uranex stopped signing foreign contracts altogether.[57] Club allegiance began to fray. By 1974, when the US AEC announced a gradual lifting of its ban on foreign uranium, only RTZ and the Canadian companies were presenting new offers for club approval.[58] During that year, American utilities contracted for more than 25,000 tons of foreign uranium.[59] By the end of the year, club members agreed with Nuexco: Sellers called the shots.

During the years when the cartel was operating, members miraculously managed to avoid accusations of monopolistic behavior while shutting out middlemen—most notably Westinghouse. In its eagerness to corner the reactor market in the 1970s, the giant US manufacturer had guaranteed electric utilities a fuel supply at $8 to $10 a pound. Speculating that prices would decrease, Westinghouse hadn't bought the uranium oxide yet. Cartel producers had been furious at this maneuver, which reduced direct demand and left mines to carry the cost of storing yellowcake during lean times. Cartel actions specifically targeted Westinghouse with bids that escalated prices for both short- and long-term contracts. Once it became clear that it couldn't make good on its promises, Westinghouse defaulted on fuel contracts for 49 reactors in September 1975.[60]

By October it was clear that rigged bidding had become superfluous. The club ceased operations.[61] The market no longer had to be "orderly" to exist. It could be constituted—and driven—by the competitive practices the club dreamed about. But members didn't want to disband altogether.

To maintain commonalities of purpose, they established the Uranium Institute—a name that conveyed academic distance. The new organization proclaimed itself to be a knowledge-producing institution dedicated to researching markets, publishing information, and hosting conferences open to all.

The Uranium Institute's masthead revealed the meridians of uranium's technopolitical geography. The UI's members were individual firms, but the masthead categorized them by nation, conveying the diplomatic impartiality of an international organization. South African firms were members in their own right, but other Africa-based companies were represented by their European parent corporations. The few buyers (admitted to avoid accusations of cartelization) were vastly outnumbered by producers. The UI's yearly meetings became forums for discussing the market's existence, boundaries, players, prices, and agency. As the UI gathered steam, members had grounds for satisfaction. Nuexco's price—happily used as an indicator even though it still bore little relation to contract prices—was $26 per pound in August 1975, and rose to a record $35 in December.

MARKET MORALITY

Any jubilation over Westinghouse's default evaporated in August 1976 when Friends of the Earth Australia surreptitiously obtained documents from the Melbourne offices of an RTZ subsidiary. Seeking ammunition for the political battle over Australian uranium, FOE activists realized that their cache also testified to a secret history of price fixing. The evidence rapidly made its way to the US Department of Justice and to the US Congress. Deeply embroiled in litigation over its fuel default, Westinghouse cried foul, charged the (now defunct) cartel with price fixing, and filed anti-trust suits against cartel members and their American affiliates.[62] The affair turned into the largest legal case in the history of the nuclear industry.

In November 1976, the US House of Representatives held hearings on "allegations that uranium prices and markets have been influenced by a foreign producer's cartel."[63] The testimony focused on prices. What was the connection between US prices and foreign prices? Had the sudden price rise been caused by cartel-like collusion? Was such collusion prosecutable under anti-trust legislation? Was there an ethical difference between price ceilings imposed by US government policies and floor prices imposed by the foreign producers, or between US protectionism and foreign price fixing?

Witnesses and congressional representatives disagreed about the answers. Regardless of perspective, however, most testimonies implicitly accepted Nuexco's exchange value as the true price of uranium, whether in the guise of "the American price," "the domestic price," or "the current price of yellowcake." The exchange value became the barometer for gauging the "fairness" of foreign prices; the latter's dramatic rise constituted evidence of suspicious activities. Crucial to Westinghouse's complaint was a contrast between a "world market price" *fixed* by producers and a Nuexco price that *reflected* "market conditions."

Testifying first, George White, president of Nuexco, argued (surprisingly) that the cartel had affected prices little. Still, nothing could have satisfied White more than the assumption that his was the one true price. Nuexco had first reported on the emergence of "world market price" contracts in 1974, asserting that the methods for determining the world price included "use of Nuexco prices." It continued to define a *spot* price, suggesting that the *short*-term figure had *long*-term value. It also intimated that price itself had agency. Rising prices had begun "performing their classic functions of bringing supply more into line with demand, eliminating year to year anomalies, increasing fluidity, and broadening the market."[64] This, White told Congress, was simply the "normal economics of price" at work.[65]

The hearings did not resolve the matter of cause and effect. Most analysts have since concluded that price increases in the mid 1970s were overdetermined, the result of Australian politics, US enrichment policies, the OPEC oil price shock, general inflation, and (to a limited extent) the cartel's actions.[66] For our purposes, the most striking point is that debates over the cartel—and especially over its effect on prices—*assumed* the existence of "the uranium market." Singling out bad market behavior implied a definition of good market behavior: open competition unregulated by governments and unaffected by agreements among sellers, a "happy" state of affairs in which "supply and demand [would] reach equilibrium."[67]

This rendition separated politics from economics, saddling the former with blame for price instability. Such narratives, of course, were themselves political claims. Congressional representatives liked to gesture toward free and fair markets. Economists preferred idealized models to messy realities and liked to analyze data in terms of conformity to model predictions. Corporations sought to conduct trade on their own terms. Any sense of irony was lost in the eagerness to assert the inherent morality of "the market."

Market morality required such assiduous affirmation in part because despite all the uproar "the uranium market" remained a shadowy entity. Beyond the halls of Congress and courts of law, buyers and sellers of uranium in the capitalist world continued to wrestle with the terms—and the fact—of commodification. One French commentator put it succinctly: Was uranium "like cattle, grain, and soy, a banal commodity that can be the object of speculation and exchange?" Buyers and sellers, he observed, remained "skeptical. They point to the uncertain character of the uranium market, whose mechanisms remain confused even for traditional transactions like spot or long-term contracts. How can one . . . envisage a futures market when there isn't even a real market for cash transactions?"[68]

According to a British commodities analyst, the market in nuclear fuels was "inherently less stable than most other commodities markets." This was due to the problems that arose from "confrontation between a technically determined monopoly and a technically determined monopsony. However many different producers there may be, their product must be sold to the electrical generation stations apart from a small military market. . . . There are no alternative uses and no substitute fuels which offer a stabilizing escape hold when industry supply gets badly out of balance with demand as happens in other markets."[69] The standard theories and mechanisms of market economics had difficulty accounting for uranium.

For buyers and sellers alike, the biggest roadblock to commodification was nuclear exceptionalism, which made it difficult to separate the politics of uranium from the economics. Investors familiar with other markets saw this as "a new factor in commodity investments." The lament was common, if untrue—one need only think of commodity chains structured by colonial states. But the orthodox view was that uranium's "overwhelming strategic importance" encouraged endless government intervention. One commodities analyst rehearsed the litany:

Governments anxious to secure supplies subsidize or sponsor exploration, mining, and fuel manufacture. Governments in fuel rich countries regulate the form and tonnage of exports. Government research programs in many countries determine the form of reactors to be used. Government owned and regulated power utilities determine the reactor building program. Government environmental and safety regulations determine the sighting [sic] and indirectly, perhaps, the total number of power stations permitted.[70]

This analyst was furious at "government revision of reactor orders" and the recent "prohibition of export sales . . . after mine development has begun" in Australia, which arose from Aboriginal land debates, and in Canada, where the reactors that made the fuel used in India's "peaceful" nuclear explosion were built. Claiming that these and other "sudden changes in regulation"—especially more stringent occupational health standards—had "clear and quantifiable consequences" on the industry's ability to plan ahead, the analyst insisted that industry had the right to make governments pay for changes they imposed.[71]

The problem of price determination remained central for Nuexco and for its rivals, most notably the Nuclear Assurance Corporation, a "nuclear services" firm in Atlanta that created the optimistically named World Nuclear Fuel Market in 1974 to compete with the Exchange. As an alternative to Nuexco's exchange value, the WNFM offered up a computer model it called the "Uranium Price Information System." Presenting a prototype at a 1976 conference, a system designer tried to explain the difference:

I'm not really sure and I'm not sure anyone here really is sure how NUEXCO arrives at their price . . . if they seek data directly and this is what they publish, it's one man's opinion—NUEXCO's. What we are proposing is different. The amount of information that this system would have would be collected from a large number of participants so that the number of data points that you would have in any given month wouldn't be one or two or three; they would be forty or fifty. . . .[72]

Even sympathetic audience members remained skeptical, though. "You will have big difficulties to define a world market price from this different data. . . . There exists no real market and you have no sort of standardized price."[73] Designers insisted that they didn't intend to define "the" world market price, but merely to provide price information that utilities could draw upon in their contract negotiations with producers: "We don't feel we have the divine guidance of NUEXCO to determine one price for

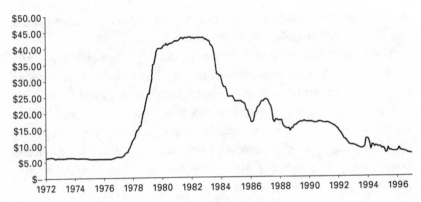

Nuexco spot market price in nominal dollars per pound of uranium oxide (U_3O_8). (data provided by TradeTech; graph prepared by Dan Hirschman)

the world. We don't walk on water. But perhaps what we can do is provide the kind of information that knowledgeable and successful fuel managers can apply."[74]

The WNFM's information service enrolled plenty of customers but didn't supersede Nuexco. Its complex algorithms could not dislodge the seductive simplicity of Nuexco's single number. The "exchange value" became *the* "spot price," giving it a privileged place in fantasies about profit, in reports on "the market," and in attempts to attract large investors. When it plummeted, producers worried. When it rose, they celebrated.

And, as we'll see in upcoming chapters, the performative power of the spot price even extended to the largest producers: those with their own marketing divisions, those who did not use Nuexco's brokerage services, those who operated across the African continent.

CONCLUSION

As an object of knowledge and a rubric for action, "the uranium market" was far from stabilized in the mid 1970s. But it was much more real than it had been in the mid 1960s. The instruments, organizations, and practices that had produced the uranium market, and now constituted it, included the OECD and its Red Books; Nuexco and its "spot price"; other brokers and their alternative price models; the cartel, the world market price contracts it had instituted, and the Uranium Institute that succeeded it; and the forecasts, price models, reserve calculations, and other market-making instruments that propelled those institutions.

The proliferation of market devices in the late 1960s and early 1970s made it possible to conceptualize uranium as an ordinary commodity. Rather than occurring in a single stroke, banalization gained momentum and power by being distributed in multiple devices. Producers, brokers, geologists, plant operators, and utilities sought to make uranium ordinary by identifying "purely" economic properties categorized by extraction costs, price, and so on, while developing "market" mechanisms to describe and circumscribe those properties.

Banalization was the key to developing commercial nuclear power and dissociating uranium from its closely guarded life as fuel for weapons. Putting some nuclear things "in the market" made them less exceptional, legitimating profits and distancing them to some extent from government purview. Market devices smoothed trade by making uranium uniform, leveling geographical lumps, producing predictable profit, and stripping the material of nuclearity. Through their content, functionality, and claims made on their behalf, market devices implicitly served to distance (certain kinds of) politics from the uranium trade.

Commodifying uranium as an ordinary thing required ongoing effort, however. Precisely because banalization was distributed and not carried by a single mechanism, it required many devices that worked together. Still, uranium often threatened to exceed the boundaries of banality. Nuclearity, as it turned out, was not the only threat to commodification. Nor did it necessarily work alone. There was also the revenge of empire.

Commodities, after all, come from places. Provenance matters in their circulation—not merely for reasons of economics, geography, or transport but also for reasons of politics. In the era of decolonization, commodities and goods from egregious imperial holdouts could produce powerful moral outrage—enough to ensure that uranium from southern Africa would not be the same commodity as uranium from elsewhere.

In 1951, simmering discontent over British encroachment on Iranian sovereignty came to a boil. The Anglo-Iranian Oil Company refused to hand over half of its profits to the state, as Aramco had done in Saudi Arabia. After Prime Minister Mohammad Mossadegh and his parliament nationalized the company, the British did their best to block sales of Iranian oil. As internal tensions rose between Mossadegh and his conservative opponents (including the young Shah), Britain appealed to the US for help. The Americans were happy to oblige, ousting Mossadegh in a 1953 coup backed by the CIA.

The episode laid bare the mechanisms used by the bastions of industrial capitalism to control access to raw materials. In the 1940s, Argentinean economist Raúl Prebisch argued that the structure of the world economy was to blame for the collapse of commodity prices that had devastated Latin American exporters during the Great Depression. Colonialism had divided the world into an industrialized "center" and a raw-materials-producing "periphery." Powerful multinational corporations controlled raw materials in poor countries, colluded through private cartels, and pocketed the profits. To break the cycle of dependence, poor countries needed to manage exports using tariffs or nationalization, create industries that enhanced the value of raw materials, and manufacture local goods to replace high-priced imports. Prebisch also advocated for public commodity cartels in which poor countries banded together to drive up export prices.

From Havana to Cairo, Prebisch's ideas resonated with national leaders seeking a global political alternative to the bipolar Cold War order. The final communiqué of the 1955 Afro-Asian conference in Bandung invoked the "vital need of stabilizing commodity trade" and emphasized "respect for the sovereignty and territorial integrity of all nations."[1] The "Bandung spirit" carried forward into the creation of the Non-Aligned Movement in 1961 and the UN Conference on Trade and Development in 1964.[2] Embracing "Third World" identity as a source of solidarity and leverage, "Group of 77" nations used the UN to articulate and legitimate their demands for (techno)political sovereignty in a global context.

Governance of natural resources occupied a central place in Third World demands. What did the principles of self-determination mean when it came to "nature" rather than people? Should newly independent states obtain *tabula rasa* and gain full control over resources extracted by former colonial powers? Or should treaties signed during colonial rule remain binding even after the end of empire?[3] Such questions reflected competing claims to global moral order, pitting the capitalist ethic of contractual honor against the anticolonial ethic of self-determination.

In the 1970s a wave of nationalizations by newly independent states, coupled with OPEC's growing muscle, fueled capitalist anxieties about privileged access to raw materials. For their part, Third World countries worried about balancing foreign investment incentives with curbs on corporate power. In April 1974, a month-long session of the UN General Assembly debated problems of developing and governing raw materials, adopting the New International Economic Order (NIEO).[4] Among other provisions, the declaration asserted "full permanent sovereignty of every State over its natural resources and all economic activities . . . including the right to nationalization or transfer of ownership to its nationals, this right being an expression of the full sovereignty of the State. No State may be subjected to economic, political, or any other type of coercion to prevent the free and full exercise of this inalienable right."[5]

The UN's Group of 77 heralded this as a transformative moment, proof that "the Third World is no fiction. It is a contemporary reality. It is a force, a responsible force." But the US and most West European countries denied that the resolution reflected consensus: "To label some of these highly controversial conclusions as 'agreed' is not only idle; it is self-deceiving."[6] In their eyes, a majority vote did not give the NIEO meaning in the absence of details about, for example, compensation for national expropriations. The bickering continued with resolutions and declarations, committees and conferences, charters and action programs.

Apartheid also occupied the UN's agenda. Pressure to sanction South Africa mounted after the March 1960 Sharpeville massacre, in which police fired into a crowd of peaceful protestors. In April, a Security Council resolution called on South Africa to abandon apartheid. As sixteen new African countries entered the UN, calls for stronger action grew. In 1962 the General Assembly passed a non-binding resolution requesting member states to break off diplomatic and trade relations with South Africa. While many African, Asian, and Communist bloc nations imposed trade boycotts, most Western countries did not comply. The newly independent African nations continued to flex their diplomatic muscles, though, leading European and American policy makers to support more concrete action. In 1963 the Security Council approved a voluntary embargo on sales of arms and ammunition to South Africa.

A defiant South African government reacted by expanding the technopolitical edifice of apartheid, articulating a drive for a nationalist modernism that appealed to progress while maintaining a distinctive South African identity.[7] Under the leadership of Hendrik Verwoerd, who served as the minister for native affairs before becoming prime minister in 1958, "grand apartheid" proclaimed a policy of "separate but equal" development for blacks and whites. "Equality" meant gerrymandered "homelands" (the so-called Bantustans) that relegated black Africans to pseudo-sovereign ethnic enclaves, reserving the majority of the country's territory, resources, and infrastructure for whites. It resulted in surveillance technologies that tracked black Africans as they commuted to work in white areas. It fostered architecture that segregated races at intimate levels, designating separate bathrooms, entrances, and living quarters for whites and blacks.

Apartheid elites saw technology as an apolitical strategy that turned racial segregation into a benign method for managing development. In a similar vein, state and business leaders insisted that politics had no place in "the market." It was inappropriate to let ongoing debates—especially on matters of national sovereignty—affect politically neutral business decisions. Western investors and customers were "just doing business." As tools for depoliticization, uranium's market devices fit this logic perfectly.

Meanwhile, ongoing colonial rule in South West Africa fueled international pressure for broad sanctions against the apartheid regime. In 1966, the UN General Assembly ended South Africa's 1919 League of Nations mandate to govern the territory. In 1967, it created the UN Council for Namibia (UNCN) to administer the territory. In 1968, South West Africa became officially known as Namibia. The UN Security Council endorsed these actions in 1969, calling for the immediate withdrawal of South African occupation, a demobilization that wouldn't get started until 1988.

As it became clear that Pretoria wouldn't budge on the initial UN demands, the politics of the liberation struggle became enmeshed with the politics of natural resource sovereignty. Taking a cue from the NIEO, the United Nations Council for Namibia issued its first decree in September 1974, prohibiting the extraction and distribution of any natural resource without explicit UNCN permission. The decree called for the seizure of all illegally exported material and warned that violators could be held liable for damages. Decree No. 1, as it became known, placed the Rössing uranium mine squarely in the sights of the Namibian liberation struggle.

The South West Africa People's Organization, the nationalist group that came to dominate the Namibian liberation struggle, took up Rössing as an emblem of colonialism. The UN declaration that SWAPO acted as the "sole authentic representative" of the Namibian people served the organization well in its internal struggle for leadership and its quest for European and American anti-apartheid allies. Targeting Rössing helped SWAPO consolidate its authority among international activists, not the least because uranium helped forge political alliances within the anti-nuclear movement.

Invoking Decree No. 1, SWAPO and its allies declared the mine illegal.[8] Just as Rio Tinto Zinc strove to make uranium a banal commodity, therefore, the Namibian liberation struggle worked to make this *particular* uranium exceptional by highlighting its problematic provenance.

Despite the best efforts of its promoters, uranium from southern Africa refused to stay ordinary. As a point of collision between moral econo-mies—of global nuclearity, of Cold War capitalism, of anti-colonial strug-gle—it was simply too prominent. When apartheid South Africa became an object of international opprobrium, the politics of place complicated commodification efforts. In the era of imperial collapse, the unabashed colonialism of the South African state grew increasingly anomalous. South African politics of racial exceptionalism became entangled with the global politics of nuclear exceptionalism. Managing these entanglements was a major task for the market devices that served the sale of uranium from southern Africa.

The cooperation between the state and the mining industry required to produce uranium in South Africa frequently devolved into conflict. Many of the deep shafts dug in the Witwatersrand ridge for the purpose of mining gold also contained uranium, which could be refined from gold tailings. The mining companies owned the gold, but the state retained title to the uranium and held final approval over sales. Though earning profit, the companies chafed under the heavy hand of the state's Atomic Energy Board. In 1967, ownership of uranium ore shifted to private indus-try and allowed companies to market their uranium through the newly created Nuclear Fuels Corporation (NUFCOR). Final approval of sales contracts remained with AEB officials, however, and their views of South Africa's uranium industry differed significantly from those of mine executives.

on facing page: A portion of the Valindaba enrichment plant. (Atomic Energy Corporation of South Africa)

The conflicts often played out as tensions between banality and exceptionalism. The mining industry wanted to treat uranium as an ordinary commodity and to remove the exceptional secrecy imposed by the state, allowing it to control the conditions of sale and to base decisions about refining capacity on profit potential. For AEB officials, uranium held potent value in domestic, regional, and international politics. Turning uranium into an ordinary commodity could bolster South Africa's increasingly precarious position in international trade networks. In parrying the threat of sanctions, apartheid leaders frequently invoked "the market" as a neutral space in which politics had no place, thus deploying uranium's market devices as powerful political tools.

But technopolitical ambitions ran even deeper. Why stop with mining yellowcake, a banal raw material produced by other African nations? South Africa should develop its own enrichment process to sell fuel-ready uranium abroad and supply its own reactors. This would put the apartheid nation in exclusive company with the few countries that had the engineering know-how and industrial capacity to enrich uranium. It would consolidate the technopolitical power of South Africa's nuclear experts at home, and would make them a force to be reckoned with abroad. Enrichment was also the key to nuclear weapons development. South Africa's secret atomic bomb program contributed mightily to the AEB's desire to retain a firm grip on uranium production, keeping AEB experts heavily invested in upholding uranium's nuclear exceptionalism, the source of their power within the state.

In the last section of this chapter, we'll see that the politics of place also disrupted British banalization of Namibian uranium. While Rio Tinto Zinc operated the Rössing mine in Namibia, it maintained a revolving door between its boardroom and the upper echelons of British ministries. RTZ was therefore caught off guard when a Foreign Office bureaucrat raised a ruckus about several thousand tons of Rössing uranium destined for British atomic weapons. Uranium also became repoliticized by the specter of an apartheid-fueled atomic bomb. Countering anti-colonial moral outrage, Rössing's defenders invoked the ethics of capitalism and insisted that uranium was an ordinary commodity. Opponents of the mining, who had little hope of stopping the contracts, grafted the politics of place onto the politics of nuclear exceptionalism, insisting that Namibian uranium was not a banal commodity.

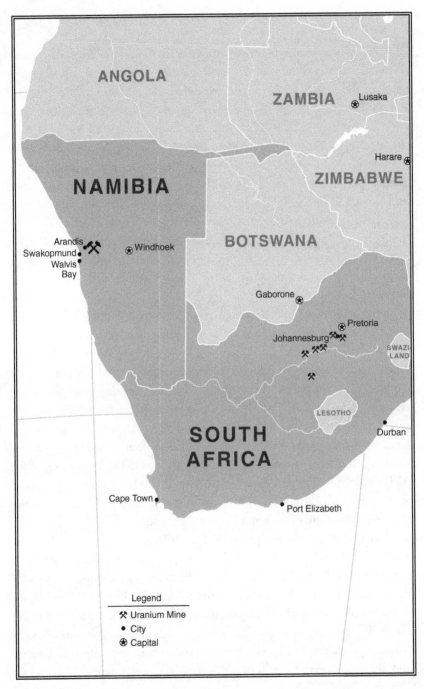

South Africa and Namibia. (map prepared by Evan Hensleigh)

PROCESSING PROFIT

In South Africa, uranium's identity suffered from the vicissitudes of domestic struggles over institutional power and the apartheid state's changing place in the world. At first, state ownership of uranium gave the Atomic Energy Board the "right to acquire or dispose" of the stuff and conclude sales contracts on behalf of mining companies, which all sent ore to a single plant for processing. Galled by the AEB's secrecy, the mining industry argued that a robust sales effort necessitated treating uranium like "a normal export commodity" and publishing production statistics and other information.[9] For the AEB, however, uranium helped to constitute national nuclearity. Its secrecy provisions were key to its status as an exceptional thing and to the AEB's power within the state.[10] These institutional conflicts were entangled with broader political tensions between Afrikaner elites who dominated the AEB and English-speaking elites who headed the mining industry.

Faced with the AEB's profound reluctance to relinquish control over uranium exports, the mining industry argued that separating state and market was critical for international sales. To act as a commodity in a capitalist economic system, South African uranium needed a brand identity separate from the state.[11] In 1967, this line of thinking resulted in the creation of NUFCOR, which took over the state's uranium oxide plant and sold its product on behalf of the mining companies. Contracts still required formal government approval, however, leaving the AEB—as the state's atomic agent—with final oversight over uranium transactions. The AEB's powerful chairman, Ampie Roux, soon made clear that he wanted to extend that oversight to RTZ's uranium ventures at Rössing and the Palabora copper-uranium mine in South Africa.

Squaring off against the powerful British-based company RTZ, Roux worried that "unfettered control by a foreign company of so large and significant a deposit might prove to be detrimental to the interests of South West Africa and the Republic of South Africa."[12] The AEB insisted on involving the Industrial Development Corporation (IDC), a South African parastatal company, in Rössing's financing and governing structure. Nevertheless, as international pressure to end the Namibian occupation intensified, the risk emerged that Rössing's precious product might escape South African control. Tightening South African oversight, Roux instructed RTZ and NUFCOR to work out a "single channel marketing" arrangement whereby the South African company would negotiate contracts on behalf of the British corporation.

Corporate executives dutifully met with Roux to discuss the matter in August 1969. But neither RTZ nor NUFCOR liked the idea. RTZ planned to raise funds from "international financiers" who required as collateral "bankable long term contracts" that NUFCOR couldn't provide. "The bankers will need confirmation that the Government of the Republic of South Africa will permit Rössing to operate in the most efficient manner, technically and commercially, to enable it to achieve the maximum benefit for its shareholders and for South Africa"—which, through the IDC, was also a shareholder. Rössing's success depended on economies of scale. Its connections to RTZ's global network enabled it to secure contracts that NUFCOR couldn't. Besides, NUFCOR already had a full plate.

Executives from both companies soothed Roux with assurances that "on the most authoritative predictions Free World supply and demand should be approximately in equilibrium by 1975/1976." Rössing, unable to begin production before 1975, would not compete with NUFCOR for at least five years. By that time, there would be plenty of demand to go around. In the meantime, to ensure "that an orderly market consistent with South Africa's best interests is maintained," the two companies agreed to remain in close contact to "promote the growth of a strong industry."[13] Even without marketing all the uranium under its territorial control through a single channel, South Africa would remain a world leader. The AEB accepted this argument. The following year, a similar argument persuaded the South African government to play a prominent role in the cartel.

The AEB would later reprise efforts to control Rössing sales. Meanwhile, it focused on the later stages of nuclear fuel production. Because South Africa sold ore as uranium oxide, customers (or NUFCOR on their behalf) had to arrange for the conversion of oxide first to uranium tetrafluoride (UF_4) then separately to uranium hexafluoride (UF_6, or "hex"). The hex served as feedstock for the enrichment process, which took place in a third plant. British and American facilities typically handled these stages. But US policies regarding foreign ore were fickle, as we saw in chapter 2. And the British Labor government couldn't be trusted to refrain from embargoing South African material. Perhaps NUFCOR could switch to a conversion plant run by the French, who had proved friendly to South Africa in the nuclear domain?

Even better, NUFCOR could build a hexafluoride plant on its home turf. After all, only colonized, backward nations produced just raw materials, as apartheid legislators liked to proclaim. South Africa had sped ahead of the rest of the continent thanks largely to its highly developed metallurgical industry.[14] Manufacturing enriched uranium was practically a patriotic duty. The esteemed AEB metallurgists affirmed that selling a more processed product would yield greater profits.

NUFCOR executives, however, knew that profits did not flow inevitably from increased processing. Conversion plants entailed high capital costs. Profits depended on economies of scale, uranium price trends, customers, ore grade, efficient removal of ore impurities, and much more. On an exploratory visit to the US in 1970, NUFCOR's chairman told an American colleague that "he personally felt his government was foolish in wanting to build such facilities, since such services could be bought cheaper elsewhere."[15] Before committing to a UF_6 plant, NUFCOR commissioned a series of feasibility studies and market forecasts.[16]

AEB officials squirmed.

"SOUTH AFRICAN CONDITIONS"

The Atomic Energy Board had hoped to lure industry with the language of profit, but grander things were at stake. AEB scientists had begun research into uranium enrichment in the early 1960s. The secret project had the blessing of the grand architect of apartheid, Prime Minister Verwoerd, and was funded directly by the state.[17] If the project succeeded, it would need a South African hex plant to produce the feed. Otherwise the AEB would have to ship NUFCOR ore abroad for conversion before bringing it back home for enrichment.

One motivation for homegrown enrichment was to reduce dependence on Americans who warned that they wouldn't resupply enriched uranium for the AEB's SAFARI research reactor until South African uranium sales followed "a safeguards policy which is acceptable to the USA." South Africans resented the US holding their research hostage by demanding oversight of their uranium sales. Exploring options, the AEB approached the French CEA, who had been "positive beyond expectations" about supplying SAFARI with enriched fuel.[18] In the long run, however, only a South African plant could guarantee enriched uranium supplies.[19]

In 1970 it became clear that technopolitical prestige provided another important motivation for pursuing enrichment. The AEB, feeling confident about its enrichment project after several years of research and development, peeled off a layer of secrecy. In a July speech to Parliament, Prime Minister B. J. Vorster—"speaking in English for the benefit of foreign visitors and pressmen"[20]—pronounced the project to be an "obvious" step in the (white, industrial) history of the nation.[21] Enrichment would allow South Africa to market uranium more profitably, and eventually supply its own nuclear power program. Vorster lauded South African scientists for bolstering "the prestige of their country. In the past they have made lasting contributions to science, but perhaps the achievement that I am announcing today is unequalled in the history of our country." He went on:

The South African process, which is unique in its concept, is presently developed to the stage where it is estimated that *under South African conditions*, a large scale plant can be competitive with existing plants in the West. . . . South Africa does not intend to withhold the considerable advantages inherent in this development from the world community. We are therefore prepared to collaborate in the exploitation of this process with any non-communist country(ies) desiring to do so, but subject to the conclusion of an agreement safeguarding our interests. However, I must emphasize that our sole objective in the further development and application of the process would be to promote the peaceful application of nuclear energy— only then can it be to our benefit and that of mankind.[22]

Emphasis on the "unique" character of the enrichment process quickly pervaded official discourse, affirming national(ist) technological prowess. The assertion that a large-scale plant would prove competitive "under South African conditions" encoded two things. First, the South African process closely resembled a German "jet-nozzle" enrichment technique that was unprofitable in Europe because of high energy costs. Hence the second bit of code: "South African conditions" referred to cheap energy from cheap coal mined by cheap black labor. Official descriptions were replete with references to these "conditions," discreetly depoliticizing them as unexceptional traits of industrial development rather than products of racial oppression.

The South African enrichment program derived legitimacy from "existing plants in the West" just as much as it competed with them. It was only a few years behind pilot plants in France and the Netherlands that heralded the end of America's monopoly on enrichment. And it was handily beating Australia, which threatened to become a major rival in the ore market but had only just begun to explore enrichment possibilities. These wide-ranging efforts to commercialize enrichment services provided an economic frame for South Africa's purely "peaceful" efforts.

Establishing the plausibility of peaceful intent was important because South African diplomats, citing concerns about commercial confidentiality, had not signed the newly ratified Nuclear Nonproliferation Treaty. They expressed apprehension that a new safeguards regime would apply to uranium mines and, as they told their British counterparts, affect "commercial security at the gold mines from which uranium is extracted as a by-product."[23] International inspectors might also compromise the confidentiality of South Africa's enrichment technology. Indeed, the word "unique" in Vorster's speech had caught international attention. British observers gathered enough credible intelligence to suggest that Vorster had not made idle claims.[24] As one official put it, "Britain's nuclear experts certainly do not pooh-pooh the idea that South Africa have come up with something new."[25]

Observers couldn't help wondering whether South African confidentiality camouflaged nefarious intentions. A 1972 report from the British Ministry of Defense mused that South Africa's "remarkably secretive" stance and the project's "tight security system . . . in itself might be regarded as an indicator of military motivation." But, the report speculated, the secrecy might also derive from "South African sensitivity about the loss of what to them is commercially valuable information," given that peaceful nuclear power ambitions demanded significant increases in enriched fuel. South Africa would be "well placed to take advantage of this trend if she can fully develop an economically competitive enrichment technique." Although the scale of work and funding seemed excessive for a pilot plant, raising "the immediate suspicion" that South Africa might be aiming at weapons-grade enrichment, the ministry could not discern the plant's true purpose without knowing further technical details. It concluded that Britain should attempt to learn more, if for no other reason than to assess "the threat to UK commercial interests which could be posed by the South African project in the longer term."[26] Apparently, commercial competition was as threatening as military potential.

Focusing on uniqueness and the problems it posed for inspections, the South Africans implied that they had something new to offer, something worth protecting, something tempting to investors. Invoking "South African conditions" helped make the case that a plant anywhere else wouldn't be as cost-effective. This reshaping of racism into technical data attempted to persuade the world that South Africa *as a place* was crucial to making the technology work—an important enticement, since the astronomical costs of scaling up a pilot plant to commercial status necessitated foreign investment. "If South Africa tried to go it alone," the enrichment project's director remarked on one occasion, "the demands on financial resources and supplies of high calibre man power and materials would be such as to seriously interfere with other important national developments."[27]

Emphasizing "South African conditions," AEB chairman Ampie Roux sought investors among European atomic agencies, manufacturing companies, and utilities. The initial enthusiasm of some in the British nuclear community was quashed by the Anglo-Dutch-German enrichment project already underway as well as by the broad political opposition to investing in South Africa. The German firm Steag and the French CEA, however, had no such qualms. Separately, each embarked on secret negotiations with South Africa's recently constituted Uranium Enrichment Corporation (UCOR). Steag contemplated an indefinite collaboration on a commercial-scale plant using the "unique" South African process. The CEA preferred a short-term collaboration on a pilot plant. While the chosen European partner would provide substantial know-how and equipment, South Africa would provide uranium at preferential prices. Where the hex feed for enrichment would come from remained an open question.[28]

Meanwhile, NUFCOR executives had begun to see the AEB's pressure to build a hex plant in a new light. Clearly the AEB had more in mind than simply selling uranium in hexafluoride form on the international market. Profuse prose about the "prestige" of enrichment made it glaringly obvious that AEB officials wanted hex at *any* cost.[29] Roux's lengthy speeches invoked physics superstars like Lise Meitner, Enrico Fermi, and Niels Bohr to situate South African enrichment in a direct lineage with the greatest hits of the nuclear age. Roux waxed eloquent about how the plant would "usher in a new era in the industrial and economic development and, what is more, in the prestige of our beloved country, South

Africa."[30] Enriching uranium would protect the country's global standing: "South Africa's position in international affairs and its prominent status as a foundational member of the International Atomic Energy Agency are very largely due to the fact that this country is one of the top three uranium producers in the world."[31] This standing absolutely had to be preserved because the IAEA "was perhaps the last international body where South Africa was permitted to make a contribution and where South Africa's viewpoint was given respectful attention."[32]

Using a logic in which several fantasies of depoliticization intersected with each other, Roux proposed enrichment as a palliative for South Africa's increasing isolation. First was the fantasy of the IAEA as a purely technical institution devoid of postcolonial politics. South Africa had long labored to reify this vision through its work on the agency's governing board.[33] Second was the fantasy of the market as an apolitical space of economic exchange where race politics had no place. This vision pervaded apartheid state discourse and offered South Africa's trading partners a rationale that legitimated their brisk business. In Roux's convoluted logic, technical and economic success made possible by depoliticization led ultimately to political prestige that would refract status and power onto the orchestrators, the AEB and UCOR.

NUFCOR executives remained skeptical about the hex plant. A scientist rather than a businessman, Roux based his seductive assertions about the global competitiveness of uranium enriched in South Africa on idealized projections. Unable even to predict the cost of making feed material, NUFCOR had little reason to believe the AEB's estimates of the cost of enrichment. Further undermining NUFCOR confidence, the AEB repeatedly requested research funding from the mining industry. Roux's warning that South Africa's refusal to sign the NPT might "harden" international attitudes and block access to foreign hex suggested that the whole endeavor would go forward regardless of cost. Adding this up, NUFCOR refused to take the plunge into a commercial hex plant. It would provide some capital for a pilot plant, but the AEB would have to erect and operate the plant itself.[34]

Unsurprisingly, relations between the AEB and the mining industry continued to deteriorate. Thickening its veil of secrecy, the AEB provided no estimates for the feed needed by the hex and enrichment plants, making it difficult for the industry to calculate the feasibility of producing more uranium oxide.[35] In retaliation, the industry reduced contributions to the AEB research budget and withheld information because of concerns about commercial confidentiality.[36] Roux grew livid, insisting "that in terms of the Atomic Energy Act . . . all information on uranium research should be disclosed to the Board."[37]

Eager for numbers that supported his case, Roux demanded to see the industry's market forecasts. NUFCOR chairman A. W. S. Schumann retorted that "the whole subject of uranium marketing was very much more tentative than the Board might think." AEB metallurgist R. E. Robinson chimed in: he needed *all* of NUFCOR's projections to establish an econometric model to predict uranium demand. Surely, Schumann replied testily, market projections went beyond the scope of a metallurgical laboratory? On the contrary, Robinson affirmed, such knowledge was "essential . . . in formulating [a] research programme."[38] NUFCOR relented and handed over projections for uranium demand through 1985. These included estimates of how much of the market South Africa could realistically expect to capture given the perpetually unmentioned problems posed by apartheid.[39]

Yet Roux's anxiety about the status of Namibia and the involvement of European partners continued to mount. As long as the Atomic Energy Act governed Namibian uranium, the AEB could pressure Rössing to write South African enrichment into its contracts and guarantee sufficient yellowcake feed to run a plant.[40] But if Namibian yellowcake went elsewhere, feedstock would have to come wholly from South African mines, which were unlikely to have sufficient capacity in the near future to justify an enrichment plant. The French had pulled out, and negotiations with Steag over its capital investment and cooperation were growing uncomfortable. The draft agreement with Steag specified that each party would provide its own feedstock for conversion and enrichment. Yet the German company made it clear that it needed to purchase its share of uranium oxide from South Africa or Namibia in order to save on transport costs— otherwise the deal was off. Roux fretted that "if they cannot purchase a share of the feedstock locally or if they are permitted to buy supplies without the requirement that these be enriched in South Africa, they may withdraw from the negotiations." Roux wanted yellowcake sales and uranium enrichment to become inseparable, so that anyone buying ore from southern Africa would have to enrich it there too.[41]

If Roux got his way, the AEB and UCOR would have even more power over industry operations. This prospect, however, made NUFCOR unhappy. Fed up with NUFCOR's recalcitrance, Roux complained to the Minister of Mines, garnering the attention of Schumann, who wrote Roux at length detailing the efforts the mining industry was making to increase oxide production.[42] Apparently the increase wasn't enough for Steag, which withdrew from the negotiations because it had no reason to collaborate on enrichment if South Africa couldn't also provide uranium. The German company insisted that its withdrawal was not related to the recent scandal in which ANC activists stole documents from the South African embassy in Bonn and leaked the secret collaboration to the press.

Whatever the case, Steag's withdrawal in 1975 effectively ended the commercial-scale enrichment project. Subsequent attempts to find funding failed, including an agreement whereby Iran would provide capital in return for enriched uranium, a prospect that perished in the 1978 Iranian revolution. Officially, the pilot plant at the AEB's Valindaba site continued as an experimental facility before joining with a second facility to produce enough fuel for the two nuclear reactors that South Africa had purchased from France. In practice, however, the two plants devoted themselves primarily to producing highly enriched uranium for apartheid's fledgling nuclear weapons program.[43]

RÖSSING AND THE POLITICS OF PLACE

Some 1,700 kilometers northwest of Valindaba, the Rössing uranium mine was experiencing its own political problems. Located in the Namib desert, 70 kilometers inland from the coastal town of Swakopmund, the deposit had been known to geologists for decades. Because of the extremely low grade of the ore, extracting uranium required mining and milling vast quantities of rock. This demanded a capital outlay that few entities—state or corporate—could even contemplate. In 1966, the British-based mining multinational RTZ took the plunge.[44]

To secure loans for mine development, RTZ had to prove that it could sell the uranium. Only substantial advance contracts would do the trick. RTZ turned first to the United Kingdom Atomic Energy Authority (UKAEA), which signed a long-term contract in 1968 for 6,000 tons of yellowcake.[45] Subsequent contracts with German utilities didn't push RTZ past the threshold of profitability, so UKAEA officials, eager to ensure the mine's viability, happily filled the gap with a second contract for 1,500 tons. Because RTZ was a British company, Rössing was as close as the UK would come to controlling its own uranium supply.

UKAEA officials were seduced not only by privileged access to large quantities of yellowcake, but also by RTZ's innovative pricing scheme. In any given year, either party could declare that the contract price was "out of line with current world prices for large long-term contracts" and renegotiate with the other for a "market price." If they failed to agree, "there is provision for arbitration." One official observed that "the practicability of meaningful arbitration has increased due to the comparatively large number of quotations for uranium now being made and the availability of consultants who give professional advice."[46] As a contract clause, the "world market price" derived credibility from its similarity to oil pricing. As a numerical value, it acquired reality by being subject to ordinary mechanisms for dispute resolution.

RTZ's market price clause was designed to navigate the volatilities engendered by nuclear exceptionalism. Unpredictable changes in foreign government policies particularly concerned UKAEA planners. For example, Canada had announced that *all* new uranium sales must go toward "peaceful purposes." These restrictions applied to the UK, despite the fact that Canada had previously supplied uranium for British nuclear weapons. UKAEA officials complained that Canada was "unreliable," its prices skewed by US policies. The South African government, by contrast, seemed happy to waive "end-use" restrictions for Britain as long as Namibia remained governed by South Africa and its atomic energy legislation.[47] All this made Rössing an attractive supplier.

As UN condemnation of South Africa intensified, the Rössing contracts came under fire from unexpected directions. In 1969–70, officials at the UK Foreign and Commonwealth Office looked closely at Britain's uranium suppliers. Combing through the paper trail, Barbara Rogers, newly appointed to the Southern Africa desk, noticed inconsistencies in claims about the provenance of RTZ's uranium. It appeared that the Foreign Secretary had approved the RTZ contracts thinking that the primary source would be the Rio Algom mine in Canada, with Rössing serving as backup if Canadian production faltered. But the reverse was true. RTZ—and probably the UKAEA as well—actually intended Rössing to serve as the prime supplier. Disturbed by this deception and its tacit endorsement of South African colonial occupation, Rogers began writing outraged memoranda that reverberated through Britain's Ministry of Technology (MinTech), the UKAEA, the prime minister's office, and the Foreign and Commonwealth Office.[48]

As tensions mounted, market devices were deployed to defend the contracts against anti-colonial outrage. "When we secured Ministerial authority to conclude this contract," one MinTech official readily admitted, "the paper to Ministers unfortunately did not refer specifically to South West Africa."[49] He downplayed this omission with imperatives of capitalist discipline: RTZ hadn't wanted to pinpoint the location of the mine for fear of wanton financial speculation.[50] "It is fair to say," the Minister of Technology himself subsequently wrote, "that in 1968 SWA . . . was not such a sensitive political issue as it became in 1969."[51] Because the contracts had been signed before the 1969 Security Council resolution demanding South African withdrawal, hefty cancellation fees would apply if the UKAEA reneged.[52] Such fees would hurt the new corporation (British Nuclear Fuels Limited) that had taken over fuel management from the UKAEA.

If it became public knowledge that Britain wasn't purchasing from southern Africa, "BNFL would lose its bargaining power and almost inevitably Canadian prices would increase in future contracts," undermining nuclear power development in Britain and impairing "BNFL's competitiveness with overseas rivals."[53] From a cost perspective, one of the *best* things about the contracts was the mine's location. With no trace of embarrassment, UKAEA officials wrote that "the new mine would be open cast involving a minimum of skilled labour and thus relatively immune from the trends of labour escalation"—in contrast to Canada, where labor costs were skyrocketing.[54] The irony of making such an argument within the corridors of a Labour Party government seemed entirely lost.

National defense interests, commercial considerations, and capitalist ethics thus militated in favor of honoring the contracts. The Cabinet agreed: BNFL should "proceed with the existing contract, and accept the political difficulties which might ensue." True, the timing was "doubly unfortunate" since it coincided with a moment when "both the Rhodesian issue and the activities of the ad hoc Committee on South West Africa were attracting attention in the United Nations."[55] If the contracts couldn't be kept secret, bad publicity could be minimized by managing the *timing* of the revelations. The UN had recently asked member states to report on their commercial, industrial, and financial activities in Namibia. Officials suggested delaying Britain's declaration until after the upcoming national elections, which already looked tenuous for Labour. All this meant close coordination with RTZ, but that was easy enough thanks to the close ties that company executives maintained with high-level state officials.

RTZ's chairman, Sir Val Duncan, even thought he could help temper adverse international reactions by portraying his company as a progressive force for black Africans. He sent Under-Secretary of State Sir Denis Greenhill the text of a presentation on how RTZ had tried "to break down the strict application of apartheid"[56] at its Palabora uranium/copper mine in South Africa:

Of course we have always been aware of the controversial nature of South African policy on apartheid and so forth, but we set out to do something to benefit all races, and I think it is no exaggeration to say that Palabora today has completely succeeded in elevating the standard of living of all races working in that complex. . . . [W]e are the only company employing 100 ton motor trucks in Southern Africa today. No white man has ever driven a 100 ton truck in South Africa, except as the instructor, because we went straight to black men and today we have a sort of "corps d'élite" of black men who operate this and other complex equipment, to our complete satisfaction. The living conditions of Africans at Palabora, who have their families with them, are very considerably in advance of the old idea of compound labor, and everyone of us is quite proud of the standards which we have set for all races. . . . It may interest you to know that our labor productivity at Palabora is higher per man than it is in the southwest of the United States.

"Development," Sir Val insisted, not only offered moral compensation for capitalist activity under apartheid; it was good for profits. The same would hold true at Rössing, which would "employ all races on a standard of living rather higher than they have hitherto enjoyed, and very markedly so in the case of the black men. We feel that any attempt on the part of the United Nations or others to curtail or discourage developments of the kind we have in mind would merely frustrate our intentions to raise the indigenous standard of living." Meanwhile, Sir Val reminded Sir Denis, RTZ's sales team was currently in Japan finalizing some very important contracts. "This sale is crucial to the successful financing of Rössing," he wrote, urging that Britain delay its declaration to the UN "until we have had a reasonable chance to conclude our broad arrangements with Japan."[57]

The Foreign Office doubted that UN audiences would find Sir Val's reasoning persuasive. Improving the apartheid norm hardly signified the end of racial disparities. Sir Val's description was reminiscent of colonial "civilizing mission" ideology. He had even published a report on the Diplomatic Service whose "major recommendation," according to Barbara Rogers, "was that the British foreign service should be drastically over-hauled so as to concentrate representation in areas of interest to British multi-national business interests."[58] The Foreign Office knew that Sir Val's arguments, though "attractive . . . to many Europeans," would "cut no ice with African political nationalism (which they largely ignore)."[59] Officials chose not to tell Britain's UN ambassador about Sir Val's letter, knowing that the prospect of RTZ's influencing British diplomatic strategy would "cause him special concern."[60]

The Foreign Office decided to time Britain's disclosure to fit RTZ's needs. Britain wasn't alone in worrying about UN reactions to Rössing contracts. In September 1970, the Japanese asked RTZ to keep their con-tracts secret "until they were safely elected to the Security Council at the end of the year."[61] In any event, Labour lost the 1970 elections to the Conservatives. Frustrated by failure, Rogers left the Foreign Office and joined the anti-apartheid movement. (We will encounter her again in chapter 5.) Of her efforts to suspend the Namibian contracts from within the halls of government, she later testified:

I wrote endless memoranda, had endless arguments with my colleagues and had endless exchanges of memoranda with the legal advisers. I finally reached the conclusion that I had done everything I could and there was absolutely nothing left for me to do. Even where it was acknowledged that I was correct on . . . a legal point, nothing happened to implement the point I had made. I left the Foreign Office on a general disagreement over southern African policy, of which this [Namibian uranium] was a major part.[62]

Contracts proceeded peacefully enough until 1974, when the UNCN issued Decree No. 1, prohibiting extraction of natural resources from Namibian territory. Inflation and anxieties engendered by the 1973 oil crisis were already making it difficult for the Conservatives. Labour regained

Sean MacBride (UN Commissioner for Namibia), Sam Nujoma (president of
SWAPO), and Theo-Ben Gurirab (representative of SWAPO in New York), 1976.
(UN photo by Saw Lwin; courtesy of African Activist Archive)

control in 1974, promising a tougher stance toward South Africa and a
revision of the uranium supply arrangements. Pressure to make good on
this promise and to cancel the contracts mounted in Parliament, in the
press, and from anti-apartheid groups.

Advocates for the contracts—led by BNFL—again defended a platform
of pragmatic capitalist ethics. Cancellation would engender mistrust, and
"the existence of a producer's cartel . . . could lead to difficulty in securing
alternative supplies." Cartel producers "might be unwilling to supply a
customer who had broken his contract with another member"[63]—not the
least if the aggrieved party was cartel leader RTZ.

As political pressure intensified, the rhetoric broadened to place at stake
the global future of UK capital and the well-being of the British public:
"Apart from the direct consequences for them and their customers of a
cancellation of contracts, RTZ fear that this could also have a seriously
adverse effect on their reputation (and that of other British-based international corporation) and consequently on their future ability to raise
capital."[64] Because the price of uranium (the Nuexco price) had risen

steadily since the contracts were signed, replacing the Rössing contracts would increase costs by 20 to 30 million pounds, "which would have to be recovered from the electricity consumer or from the tax payer."[65] Finally, Rössing had agreed to deliver its uranium in oxide form. Other producers were insisting on delivering processed uranium. "Conversion to hexafluoride would not then be undertaken in the UK. This would add about £5 million to foreign currency expenditure."[66] Maintaining the colonial tradition of importing raw materials thus had significant financial advantages.[67]

Many at the Foreign Office felt that Decree No. 1 would have little effect once the initial publicity blew over. One official reassured the Department of Energy in October 1974 that the UNCN was "an extremist body" without much "Western" support.[68] The provisions of the decree were "drafted in extreme language and if implemented would amount to sanctions. We think it unlikely that other governments will feel bound to implement them."[69] RTZ, for its part, reassured investors that it would proceed with business as usual. At the annual shareholders meeting in 1975, Sir Val haughtily declared "I am not prepared to fail to deliver to the United Kingdom and others under a contract solemnly entered into for the provision of uranium from South West Africa. I am therefore not prepared to take any notice of what the United Nations say about that."[70]

Meanwhile, the British government formally rejected the legitimacy of the UNCN. Decree No. 1, it argued, was a back-handed means of obtaining sanctions. Other industrialized nations followed suit, most notably Japan. Poised to become Rössing's largest customer, the Asian nation allowed its utilities to keep their contracts and contented itself with making donations to the Namibia fund.[71]

RTZ, BNFL, and their allies thus pitted the moral economy of capitalism against the moral opprobrium of anti-colonialism. Sir Val's logic followed the arguments of most industrialized nations and multinational corporations concerning the meaning of sovereignty after decolonization. As the fundamental covenant of capitalism, the contract formed the foundation of social order. Its maintenance was critical in the struggle against communism. It could not be broken. In Britain's Labour government, this logic overran the ideological high ground of anti-apartheid policy.

Interestingly, the polarization between postcolonial sovereignty and transnational corporate interests sometimes hid more complex exchanges.

Throughout the liberation struggle, for example, SWAPO strongly condemned the uranium contracts in public. But as early as 1976, SWAPO signaled that its private posture might differ. Eager to ensure that "SWAPO and RTZ develop a reasonable relationship in advance of independence," the Foreign Office's new southern Africa specialist, Martin Reith, hosted a luncheon to grease the wheels. Guests included Peter Katjavivi, SWAPO's representative in Europe, and Frances Vale, RTZ's political advisor. The outcome surpassed Reith's hopes. "After conviviality had developed," Katjavivi "lightly derided the British Government for acting solely in its own interest over Namibia." Reith replied that Britain was using "what credit [it] had with the South Africans in order to get them out," but that "SWAPO's threats about uranium were not much help." At which point, Reith reported, "Mr. Katjavivi said not to worry. We should realize that SWAPO had to take a certain public line on Rössing. 'We have to create an atmosphere of insecurity.' He went on to say, looking at Miss Vale and myself and emphasizing that his remarks were off the record, that a SWAPO Government would not disturb RTZ's position in Namibia." Later in the course of the lunch, apparently, Katjavivi "went on to speak slightingly of the UN Council and [its chairman] Mr. MacBride, saying that they were 'irrelevant' to Namibia."[72] These surprising words contradicted SWAPO's public position and the stance taken by its president, Sam Nujoma, a year earlier while meeting with Foreign Secretary James Callaghan.[73] Frances Vale told her chairman that Katjavivi's remarks were "a clear signal to RTZ that operations will be allowed to go on." Reith told his superiors that they provided "as solid confirmation as we could ever expect to get that our uranium supplies are likely to be reasonably safe with SWAPO."[74] Callaghan reported the lunch conversation to his prime minister, adding that Katjavivi was "a professional who would not speak thus unless quite sure that he accurately reflected SWAPO policy."[75] But the comments were not for attribution: Katjavivi made clear that he would deny them if necessary.[76] This put British government officials "in a difficult position," able to justify pragmatism to themselves but lacking ammunition to counter public pressure to cancel the contracts.[77]

The British Foreign Office clung to arguments centered on upholding the moral economy of the capitalist order. They suggested that Rössing could shift "apartheid tendencies" by engaging in enlightened hiring and promotion practices.[78] Through contracts and better labor management,

capitalist ethics could overcome colonial occupation. At least in government corridors, such arguments upheld the conviction that uranium could
be treated as an ordinary commodity.

Namibian uranium refused to remain banal, however. In 1976, when
the news broke that South Africa and Israel had signed a broad cooperation pact, speculation that they would collaborate to build nuclear weapons
ran rampant. Uranium's nuclear peculiarities once again became a subject
of intense contestation. In London, Farouk Helmy from the Egyptian
embassy expressed concern to Martin Reith that "RTZ was knowingly
assisting [weapons development] by agreeing to supply uranium." Reith
replied that RTZ's people "were politically conscious. Of course, they
would have no control over the end use of uranium which they supplied,
but as a personal view I found it inconceivable that they were knowingly
involved in such business." Furthermore, even if RTZ refrained from supplying uranium to Israel, South Africa had plenty of other uranium sources
that it could sell.[79]

Informed by Reith of Helmy's query, RTZ's Frances Vale confirmed
that Rössing "had no contracts to supply uranium to Israel either directly
or through South Africa. . . . But she thought the Egyptian Embassy call
was significant, because RTZ was hoping to sign some valuable contracts
with Arab countries soon."[80] Reith forwarded the information to the
British Chancery in Cairo, commenting that "no Arab country has taken
an active part in the campaign against RTZ's activities in Namibia or the
British contracts, and we would not wish them to begin. If you get any
inkling that the Arabs remain concerned at the allegation that RTZ might
supply uranium from Rössing to Israel via South Africa, for the purpose
of developing a nuclear warfare potential, it would be useful to scotch the
rumour."[81]

Meanwhile, Tanzania and the Organization of African Unity also
expressed concern about the implications of the Israeli pact for a potential
South African nuclear weapon. Reith found this anxiety harder to allay:
"I do not think we would be able to set at rest black Africa's apprehen-
sions about South African nuclear potential even if we thought it worth
trying. But I do think we should do something to take RTZ out from
among their targets."[82] South Africa, he insisted again to anyone who
would listen, had access to its own uranium. It didn't need Rössing's

product to build a bomb. The British company should be left to hawk its wares in peace.

Martin Reith was right that observers in "black Africa" and elsewhere wouldn't be easily assuaged by reassurances about South Africa's nuclear intentions. By the mid 1970s, the prospect of a South African atomic bomb became a significant target for ANC action. As we'll see in chapter 5, the global anti-apartheid movement would find nuclearity an exceptionally useful weapon in the battle for the liberation of Namibia and South Africa.

CONCLUSION

In both South Africa and Britain, ambiguities in uranium's commodity status fueled struggles over state power and the relationship between public and private capital. Controlled by a British company, the Rössing mine offered the UK the next best thing to a national fuel source, enabling it to avoid the political and financial costs of nuclear exceptionalism by using an alternative to North American supplies. RTZ's deployment of market devices promoted a view of uranium as an ordinary commodity. This view fit neatly into British efforts to develop commercial nuclear power by separating fuel production (via BNFL) from the more research-oriented UKAEA.

Of course, Rössing was not on British soil. As opposition to South Africa's occupation of Namibia gathered steam, public debates pitted the moral imperatives of anti-colonial resistance against those of capitalist business ethics, tying the latter to British national interest. Contracts were the foundation of capitalism. The global credibility of British business and the pocketbooks of British taxpayers depended on honoring them. Morally comforted by RTZ's public promises of racial uplift and SWAPO's private assurances of future cooperation, British officials increasingly settled on the view that uranium was an ordinary commodity.

Nevertheless, uranium again became controversial, not for its ontology but for its origin. The prospect of collaboration between South Africa and Israel on nuclear weapons development threatened uranium's comfortable commercial identity by re-entangling the politics of place with the technopolitics of exceptionalism. Conflicts over Namibian uranium paralleled ongoing debates over resource sovereignty between rich and poor nations, former colonizers and newly independent states. On that international stage, what mattered most was where uranium came from.

Within South Africa, though, the conflict was about the *stuff*. Uranium's identity and its material form, whether as ore or as highly processed feedstock, shaped the balance of power within the apartheid state and affected South Africa's political and economic relationships with the world. Elites sought international legitimacy via "the market," which they carefully construed as an apolitical space of economic exchange. The AEB tapped into this discourse when it promoted its vision of a commercially competitive enrichment plant. But AEB leaders wanted it both ways, simultaneously seeing the enrichment plant as a contribution to South Africa's increasingly exceptional nuclearity. By contrast, the mining industry, led by NUFCOR, resisted efforts to encumber commercialism with exceptionalism because it sought more room to maneuver, both within and outside South Africa.

Clearly, commodification is not the same everywhere. It is not enough to observe the emergence of market devices. We must also trace their deployment, their uses, their transformations, their failures. And we must do this in different places.

In February 1960, France exploded its first atomic bomb in the Algerian Sahara. Emotional renderings in the metropolitan press extolled the renewal of French "radiance." By becoming the world's fourth nuclear weapons power, France had secured its place in the new global order. The subtext—never explicit in these glowing reports—was the crumbling empire. Also absent: the test took place during the height of Algeria's war for independence. Everyone knew about the war, so why belabor the point? Unmentioned, too, because no one could guess its future significance: the explosion occurred just a few hundred kilometers north of the Nigérien deposits, the largest uranium reserves in Africa.

To put the explosions in their African context, let's back up a bit. After World War II, a limited number of Africans who had received Western education (and who came to be known as *évolués*, or "evolved ones") were granted the right to vote. Their parliamentary deputies went off to Paris, where they helped to promulgate significant changes to French rule, ending forced labor and granting Africans citizenship in a greater "French Union." Reforms in working conditions and benefits for wage labor followed, along with ambitious development plans for modern infrastructures in rural and urban areas. As reforms multiplied, however, so did the costs of colonial administration. Empire was becoming unaffordable.[1]

In 1956, hoping to lift the burden from metropolitan taxpayers, the French government passed the *loi cadre*, a law that made African territories responsible for funding their own administrative services and infrastructure development. In a 1958 referendum called by President Charles de Gaulle, most territories voted for autonomy within the French Union rather than total independence. In 1960, most of French Africa formally achieved

independence.[2] But African and French elites soon learned that "auton-omy" carried significant costs, both political and financial.

In negotiating the terms of decolonization, France signed a series of defense and raw materials accords with Gabon, Niger, and other newly independent African nations. These treaties gave the former colonizer privileged access to uranium and other strategic raw materials, prohibiting exports that ran counter to French military interests. For African leaders, the tradeoff gave them military security, guaranteed product markets, and offered development aid. Many of the treaties had a secret appendix that signaled the limits of independence: France would help defend the new leaders not only against external threats, but also against coups d'état.[3]

French and African political elites wanted to preserve privileged rela-tionships during this transition. Most of the new states, for example, kept as their currency the FCFA, which had been created in 1945 as the "franc of the French Colonies of Africa" before becoming the "franc of the French Community of Africa" and finally the "franc of the African Finan-cial Community." Because its exchange rate was pegged to the French franc, the FCFA gave African states financial stability—with the caveat that half of their export earnings had to be deposited in the French treasury. This made FCFA states vulnerable to the vagaries of the French economy. When the value of the French franc dropped, so did that of the FCFA, resulting in higher prices for non-French goods (and privileged markets for French goods) in the FCFA states.

The French Ministry of Cooperation, created by de Gaulle in 1959, offered another formal framework for continuing privilege. For four decades the Ministry of Cooperation remained structurally separate from the Ministry for Foreign Affairs as it managed development aid to Fran-cophone Africa. It stood as a clear symbol that the French state treated former African colonies differently from other nation-states.

Most significant of all in maintaining this "special relationship" was the mysterious and powerful Jacques Foccart. As a member of the French Resistance, Foccart had developed a fierce and loyalty to its leader, General Charles de Gaulle. Known to some as *le sphinx* and to others as *l'homme de l'ombre* (the man in the shadows), Foccart served as the French presi-dent's personal "Monsieur Afrique" from 1958 to 1974.

on facing page: Scenes from French atomic bomb testing in Reggane, Algeria, 1960. (Keystone France/Getty Images, used with permission)

Gabonese president Léon Mba and French president Charles de Gaulle shake hands; Jacques Foccart, the "shadow man," surveys from the side. (STF/AFP/Getty Images, used with permission)

By the time he helped engineer de Gaulle's return to power in 1958, Foccart had developed a far-reaching network of "friends" across Francophone Africa, including African political elites, industrialists, colonial and military officers, French secret service agents, and a smattering of mercenaries who prowled the continent for lucrative jobs in arms trafficking and other enterprises. Foccart felt that maintaining these relationships would ensure the continuity of French interests in an independent Africa. Persuaded by this argument, President de Gaulle created a special office for African affairs housed directly in the Elysée presidential palace, giving Foccart unparalleled access to executive authority and protecting him from parliamentary and ministerial oversight.

In essence, Jacques Foccart directed French policy in Africa. The pronouncements of a succession of ministers speaking in the official voice of diplomacy meant little. Foccart dispensed the favors, collected the information, paid the bribes, and approved the occasional political assassination, all the while sheltering the president from unsavory events. The web of people, money, and businesses that Foccart and his allies spun from his Elysée quarters would eventually be called *la Françafrique* by its critics. Intended to denote the corrupt relationships that bound France to its former sub-Saharan colonies, the term is also a pun on *France-à-fric*, in which *fric* is slang for money. But Foccart, knowing better than anyone that accepting money personally could create unwelcome obligations, refused all direct payment for his services, instead living comfortably off his private import-export business.[4]

Foccart was on close terms with nearly all of the presidents of Francophone Africa, some of whom he had personally shepherded to power. When African rulers fretted about coups, Foccart offered them secret guarantees in the form of undated requests for French military intervention. Rulers signed these in advance and securely lodged them in places where Foccart could reach them if the time came. The first opportunity to make good on such promises came in Gabon.

In the 1950s, Léon Mba, who would become Gabon's first president, had wanted Gabon to become a French *département* (like the Caribbean islands of Martinique and Guadeloupe) rather than an independent nation. Foccart shrewdly interpreted the request, which de Gaulle refused to grant, as a sign of Mba's dependence on France for political survival. This made Gabon an excellent hub for France's activities in Africa, licit and illicit.

The discovery of oil in the Gulf of Guinea soon after Gabon's independence further increased the nation's strategic significance. When a coup
d'état attempted to dislodge President Mba in 1964, Foccart didn't hesitate.
Demonstrating how *la Françafrique* could help ruling elites, France sent
troops to quash the coup and rescue Mba from his captors.[5]

Not long after his restoration, Mba was diagnosed with terminal cancer.
Foccart and his friend Maurice Delauney, France's ambassador to Gabon,
stepped in again to nurture (financially and otherwise) the political for-
tunes of a young man named Albert-Bernard Bongo, whom they deemed
a loyal Gaullist. As Mba lay on his deathbed, Delauney persuaded him to
install Bongo—who by then had held several ministerial posts—as vice-
president. When Mba died, in 1967, Bongo, 31, became one of the young-
est heads of state in the world. Backed by the constant presence of French
troops, Bongo remained president of Gabon until his death in 2009.

By and large, Bongo (re)paid the French political elite handsomely.
Money flowed from oil production to Bongo and from Bongo to French
politicians in an endless, murky cycle of patronage. But the Gabonese
president wasn't quite the puppet that many made him out to be. There
were significant moments of tension with France during Bongo's long
reign. And some of those involved uranium.

Uranium mines require massive capital investment and complex technologies, neither of which were readily within reach for the first postcolonial leaders of Gabon and Niger. The 1960 defense accords with France acknowledged this state of affairs. Allowing France to retain effective control over certain raw materials seemed like a pragmatic arrangement, one whose implications could be deferred pending more pressing matters of governance. So what—if anything—did natural resource sovereignty mean in la Françafrique?

UN decrees, resolutions, and charters made it possible to invoke international agreements as threats or as sources of legitimacy.[6] The concrete conundrums of sovereignty, however, did not take the form of texts and accords. Rocks and fluids and plants didn't become *resources* until technological processes identified, extracted, refined, and exported them—all of which shaped their "value." And resources didn't become *commodities* without trucks, ships, airplanes, processing plants, accounting systems, and myriad market devices that enabled their commercial circulation. As I argue in this chapter, sovereignty was distributed through the technopolitical systems that turned nature into resources, and resources into commodities.

In the late 1960s, leaders of Gabon and Niger began to pursue three main ways of expressing postcolonial sovereignty over their uranium mines. The first had to do with the "development" scope of mining operations: the financial, social, and infrastructural dimensions of building mine

on facing page: President Omar Bongo arrives in Mounana to inaugurate the COMUF's new yellowcake plant in 1982, with Maurice Delauney in sunglasses behind him. (photo courtesy of COMUF) capacity. In the early 1970s, Gabon's President Omar Bongo and Niger's President Hamani Diori began to press the French for more and better development projects around the uranium mines. These demands were usually met, and in Gabon in particular the results were touted by Bongo as emblems of modern nationhood.

The second, much trickier expression of postcolonial sovereignty had to do with value. As we have seen, mining companies, reactor manufacturers, utilities, and atomic energy agencies all wanted to put a price on uranium. Invoking the market and its mechanisms, they sought to de-exceptionalize nuclear trade and skim the politics off from the economics. But just as uranium had value beyond its price, this chapter argues that *price* had meaning beyond the company ledger. Price could express sovereignty in the struggle over the meaning of national independence in Francophone Africa. Who should rightfully determine the price of African uranium? What should go into that determination?

Inspired by the 1973 oil crisis, Bongo and Diori began to demand a role in setting the price of their uranium. Countering French efforts to control the terms of trade, Bongo insisted that uranium was an ordinary commodity whose price Gabon should set. Diori insisted that uranium's nuclearity gave it exceptional value that carried its own price tag. In the boundary between banality and exceptionalism, I argue, Bongo and Diori each sought a separate foothold in the uranium market, and on the sovereignty that it could help them enact.

This led to a third question: Who could sell the uranium, and to whom could it be sold? Gabonese and Nigérien efforts to express sovereignty through price met with limited success. The imbalance of power between France and its African interlocutors was simply too great. France's refusal to put a price on the nuclearity of uranium led Bongo and the new Nigérien president, Seyni Kountché (who deposed Diori in a 1974 coup), to shift focus. Both Gabon and Niger had acquired substantial stakes (around a quarter to a third) in their states' uranium companies. If uranium was truly a banal commodity, state officials began to argue, host countries should take charge of selling their shares. Bongo, who sought an alliance with Iran on matters related to oil and OPEC, summarily offered to sell a small lot of uranium to the Shah. Kountché turned to Niger's difficult but rich neighbor, Libyan leader Muammar Qaddafi, who happily acted as a broker for other countries interested in uranium. (Saddam Hussein

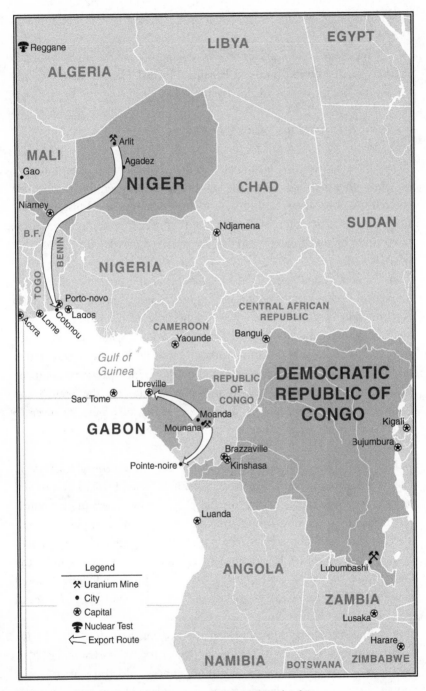

Niger, Gabon, and D. R. Congo. (map by Evan Hensleigh)

didn't obtain yellowcake from Niger in 2002, but he did get some in 1981.)

The technopolitical distribution of sovereignty—and the shifting boundary between exceptionalism and banality—turned France's refusal to pay a premium for uranium's nuclearity into risky business.

GABON: A STRANGELY NUCLEAR PLACE

The defense accords that inaugurated *la Françafrique* lumped uranium with oil and other strategic materials. In practice, however, each substance entered the postcolonial period with its own technopolitical history, infrastructure, and trade circuits. During the first decade after decolonization, the technopolitics of uranium in Gabon remained separate from those that governed oil.

As French technologists liked to say, both oil and uranium contributed to France's "energy independence." Engineers and planners exhibited no compunction over the neo-colonial implications of the phrase. Independence, in this formulation, signaled freedom from US nuclear dominance and Middle Eastern petroleum sources. It quickly became apparent that oil from the Gulf of Guinea—extracted by the French parastatal company Elf-Gabon—was abundant and immensely lucrative for the French and Gabonese state treasuries. Plus there were plenty of profits left over to fund Foccart's covert activities and to line the bank accounts of Gabonese elites.

Uranium production wasn't about the profits, however—at least not at first. It was about power and prestige. Uranium fueled France's bombs and reactors—nuclear things meant to replace empire by generating geopolitical "radiance" as the new means and emblem for projecting power.[7] Their material and symbolic significance meant that France's Commissariat à l'Énergie Atomique (CEA) enjoyed tremendous autonomy within the French state as well as in Gabon. Foccart and his friends long ignored uranium, letting the CEA negotiate the political, economic, and social value of Gabonese uranium.

The end of empire shimmered on the horizon when CEA geologists found uranium in eastern Gabon in the late 1950s. Earlier in that decade, the CEA had operated uranium mines in Madagascar; it was a simple matter to retain that oversight after independence, as we'll see in chapter 7. In the Gabon of the late 1950s, when the defense accords had not yet been conceived let alone signed, planners worried about the ease with which an enterprise run by a French state agency could be appropriated by a newly independent state. The CEA joined forces with Mokta, a private mining company with considerable colonial experience, to form the Compagnie des Mines d'Uranium de Franceville (COMUF) and run the uranium site, which they dubbed Mounana. Decisions concerning mining and milling triangulated the CEA's fuel needs, its expertise with uranium mining and processing, and Mokta's experience running profitable mines.

The COMUF sent its first lot of uranium oxide to France in 1961, the year after Gabon's independence. Already, though, Gabonese uranium seemed expensive, particularly relative to the rock-bottom prices offered by South Africa.[8] What justified ongoing production?

In France, COMUF uranium could be readily construed as French. Produced by a French-owned company, the stuff contributed to France's long-term security. This mattered not only for France's atomic weapons program, but also for its ability to secure a privileged position as a global player in a uranium market projected to take off in the 1970s. In anticipation of that growth, CEA's director of mines, Jacques Mabile, made an apparently counterintuitive choice: he decided to cut metropolitan exploration efforts while expanding the agency's overseas prospecting programs. Uranium in metropolitan soil wasn't going anywhere else, he reasoned, and laying claim to overseas deposits would help France establish a global market presence. Besides, Gabonese uranium was good for metropolitan employment. The CEA had specifically redesigned its Gueugnon yellowcake plant to suit the processing requirements of COMUF ore.[9] All these factors justified the high price of COMUF uranium in France.

In Gabon, however, arguments based on "French national security," "French global market strength," and "French employment" smelled strongly of colonialism. There, the words of the day were "development" and "modernization," the means by which Gabon would manifest autonomy, improve its infrastructure, attract foreign capital, and enact nationhood. If the French wanted to keep their privileges in Gabon, they would have to get on board—at least nominally—by reframing "French security of supply" as "Gabonese development." The 1964 coup against Mba made it particularly clear that the French presence in Gabon required careful stage management.

"Development" presented a solution to the postcolonial problem of natural resource sovereignty. The uranium would remain in French hands, but the COMUF promised the Gabonese state a modern infrastructure to support production: housing, electricity, running water, roads, technical training, and jobs. The CEA also deployed the discourse of development to reshape the outlook of its French employees. A European man, the agency warned, might well see himself as "a representative of that very progress that leads to a better use of natural resources" and feel that he represented "a civilizing nation that had brought numerous benefits to less technologically advanced peoples." But he had to remember that technical expertise was a vehicle for uplifting Africans, not a justification for arrogance or violence.[10] France's relationship with African countries was now one of "cooperation."

Development discourse thus offered a powerful moral script. Its allure derived not only from its promises, but also from its flexibility.[11] It could propel people like the social worker who taught (French) home economics and social manners to the (Gabonese) wives of COMUF employees, or the doctor who insisted that the mine hospital be open to everyone in the region. But its teleological thrust could also be marshaled by Gabonese state officials to make demands of the company. In the early 1970s, just such an opportunity opened up.

By 1971, the year of the uranium cartel, the COMUF faced serious financial difficulties. Construction of nuclear power plants had stalled in France, and uranium prices had hit bottom throughout the capitalist world. Although the cartel might eventually force a price increase, low cash flow severely threatened the COMUF's expansion plans, which included prospecting, adding underground shafts, and expanding milling capacity. Management also hoped to build a yellowcake plant on site, saving production and transport costs and giving future (non-French) customers more flexibility.[12] In the long run, such expansion would secure the COMUF's position as a commercial uranium supplier. In the short run, however, financing appeared nearly impossible.[13]

Trimming any one of the COMUF's projects posed problems. Digging more underground shafts and building new plants took years. If construction didn't start soon, the company might not profit from the projected demand increase.[14] Provisionally curtailing production and laying off personnel offered temporary solutions but carried direct costs in the loss of investment in training and indirect costs through local upheaval and state displeasure.[15] As things stood, the Gabonese state was already holding the COMUF to its "development" claims by demanding upgrades in workers' living quarters and mandating wage increases.[16]

The company first tried to solve its cash flow problems by finding new customers and hence demonstrating the viability of expansion plans to financiers. But then a nuclear freak of nature intervened. Tests revealed that the newly opened Oklo deposit—right next to Mounana—contained *depleted* uranium. The composition of the ore *in the ground* resembled that of spent reactor fuel, as though it had undergone fission! CEA scientists hypothesized that high concentrations of uranium had led to spontaneous fission reactions some 1.7 billion years ago.

Suddenly, Gabon had become a strangely nuclear place. Speculation caused international excitement among scientists, who began referring to Oklo as a fossilized reactor. COMUF management, though, began to despair. Who would want to buy depleted uranium? Back in France, CEA scientists scrambled to find ways to make use of the stuff, and the agency finally agreed to buy it on discount.[17] Clearly, the COMUF had to get the depleted ore out of the ground before it could persuade most investors that enough good ore remained to justify expansion.

The postcolonial Gabonese state, however, seemed readier for a leap of faith than other investors.

Gabon had just undergone a series of transitions. Albert-Bernard Bongo became president in 1967, the same year that saw the Biafran secession from Nigeria and the creation of the huge French company Elf. Jacques Foccart saw the Biafran secession as a chance to weaken the Nigerian state. Elf had its eye on oil fields off the Biafran coast. Repaying Foccart and his clan for bringing him to power, Bongo made Gabon the launch pad for French military support to the secessionists. Neither the secessionists nor the French obtained the result they had hoped for, and over a million people died from starvation or combat. Elf didn't get access to the Biafran oil fields. But in the meantime, the oil fields off the Gabonese coast had proved lucrative beyond expectation. By 1970, when the Biafra conflict ended, oil revenues from Elf-Gabon were flowing into the coffers of the Gabonese state—and those of its new president.[18]

Flush with youth, money, and ambition, Bongo imagined Gabon as a modern technological nation, a powerhouse of central Africa, and a player in global oil markets. He adeptly filled state-owned media with his message of *Rénovation*, which promised to convert revenues from Gabonese natural resources into housing, schools, and hospitals. His signature project was the Transgabonais, a modern railroad intended to unify the nation by connecting the coast to the interior—specifically, the Haut-Ogooué, his province of origin. Bongo thus sought to bring his power base symbolically and materially closer to the capital, Libreville. As with large-scale technological projects elsewhere in the postcolonial world, the Transgabonais became a nationalist spectacle dramatized in the state-run media as the triumphant victory of modernization over Gabon's notoriously recalcitrant nature.[19]

The new president was equally flamboyant in positioning himself alongside the world's oil producers. In 1973, allegedly following advice from Muammar Qaddafi, he converted to Islam, made a pilgrimage to Mecca, and changed his name to El Hadj Omar Bongo. Eager to assert natural resource sovereignty as forcefully as any OPEC leader, Bongo began making declarations that challenged France's control of Gabonese oil and uranium. Displaying a new-found Third Worldism, he attended the Non-Aligned Movement's 1973 summit in Algiers and declared his support for the movement's position on natural resource sovereignty.[20]

Warning that uranium-producing countries would join forces to control prices, Bongo admonished France not to speak about "its" uranium without mentioning Gabon: "It's Gabonese uranium and it cannot be used to do anything without the Gabonese government—that is to say myself—expressing its point of view."[21] Gabon had applied for OPEC membership, and it would be "aberrant and wrong" to expect Gabon to treat uranium differently from oil. In December 1973, Bongo asked the Israeli, Italian, German, American, Spanish, and Japanese ambassadors in Libreville to search for potential customers for Gabonese uranium.[22]

Bongo thus seemed prepared to extend Gabonese sovereignty over natural resources. But not too far. Although postcolonial nationalizations were in the global air, Bongo's experience with Elf had already shown him the personal and national benefits of cooperating with French corporations. No need for expropriations when a stake in the companies would suffice. In 1974—the year of the UN Council for Namibia's Decree No.1—the Gabonese state acquired 25 percent ownership in COMUF with an infusion of capital large enough to provide a substantial advance for expansion. Without this, the COMUF would have faced severe difficulties just maintaining production levels, let alone financing expansion.[23]

Both parties framed the capital infusion as a matter of postcolonial sovereignty rather than a rescue. The COMUF vaunted good corporate citizenship in a postcolonial world. Bongo's state displayed sovereignty over its natural resources, claimed three seats on the company's board of directors, and received dividends on profits. In 1975, Bongo and his allies obtained a more personal piece of the pie by mandating that 4,000 "B" shares, which produced dividends but didn't grant seats on the board of directors, be made available for purchase by Gabonese citizens. In principle, the COMUF's Gabonese employees could claim purchase priority, but few had the money to invest. Bongo, his clan, and his ministers snapped up these shares, ensuring profits for the country's ruling elites and consolidating the president's hold on the Libreville bourgeoisie.[24]

Company executives anticipated that the state would make demands in return for its investment. But they expected these claims to keep with Bongo's displays of *Rénovation* and to concern generic matters of "development," such as improving the housing stock, intensifying the training programs, and promoting Gabonese employees to management positions.[25] Such measures could even benefit the COMUF. Gabonese managers, for instance, earned less than expatriates.

The French were caught short in December 1974, though, when they learned that Bongo not only had offered to sell Gabonese ore to Americans, but had requested a quote from Union Carbide for the construction of a yellowcake plant at Mounana. At this point, the CEA's Gueugnon plant was still converting Gabonese ore to yellowcake. The COMUF successfully rebuffed Union Carbide, and in April 1975 corporate executives met with state officials to discuss the price of Gabonese uranium.[26] If they thought they'd successfully reined in Bongo, however, they got a nasty surprise when they learned in October that El Hadj Omar Bongo had promised over 500 tons of uranium to Iran. This time, Bongo refused to back down.

PRICING NUCLEARITY IN NIGER

To understand the events in Gabon, we must travel to Niger and go back a few years to when uranium operations were just getting underway there. Soon after Hamani Diori became the first president of independent Niger, he began to contemplate the implications of the enormous uranium deposits that geologists had first glimpsed in 1957. Clearly, France considered these deposits strategically important. How quickly, Diori wondered, could uranium be made to benefit the Nigérien economy?

In Gabon, production had taken off five years after uranium's discovery. But delineating the extent and nature of deposits in the Sahara proved more challenging. Technological reinforcements arrived after the French military closed its Algerian atomic test sites and shipped some of the geological equipment south. In 1966 the CEA, having finally decided that the huge deposits were commercially viable, began drafting plans for a mixed-capital mining company on the model of the COMUF. But whereas the COMUF had been constituted during the last years of colonial rule, the CEA had to contend with Niger as an independent state. For the next two years, Diori and his advisers bargained long and hard to maximize the state's financial benefits.

Diori walked a fine line in his relationship with the former colonial power. Under the tutelage of Félix Houphouët-Boigny, president of Ivory Coast, he supported Niger's inclusion in the French Union in 1958 rather than opting for independence. Supported by the elites, Diori's party handily won legislative elections. Less than two years later, however, the French Union dissolved. In August 1960, Diori proclaimed the very independence his rivals had advocated. Like other leaders in Francophone Africa, he consulted regularly with Jacques Foccart, who on more than one occasion helped him stay in power. But Diori also pursued policies in defiance of Foccart—notably by sending arms to Nigeria following the Biafran secession.[27]

Nowhere was Diori's balancing act more delicate than in discussions over the development of Niger's uranium resources. The Nigérien president didn't like the CEA's initial proposal for the distribution of company profits. When the agency resisted his efforts to obtain a better deal, he asked Foccart to intercede. This worked at first. On Foccart's recommendation, de Gaulle personally granted a larger percentage of the profits to Niger, throwing in a special development aid package and a promise to send French troops as needed to defend the deposits against Algerian incursions.

Diori accepted de Gaulle's offer but had second thoughts a few months later after a visit to the United States. Foccart speculated that the Americans had taken Diori to visit uranium mines in Arizona and suggested that uranium would allow him to "turn Niger into an African Arizona." Diori subsequently asked that the Nigérien state receive a substantial (and free) capital stake in the company in addition to a share of profits. The CEA balked. If Niger were a shareholder, it might demand a say in corporate decisions. Dismayed, de Gaulle instructed Foccart not to let the CEA's obstinacy drive Diori into the arms of the Americans. Niger's uranium had to remain exclusively French.[28]

Foccart was only too glad to oppose a rival for de Gaulle's affection. By providing the general with the *force de frappe* and other nuclear things, the CEA had helped redefine France's global identity in the post-imperial world, giving it a special place in de Gaulle's heart and out of Foccart's grasp. The agency enjoyed institutional autonomy and wielded a technopolitical approach to power, proving itself resistant to Foccart's strengths, which lay in his network, his personal relations with African leaders, and his ability to manipulate secrecy. Trying to rein the agency in, Foccart tried and failed both to obtain a guarantee of French exclusivity over Niger's uranium and to prevent the CEA from seeking uranium outside Africa. On the latter point, de Gaulle didn't see the sense in stopping the CEA from exploring elsewhere, since nuclear radiance enabled France to reach beyond the boundaries of its former empire.[29]

Diori understood de Gaulle's obsession with nuclear things, and viewed uranium as a tool to extract exceptional concessions from France. The IAEA and various national atomic energy agencies had touted the uses of radioisotopes in agriculture and medicine. Surely Niger could benefit from such nuclear things. Why not a nuclear power plant too? Although the CEA eventually set up a lab to investigate agricultural uses of radioisotopes, a nuclear plant remained out of the question. But in this and other instances, Diori argued that uranium's unique properties—along with the 1960 defense accords—mandated that negotiations be conducted at the highest levels of diplomacy.

In 1968, conceding Diori's point, de Gaulle agreed to the creation of a commission that met regularly to discuss profits, security of supply, development, cooperation, and fiscal revenues for the two companies that mined the reserves.[30] The Somaïr mining company, set up in 1968, allocated a 20 percent stake to the Nigérien state. Two years later a second company, Cominak, was formed to mine another deposit. This time, the state received a 32 percent stake. Eventually German, Italian, and Japanese investors also bought stakes that mostly came out of French-owned shares.

Since the mines would produce no yellowcake—and therefore no profits or taxes—until the mid 1970s, Diori sought special financial contributions to develop the country's infrastructures. He requested an initial payment of one billion FCFA into a special investment fund, with similar contributions in subsequent years. This money, he specified, should not come out of France's regular foreign aid budget because that would take money away from other aid recipients. It had to be a *special* payment that reflected uranium's *exceptional* status. The French vaguely indicated that they would consider the request. For several months, though, nothing happened.

In 1969, soon after Georges Pompidou took office as president of France, Diori wrote to him, hoping to cement the understanding he'd reached with de Gaulle to treat uranium as an exceptional thing. Diori felt slighted by France's failure to appoint high-level decision makers to the interstate uranium commission, and he worried that his requests had fallen into bureaucratic cracks. Pompidou reassured Diori: "I understand your concern not to confuse our traditional aid, shared by all Francophone countries, with the special cooperation that results from our accords concerning . . . uranium mines."[31] He agreed to set up a special three-year investment program.[32]

I was refused access to archival records documenting subsequent exchanges. But it's clear from a variety of sources that Diori was only partly mollified by this concession. He decided to commission his own inquiries into the political economy of uranium. These were spearheaded by a new member of his entourage, Jacques Baulin, a former advisor to the Ivoirian president and a controversial figure in his own right.[33] Foccart's journals and Areva's official history both present Diori's ceaseless pressure to increase uranium revenues as fickle behavior resulting from the undue influence of a nefarious, self-serving advisor (namely Baulin). But

Georges Pompidou and Hamani Diori during Pompidou's visit to Niger in 1972.
(Keystone France/Getty Images, used with permission)

increasing revenue had become a matter of extreme urgency for the
country and for Diori's political survival. A late 1960s dry spell had become
a full-fledged drought, leading to famine and deep political dissatisfaction
with Diori. Revenues from uranium were keyed to sales and profits were
years away. As of 1973, Niger still hadn't received any dividends.

Diori thus had ample motivation to maintain pressure on the French.
He had also begun to mistrust Foccart. Perhaps seeing a way to circumvent
"Monsieur Afrique," Diori and his advisors engaged the CEA on technopolitical terms, attempting to figure out alternative price calculations
without drawing on CEA expertise. Baulin's Canadian friends shared
demand projections based on forecasts for nuclear power plant construction. His friends at Electricité de France (France's nationalized electric
utility, which had just won a huge institutional battle with the CEA)
shared comparisons between the costs of different forms of electricity.
From all this, Baulin learned that uranium accounted for only 3 percent
of the cost of nuclear power production. He also discovered that nuclear

plants came with bonuses for French rural communities, staggering sums when set against Niger's national budget.

Baulin concluded that the price of uranium could rise without greatly affecting the commercial competitiveness of nuclear electricity.[34] Diori's advisors assembled the findings into a proposal for increasing the price of Somaïr's uranium. The company director reacted badly. He told Diori that the sale price of uranium was fixed three years in advance, presumably eliding the role of the still-secret cartel. Worse, increases in oil and sulfur prices were cutting into company profits. Niger should therefore expect even less in the way of revenues. Scandalized, Diori complained to the French foreign affairs minister. A week later, Diori and his minister of mines received a formal apology.[35]

Diori, like Bongo, had found OPEC's 1973 price increase an inspiring example of how producers of raw materials could cooperate to shift the global balance of power. After conferring with each other in early 1974, the two African leaders demanded a tripartite meeting of France, Niger, and Gabon to negotiate the price of uranium from Francophone Africa.[36] This took place in March 1974 in Niamey, Niger's capital city.

For Bongo, uranium occupied the same conceptual frame as oil. "We don't want any more distinctions between strategic products and ordinary products," he had told the newspaper *Le Monde* in what the French embassy in Libreville deemed a "deliberately disagreeable tone." Bongo insisted that both were "commercial resources, whose price we are entirely free to determine." Gabon, like OPEC, could unilaterally increase the price of uranium. France would have to decide whether to take it or leave it.[37]

Oil-free Niger, however, didn't want uranium to become an *ordinary* commodity. Instead, Diori placed the tripartite discussions under the rubric of nuclear exceptionalism. As Baulin wrote in his memoirs, the "Nigérien thesis on the unusual character of uranium" held that "the content of uranium transcended commercialism." Diori banked on the significance of atomic bombs and reactors for France's national identity, a bet made even shrewder by the recent, massive expansion of France's nuclear power program. Diori reasoned that if Niger's uranium could fuel France's exceptional nuclearity (my term, not his), surely France could fund exceptional contributions to Niger's treasury.[38] Reframing Bongo's logic, he proposed indexing uranium to the price of oil, a move that would increase Nigérien revenues tenfold.[39]

French delegates refused to accept calculations based on comparing energy outputs from uranium and oil. Instead, they sought to *de*-nuclearize uranium by insisting on its market banality. They maintained that the only way to determine the value of uranium was to treat it like an ordinary market commodity, which meant (among other things) disassociating it from oil. Revenues to African states could conceivably increase, but only if pegged to the *international* "revalorization" of uranium. To Nigérien ears, this reasoning sounded hollow and condescending.[40] The African delegations countered that when it came to uranium, the problem of calculability transcended ordinary commercial considerations: "Outside of the calculable parameters, there are other more significant ones that are not a function of calculation, such as the economic independence of France, the guaranteed satisfaction of its energy needs, a substantial savings in foreign currency and the reinforcement of the franc zone, and finally the solidarity of the three countries which could together represent 15 percent of the world uranium market if they coordinated [their efforts]."[41]

The Nigérien delegates—evidently unaware that France had already "coordinated" its marketing with other leading producers, thereby assuming exclusive access to Nigérien and Gabonese uranium in the process—appealed to France's breed of nuclear exceptionalism. Referring to the "planetary dimensions" of uranium-related problems, they insisted that this exceptionalism itself had value that could be expressed (among other ways) in terms of market shares.

Negotiations stalled. A week after the meeting in Niamey, Pompidou died. Traveling to Paris for the memorial service, Diori met with the prime minister to request the resumption of uranium negotiations within the week. He wanted to resolve matters before his scheduled appearance at the UN General Assembly's special session on natural resource sovereignty, which was already in progress. Five days later, Diori was overthrown in a military coup led by Lieutenant-Colonel Seyni Kountché.[42]

The timing of the coup led to speculation about France's involvement. Why didn't French troops rescue Diori, as they had Léon Mba in 1964? Given the extent of Foccart's intelligence networks, surely the French knew that the coup was in the works? Foccart claimed in his diaries that he favored intervention but others did not.[43] The Nigérien military, for its part, insisted that it had planned the coup long before the uranium negotiations. Diori's opponents accused him of pandering to French capital and allowing deplorable living conditions for Nigérien mineworkers. They dismissed his efforts to stand up to the French on matters nuclear, characterizing these as no more than "demagogic declarations aimed at international opinion."[44]

Whatever the case, the country's dire economic situation, famine, mounting political corruption (including the diversion of food and other relief aid), and the repression of opposition gave plenty of cause for discontent. Most scholars agree that Diori's fall was overdetermined. The regime "fell victim to its own contradictions."[45] Many have found France's official explanation—that it couldn't coordinate a rescue operation because of the crisis caused by Pompidou's sudden death—unconvincing. Nigérien scholars conclude that France simply didn't want to help Diori. He had caused too much trouble.[46]

Kountché promised to take a harder line with the former colonial power. Rather than pushing for greater revenues from sales controlled by the French, the new president negotiated for Niger to sell—directly and independently—an amount of yellowcake proportional to its financial stake in the mining companies. Other non-French investors in the mines could do the same.

Under Kountché, the state apparently found it more lucrative to plunge into "the uranium market" directly. Reliable, accessible sources on subsequent contracts signed by Niger are scant. Most agree that customers for the Nigérien state's portion of uranium included the following countries:

• Libya—perhaps up to 1,200 tons in the early to mid 1970s. These purchases supposedly occurred between the time that Libya signed the NPT in 1968 and the time that it ratified it in 1975.[47] Some reports suggested a second sale, perhaps up to 1,500 tons, in 1980–81.[48]
• Iraq—around 300 tons in 1981.[49]
• Pakistan—around 500 tons in 1979, mostly routed secretly through Libya, with more perhaps in the mid 1980s.[50]

During the Kountché years, Niger's market had a distinct geography, one that many Western governments would find increasingly dangerous and come to characterize as a black market. Niger—like France—didn't accede to the NPT until 1992. Local and regional issues mattered far more

to its leaders than Cold War superpower politics. For example, Kountché
threatened to cut off supplies to Qaddafi in January 1981 after Libya
attempted to annex Chad.[51] But he apparently changed his mind
a few months later, famously declaring that Niger needed funds so
badly that "if the devil asks me to sell him uranium today, I will sell it
to him."[52]

Kountché's priorities were not those of the NPT heavyweights. Relative
to the intense pressures his government faced, northern nuclear anxieties
seemed distant and insignificant. For a time, France also found advantages
in partitioning responsibility for uranium sales. Coming under fire after
news coverage of sales to Libya and Pakistan, it denied involvement by
insisting that each shareholder controlled its portion of product independently of the others.[53] In some respects, then, the simmering political
tensions enabled the mines themselves to thrive.

In the end, however, there were limits to how much uranium Niger
could sell under advance contract. It began resorting to spot transactions.
When the spot price began to decline in 1981, other spot sellers could
respond by stockpiling yellowcake until the price increased again. But the
Nigérien state could not afford that strategy: it needed the cash flow.[54]
Cogéma—a French parastatal company formed in 1976 to take over the
nuclear fuel cycle from the CEA—agreed to fill the gap. Its chairman
noted self-righteously that "the best support for Niger is to not give them
artificial prices." Simply guaranteeing sales would be far more valuable.[55]
After several rough years, the two states renegotiated their arrangements,
leaving Cogéma in charge of marketing Niger's uranium.

LE PRIX AFRICAIN IN GABON

The stakes of the 1974 tripartite meetings were very different for Gabon,
which had much smaller uranium reserves than Niger and earned far
greater revenue from oil than from ore. Six weeks after the tripartite meetings in Niamey, Gabonese officials and company executives signed the
papers that sealed the state's 25 percent stake in the COMUF. As Diori's
government fell, Bongo's rule strengthened.

In June 1975, Bongo celebrated Gabon's admission to OPEC by hosting
a meeting of the oil cartel. Banners that greeted delegates in Libreville
proclaimed Gabon "a country of dialogue, tolerance, and peace."[56] Was it

during a backroom discussion that Bongo offered to sell Iran 520 tons of uranium? Available sources are mute. Clearly, however, Bongo made the offer without consulting the COMUF or Uranex (the broker that was charged with selling French-controlled uranium).

The COMUF first learned about the promise to the Shah from Edouard Alexis M'Bouy-Boutzit, one of Bongo's senior ministers and an owner of 88 of the company's "B" shares.[57] In October 1975, M'Bouy-Boutzit instructed the company to make contractual arrangements to deliver the first 100 tons immediately, with the rest coming in the next few years. "This decision," he added, "is irrevocable."[58] This time, Bongo wouldn't tolerate any opposition.

Price posed a problem. Bongo had offered Iran the same price given to the CEA, a figure increasingly referred to as the *prix africain* and viewed as a form of development aid. Upon reflection, M'Bouy-Boutzit recommended instead that Iran pay a "world market price."[59] Although it still didn't reflect actual sales, the Nuexco exchange value had acquired sufficient heft to become equated with "the world market price" in most contracts (though not all, as we'll see in chapter 5). Using it would carry symbolic weight by aligning Gabon with the US, Canada, and Australia. Two years earlier, a similar argument had failed to persuade Diori. But since then the spot price had risen dramatically until it came to exceed the CEA price. The *prix africain* would remain a reward for the CEA's loyalty and investments. New customers like the Iranians, though, would pay more.[60]

The COMUF offered the Atomic Energy Organization of Iran (AEOI) the Nuexco spot price: $48 per pound of uranium oxide. The Iranians, however, did not want a spot transaction. AEOI president Akbar Etemad explained that fueling all of Iran's planned reactors would require 40,000 tons of uranium between 1977 and 1994, and 3,000 tons per year after that. Furthermore, Etemad hinted, Iran might want to invest in mining and exploration in Gabon following this mere token of a first order. Dangling the prospects of grander arrangements, Etemad offered to pay the *prix africain*: $24 a pound. COMUF executives and state officials refused, arguing that the CEA's preferential *prix africain* "could not serve as a standard referent for new buyers."[61] But they agreed to negotiate down to $39 a pound as "a good will gesture by the Gabonese Authorities."[62] The AEOI turned the offer down flat.

Discussions dragged on for two excruciating years. The parties continued to insist on the "purely commercial" nature of the transaction, though
it's hard to imagine that either the COMUF or the Gabonese state would
have tolerated the aggravation had this really been the case. By October
1976, the AEOI had secured a bargain rate of $28 per pound, yet it repeat-
edly prolonged negotiations by introducing new requirements and conditions. By early 1977, COMUF executives and M'Bouy-Boutzit were fed
up. Did the AEOI "really intend to engage in an equitable transaction,"
the minister wondered in a letter to Bongo, or was it trying "to profit
excessively from your promise to the Shah"?[63]

Hoping that a more personal approach would help, COMUF president
Jacques Peccia-Galleto invited Etemad to dine with him in London at the
second annual meeting of the Uranium Institute in June 1977. The AEOI
had just joined the Uranium Institute, and Etemad was delivering the
banquet address. Seen from Gabon, the AEOI's membership in the Uranium
Institute was striking. The COMUF wasn't a member in its own right,
though its parent companies were. This meant that Gabon did not appear
on the UI masthead, an absence that clearly signaled the subordination of
Gabonese uranium to French control.

Iran's massive nuclear ambitions, by contrast, made it an instant powerhouse at the Uranium Institute. The AEOI had signed large uranium
supply contracts with South Africa's NUFCOR, acquired a 10 percent
share of the Rössing mine, and invested significant sums in France's
Eurodif uranium enrichment plant. Iran's desire to avoid excessive dependence on American technology aligned perfectly with the UI's *raison
d'être*, as we saw in chapter 2. In his banquet speech, Etemad lamented
the uranium enrichment policies of the US. "I genuinely hope that
the world refrains from the political manipulation of this important
resource, as this would of course have a detrimental effect on the
economics of the fuel cycle," he declared.[64] His rehearsal of the separation
between politics and commerce doubtless had most diners nodding
sagely.

Did Peccia-Galleto privately note the irony of Etemad's declaration?
Whatever the case, their tête-à-tête in London didn't help. Back in Gabon,
company and state officials began plotting to withdraw from the negotiations without sparking a diplomatic incident.[65] Then, suddenly, everything
seemed to fall into place. A contract was drafted in English and translated

into French. It included a nominal safeguards clause, even though weapons proliferation had never been a point of discussion in the negotiations.[66] But as political opposition to the Shah deepened over the course of 1978, the deal looked doubtful.[67] Etemad had resigned; some thought he might be imprisoned.[68] Gabon's ambassador in Iran couldn't get an appointment with the new AEOI officials. Bongo, according to M'Bouy-Boutzit was "very vexed."[69]

M'Bouy-Boutzit gave the Iranians a deadline of 26 October. The ultimatum passed unheeded. The Iranian revolution had killed the contract. Nevertheless, the negotiations themselves represented a shift in the relationships that bound the Gabonese state, the COMUF, the CEA, and other uranium purchasers. The negotiations enacted the meaning and limits of postcolonial Gabon's sovereignty over natural resources. Until the early 1970s, the COMUF and the French state as embodied by the CEA and Uranex had full power over the circulation of Gabonese uranium. As the decade progressed, the combined effects of oil wealth, the changing context of Françafrique networks, and global debates on natural resource sovereignty gave the Gabonese state new financial and political leverage over its uranium.

The COMUF, forced to heed Bongo's promise to the Shah, had tried to regain some control over subsequent negotiations. It invoked the "world market price" to recast the meaning of Gabonese sovereignty over uranium as a matter of "commerce." Government officials followed suit, invoking "commerce" in efforts to sell uranium outside the COMUF's orbit, most notably to the Philippines in 1977. Company executives complained that such attempts undermined their credibility, especially when offers were made at the *prix africain*. They repeatedly asked officials to refrain from issuing offers directly.

Professing ignorance, the Director of Mines claimed that he would have to "ask around Libreville to find out the origin of the interference" with company business.[70] Available archives do not reveal the results of his investigation. Did some officials feel that alternative "circuits of negotiation"[71] for uranium would express state sovereignty? Did they believe that their "B" shares entitled them to sell uranium directly? Or did they anticipate that, as with deals conducted within Françafrique networks, contracts they landed would provide juicy additions to their incomes?

Whatever the case, by 1977 the meaning of the *prix africain* had changed. After serving as a subsidy to the COMUF and a privileged rate for the CEA, the figure became a number used and produced by the Gabonese state. State officials acquired final approval over its value. Although company executives provided data concerning their production costs, price levels contracted by other producers, and so on, the Gabonese state made the final determination. The shift, subtle but meaningful, was accompanied by others as state officials became more involved in negotiations over contracts and prospecting partnerships with firms in Italy, Japan, Belgium, Germany, and South Korea.[72]

DISTRIBUTING SOVEREIGNTY IN "YAYA" BONGO'S STATE

During Gabon's first decade of independence, the COMUF operated largely outside of Foccart's network, a rarity for a French company working in Gabon. This may have stemmed from the CEA's stature in the Gaullist state and the COMUF's small size relative to Gabon's oil, manganese, and lumber companies. When Foccart finally visited the Mounana mine in January 1970, his observations focused on the mine's location in the Haut-Ogooué.[73] At the time, uranium was shipped by *téléférique* over the Gabonese border to Congo-Brazzaville's port in Pointe Noire, where it boarded a ship for France.[74] This routing irked Bongo because it subverted the nationalist meaning of these ores. The Transgabonais railroad promised a technologically resplendent connection between the Haut-Ogooué and Libreville. Pending its completion, Bongo sought other ways to bring the COMUF closer.

In 1979, Bongo obtained the appointment of his old friend Maurice Delauney to the top slot on the COMUF's board of directors. As France's longtime ambassador to Gabon, Delauney was close to Foccart and had helped to usher Bongo into the presidency. Clearly the new chief was Bongo's man. Delauney explained in his memoirs that "the president wanted me . . . to make myself available for whatever mission he might want to entrust me with."[75] The technologists charged with operating the mine, however, didn't much "appreciate" the slippery politician.[76] One engineer who served as interim director under Delauney resigned, citing "ethical reasons."[77] In coping with these tensions, COMUF managers sought to distribute sovereignty over uranium into separate symbolic and operational domains.

The launch of the mine's yellowcake plant in 1982 epitomized the delicate dance of distributed sovereignty. Perhaps spurred by Bongo's rogue request to Union Carbide, COMUF management decided in 1977–78 to build a yellowcake plant at Mounana and stop sending the ores to France for further concentration into yellowcake at the Gueugnon plant. Although the conversion into hexafluoride and subsequent enrichment would keep Gabonese uranium deeply embedded in French technological systems, making the yellowcake *in situ* would save money, mainly because Gabonese employees commanded lower salaries.[78] Ensuring that the plant had the proper cultural and political valence required careful calibration, however. Producing yellowcake locally would eliminate several hundred jobs in France, and the salary disparities between French expatriates and Gabonese nationals could look like neocolonial exploitation.

The plant inauguration—that timeless ritual of industrial grandeur— marked Gabon's ascension in the global technopolitical order from lowly supplier of raw materials to manufacturer of yellowcake rivaling that of non-African producers. Bongo himself presided over a two-day ceremony that awarded medals to employees for 21 years of service. The odd number explicitly referred to a "coming of age," marking Gabonese workers as (industrial) adults who had finally transcended the colonial rendition of Africans as children. Speakers gushed about the "Gabonization" of the personnel, the mine's magnificently modern infrastructures, and the social harmony that radiated out from the company town. The proceedings attributed the success of such modernity to Bongo himself.

Not incidentally, the proceedings were designed to be a regional per- formance of Bongo's political authority. The leader of the local branch of Bongo's Parti Démocratique Gabonais praised "Yaya Bongo's . . . policy of progressive and concerted democracy, spearhead of Gabon's economic and social development." Bongo in turn awarded Gabon's "Étoile Équatoriale" to over 40 COMUF managers and employees, giving even the French honorees a share of Gabonese nationalism. The festivities displayed the alliance between the company and the Bongo-state by subsuming the worker-citizens who, as we'll see in chapter 7, had a different interpreta- tion of their labor. The company magazine concluded grandly that the occasion marked a "turning point in history."[79]

President El Hadj Omar Bongo (front row, fourth from left) and his entourage at the opening ceremony for the COMUF's yellowcake plant, 1982. (photo courtesy of COMUF)

Outside Gabon, Director of Mines Paulin Ampamba-Gouerangué, owner of 225 "B" shares,[80] calibrated this nationalist narrative for an international audience of investors and competitors. During a presentation at the Uranium Institute's 1982 meeting, he began by laying down markers of statehood: Gabon's independence in 1960; Bongo's presidency; and some generic indicators like population, climate, and breakdown of GDP. Ampamba-Gouerangué insisted that Gabon was not a French preserve. "Many companies of various nationalities have operated and are still operating in Gabon," he affirmed, listing examples. His exhaustive geological descriptions served as a catalog of investment opportunities.[81]

Concluding his presentation, Ampamba-Gouerangué noted that uranium producers could "dispose freely of their products" on three conditions. First, they had to satisfy "national requirements" when the time came. Gabon had no nuclear power program, so this condition represented a fantasy of greater nuclearity. Second, they had to accept the state's authority over contract approval—a clear assertion of sovereignty. Third, customers had to be NPT signatories unless there was a "special agreement between the Government of Gabon and that of the country concerned."[82]

This final condition exposed ambiguities in uranium's status relative to NPT compliance and Gabon's relationship to France. Yellowcake had been explicitly excluded from IAEA definitions of nuclear materials. But unlike France or Niger, Gabon had acceded to the NPT in 1974 in yet another expression of its sovereignty. NPT adherence thus served to mark Gabon's autonomy from France, while the reference to potential "special agreements" allowed for the French exception.

In 1983, the COMUF joined the Uranium Institute, making Gabon the first postcolonial nation to appear on the UI's masthead. In the marketplace symbolically delineated by the UI, COMUF uranium became Gabonese yellowcake. Back in Gabon, the company and the state had found a comfortable middle ground. Legally, the state retained final say over contracts. In practice, the COMUF and its new parent company, Cogéma, controlled customer selection and prices. France remained the COMUF's largest customer and continued to receive a significantly better price than other customers.[83] The COMUF, for its part, invested in Bongo's pet projects and agreed to intensify training programs and hire more Gabonese into management and technical positions.[84] Available archives did not reveal whether—as in the well-documented case of Elf-Gabon[85]—any of the COMUF's investments were in reality personal payments to Bongo. Clearly, though, Delauney's presidency had enabled an accommodation between the COMUF and the Gabonese state elite.

In 1989, Michel Pecqueur took over as president of the COMUF. In contrast to Delaunay the diplomat, Pecqueur belonged to the French technopolitical elite. A graduate of the Ecole Polytechnique and a member of the Corps des Mines, he had risen rapidly through the CEA's ranks and had served as president of Cogéma before briefly switching sectors to preside over Elf. During his ascension, he had forged personal ties with Gabonese state officials, including El Hadj Omar Bongo.[86] His appointment signaled that, in the ongoing accommodation between Gabonese and French elites, the distribution of sovereignty profited both parties. The cost, as we shall see in part II, was borne by the workers.

CONCLUSION

Over the course of the 1970s, Gabonese and Nigérien leaders struggled to define uranium's role in the enactment of state power. As global debates over natural resource sovereignty grew, and as OPEC flexed its muscle, Bongo and Diori began to strain against the defense accords signed in the heat of decolonization. A decade into independence, the accords seemed as much prison as safety net. In the figures that it contributed to the development of global uranium data, France still assumed that Gabonese and Nigérien uranium, passing through conversion and enrichment plants on metropolitan soil, were its products.

Although outright nationalization of uranium companies was never an option, Bongo and Diori found leverage in matters of price—albeit in different ways and with different outcomes. Bongo promoted the *banality* of uranium so that he could make it commercially analogous to oil. Diori emphasized its *exceptionalism* and its political value to France, insisting that nuclearity commanded a premium. Their efforts met with limited success. A tremendous imbalance of power and wealth gave France the upper hand. French officials countered attempts to value Nigérien yellowcake in terms of nuclear exceptionalism by framing uranium in ordinary market terms. But their refusal to include nuclear exceptionalism in the "market" value of African uranium carried substantial risks.

France's success stemmed from the broad distribution of its uranium activities, and especially because of the work done, in conjunction with other states and companies, to de-exceptionalize the uranium business. The uranium cartel, the Red Books, global predictions of supply and demand, and France's own burgeoning nuclear power program all built on the notion that uranium could *and should* be traded like any other commodity. All this made it difficult for Gabon and Niger to gain leverage over the price of uranium. State officials had trouble finding and sustaining a customer base that did not involve French expertise and infrastructures. Ultimately, the customers that Gabonese and Nigérien state officials did find independently—Iran, the Philippines, Libya, Pakistan—were not reliable sources of income.

In Gabon, Bongo sought to use uranium to make alliances with OPEC powerhouse Iran. Because Gabon's economy ran primarily on oil, uranium mattered to Bongo more as a political chip than as a source of income, though that was certainly not the case for COMUF workers. Bongo's personal bank accounts may well have benefited under Delauney's leadership of the COMUF, though I found no direct evidence for this. Far more visible, certainly, was Bongo's use of the COMUF as an enactment of

Gabonese modernity, a spectacle of development enabled and ennobled by the big man El Hadj Omar Bongo, the benevolent father Yaya Bongo.

In Niger, Kountché (Diori's successor) used uranium to palliate regional tensions with Libya and alleviate budget constraints at home. Here, Western nuclear exceptionalism confronted postcolonial capitalism, regional politics, and the dreadful mundanity of poverty. With few other revenues, Niger could not afford to stockpile its product when prices slumped. Sales oversight reverted to France until 2007 when then-president Mamadou Tandja again demanded control, as we'll see in chapter 10. Divesting uranium of nuclearity pushed Nigérien yellowcake into a market, just not a market that was licit by NPT definitions. Yellowcake from Niger ended up in the 1998 Pakistani bomb and would have fueled a Libyan bomb had Qaddafi not renounced his program.

Just as uranium connected France to Gabon and Niger in ways shaped by the imbalances, values, and valences of *la Françafrique*, so too did uranium shape the wider world's implication in apartheid South Africa and in its colonial occupation of Namibia.

NUCLEAR FRANKENSTEIN

In February 1979, the newly formed "World Campaign against Military and Nuclear Collaboration with South Africa" hosted a high-profile seminar at the United Nations during which participants described how apartheid transcended its obvious manifestations in racial segregation and violence.[1] Embedded deep in South Africa's technological infrastructures, the apartheid system stretched beyond that country's borders, relying on US and European expertise and technology for its maintenance. In World Campaign director Abdul Minty's pithy formulation, South Africa was the "nuclear Frankenstein" of the West. Only through strict sanctions could the West hope to control—and eventually fell—its monster.[2] Thus began a global campaign that, as I have argued elsewhere, used technopolitical strategies to combat the apartheid regime.[3]

Campaigners had been building their case for some time. In 1975, members of the West German cell of the African National Congress broke into the South African embassy in Bonn and stole nine files that covered uranium enrichment negotiations between the German firm Steag and the South African parastatal company UCOR. Using these documents, the ANC argued that South Africa and West Germany were secretly cooperating to build a large-scale uranium enrichment plant based on jet nozzle technology that was developed in Germany rather than (as the AEB's Ampie Roux had insisted) in South Africa. They also claimed that this enriched uranium was destined for a South African atomic bomb.

Spreading their pamphlet *The Nuclear Conspiracy* to every embassy in Bonn, to German ministries, to members of Parliament, and to the German newspapers, the ANC challenged the AEB's claim that "South African conditions" would ensure the commercial success of jet nozzle enrichment.[4] Clearly aware that—for a Western audience worried about Cold

War nuclear proliferation—technological and economic reasoning trumped racism, ANC analysts did not point out that apartheid labor practices explained the low cost of electricity in South Africa. Instead, they used data from existing enrichment plants to argue that the capital and operational costs of the proposed jet nozzle plant would far exceed any advantage from cheap electricity. "South Africa's export of enriched uranium could only be profitable," they argued, if it "operated a virtual black market supplying enriched uranium to states which were not prepared to accept the safeguards of the Non Proliferation Treaty."

Furthermore, cheap energy from coal meant that the erection of nuclear power plants in South Africa was not economically justified. Given the AEB's obsession with secrecy, only "political and military considerations" fully explained the nuclear program:

With control of nuclear material the Pretoria regime could consolidate its military links with the western powers, while at the same time operating outside international controls, it could buy "friends" by providing nuclear materials and technology to non-signatory states. The development of nuclear weapons would be used to demonstrate to supporters in the country that the regime could stand alone in defiance of world opinion and internal opposition. The aggressive posture in Africa would be enhanced, and having already threatened to "bloody the nose" of "interfering" African Presidents, Pretoria will threaten to reduce recalcitrant or "unfriendly" African capitals. The regime could and would use its nuclear muscle to weaken boycotts, embargoes, and sanctions: it could blackmail Africa and the international community into acquiescence in its apartheid policies at home and expansion of its economic stranglehold over the continent.[5]

The ANC also challenged the AEB's nationalist nuclearity narratives. South Africa had long exchanged scientific knowledge and technology with the West. The jet nozzle design was too unusual to have been developed independently in two places; German experts must have given South Africans the design. The West's complicity in South African nuclearity subverted its own non-proliferation discourse. "All those who join hands with the Pretoria regime must," the ANC asserted, "bear the responsibility for the holocaust that must surely follow if this nuclear conspiracy is allowed to reach its aims."[6] A few months after *The Nuclear Conspiracy* appeared, the German firm Steag withdrew from its prospective collaboration with UCOR, citing South Africa's inability to guarantee sufficient uranium feed from its supply, as we saw in chapter 3.

American and Western European governments ignored the ANC's claims in public, though behind the scenes they suspected a potential South African weapons program. Presented in the language of the non-aligned movement and anti-imperialist socialism, the ANC's largely circumstantial evidence linking enrichment to bombs was easily dismissed. But in August 1977, a Soviet spy satellite photographed (and a US satellite later confirmed) a potential nuclear weapons testing ground in the Kalahari desert. That got everyone's attention. A month later, international revulsion toward apartheid heightened with the murder of Steve Biko.

These events finally enabled the UN's Group of 77 to oust South Africa—after nearly two decades of trying—from the IAEA Board of Governors. Encouraged, activists continued probing the technopolitics of South African nuclear development. In 1978 Zdenek Červenka and Barbara Rogers (whom we last encountered in chapter 3, resigning in disgust from her job at the British Foreign Office) published *The Nuclear Axis: Secret Collaboration between West Germany and South Africa*. The book analyzed South African uranium contracts, the Kalahari test site, and everything in between. It argued that transnational networks underwrote South African technology and expertise, and therefore apartheid itself.

Červenka and Rogers pointed to parallels between Nazi and nuclear holocausts, and discussed the Nazi past of the general who'd traveled secretly to Pelindaba during the enrichment negotiations. They quoted a German official saying that "in no part of Africa were blacks better cared for than in South Africa." One Steag scientist had apparently insisted that "Steag had no secrets to hide about its own enrichment process. The blackest Zulu kaffir can find out all what he wants to about it."[7] Had Germany hoped to get its own atomic bomb out of the collaboration? Červenka and Rogers pointed out that this prospect—made all the more terrifying by their examples of unrepentant racism—was explicitly prohibited by the NPT.

The mainstream international press initially dismissed *The Nuclear Axis* as the conspiratorial rantings of political extremists. But in September 1979, a year after the book's publication, a US Vela satellite detected an intense double flash of light—a telltale sign of a nuclear test—over the South Atlantic. American scientists and intelligence agencies scrutinized sensor readings and climate data. Their analysis was inconclusive. But scientists at Los Alamos who'd run the data through their own models

Poster from campaigns to end military and nuclear collaboration with South Africa. (from collection of Laka Foundation)

seemed convinced that the double flash was the result of a surface nuclear burst.[8]

Suspicion immediately fell on South Africa and its main military trading partner, Israel. Had the apartheid government managed to test a bomb after all? Had it done so in conjunction with Israel? Or had South Africa simply facilitated the one and only test of an Israeli bomb? The US government officially concluded that the flash was a meteoroid hitting the satellite. It remained unwilling to accuse South Africa of building nuclear weapons, or Israel of testing one. But many other observers were prepared to do so. The Vela incident, as it became known, piled onto the mountain of evidence suggesting an apartheid bomb program. Over the next decade, scholars, policy wonks, and activists churned out reports, books, articles, and documentaries making the case.[9]

Activists were right about many things. The apartheid state did have a nuclear weapons program. In its dying days, the apartheid government admitted to building six atomic bombs and—eager to display their reformation as responsible global actors and afraid of leaving nuclear weapons in the hands of black Africans—had the bombs dismantled in 1994. Claiming that they didn't want blueprints to fall into the hands of terrorists or other unreliable actors, they also ordered the destruction of archival records documenting the weapons program. This radical production of ignorance and invisibility, perpetuated under the guise of proliferation prevention, has left historians nibbling at the edges of evidence.

Memoirs and lingering bits of documentation cast doubt on some activist theories. It seems likely that South Africans did develop the Valindaba enrichment process from scratch, perhaps inspired by German work but executing the final design quite differently. It seems unlikely that South Africa had made enough progress toward building weapons to be able to test one in the South Atlantic in 1979, though Israel certainly had.[10] All evidence, however, strongly supports the activists' central case: from uranium production to bomb building, international technological systems were deeply implicated in South African nuclear development. To put it in our terms, the apartheid bomb was a creature of transnational technopolitics.

PLUNDER OF NAMIBIAN URANIU

"No person or entity, whether a body corporate or unincorpora may search for, prospect for, explore for, take, extract, mine, process, refine, use, sell, export, or distribute any natural resource, whether animal or mineral, situated or found to be situ within the territorial limits of Namibia..."

(Decree No. 1, adopted by the United Nations Council for Namibia on 27 Septemb

UNITED NATIONS HEARIN
NEW YORK, 7-11 JULY 198

The likelihood of an apartheid atomic bomb elevated uranium's position in the struggle for freedom in southern Africa. By forging ties with anti-nuclear activists, the anti-apartheid movement gained access to a wider set of knowledge, tools, and arguments. By itself, African uranium did not attract much interest from the European anti-nuclear movement, which preferred to set its sights on the high-tech systems closer to home. In highlighting the ore's role in an illegitimate weapons program, anti-apartheid activists sought to make southern African uranium a more compelling target to anti-nuclear activists and other opposition groups.

This chapter explores the technopolitical strategies deployed by both sides in the battle over Namibian uranium. The UN Council for Namibia's Decree No. 1 had placed natural resources of all kinds at the center of the struggle for independence. Uranium came to occupy center stage when activists traced the chains that linked yellowcake from the Rössing mine to atomic power plants and weapons elsewhere. The power of nuclear things, I argue, made uranium a more compelling avatar for the liberation struggle than, say, copper or diamonds. European allies of the Namibian liberation struggle clearly understood this.

In the first two sections of this chapter, I argue that activists sought to *nuclearize* Rössing's uranium on the world stage. Boosted by the prospect of a South African bomb, their case for the illegitimacy of Namibian uranium grew more robust and visible. They laid out their points in UN hearings and international lawsuits, European marches and rallies, US lobbying firms, and Congressional hearings. Critics also emphasized the role

on facing page: Poster for 1980 United Nations hearings on Namibian uranium. (courtesy of African Activist Archive Project) of Namibian (and South African) uranium in capitalist nuclear systems, paying special attention to the technopolitical attachments Rössing had formed in North America, Western Europe, and Japan. They portrayed these attachments as doubly immoral because of their violation of Namibian resource sovereignty and their support for the illegitimate weapons program of a racist regime. Gaining momentum throughout the 1980s, these critiques made it increasingly difficult to circulate Namibian uranium.

As with other transnationally traded goods, the distinction between licit and illicit transactions depended on one's place in the geography of nuclear things.[11] Rössing and its parent company, Rio Tinto Zinc, certainly did not deem their own activities illicit. In the third section, I demonstrate how company officials used a language of market autonomy bolstered by geographically distributed market devices, including contracts channeled through a front company in Switzerland, to maintain that Rössing uranium was an *ordinary* commodity to be governed by supply and demand. For them, politics could (and should) be bracketed as irrelevant.

As liberation struggles made its claims increasingly untenable, Rössing sought help from its partners further along the nuclear fuel chain. Activists were right in that Namibian yellowcake played a central role in the uranium market of Cold War, capitalist economies. It helped northern conversion and enrichment plants stay in business. It fueled bombs in Britain and reactors in Europe and Asia. The final section of this chapter argues that these links enabled Rössing and RTZ to counter the technopolitical claims advanced by activists with powerful strategies grounded in material transformations. As we will see, British and French conversion plants worked to give Namibian uranium a new nationality. Swiss power plants swapped fuel with facilities elsewhere that were forbidden by their governments from loading Namibian uranium into their reactor cores. RTZ and its allies, I conclude, deployed technopolitical mechanisms to reconfigure the provenance of Namibian ore, dissolving it into a uniform pool of global yellowcake.

Still, when we contemplate the considerable imbalance in resources and power between RTZ and the campaign against Namibian uranium, it's remarkable that activists made any headway at all.

AGAINST NAMIBIAN URANIUM

The struggle for Namibian independence found an extraordinary asset in Barbara Rogers. In 1972, a year after she left the Foreign Office, Rogers traveled to South Africa and Namibia to witness apartheid herself. She was particularly moved by a visit to northern Namibia, where she witnessed the strong support for SWAPO. Returning to Britain, she joined the Friends of Namibia, an offshoot of the liberation struggle, and began to write about Namibian uranium and South African nuclear ambitions, hoping to stimulate an international boycott of Rössing's yellowcake.[12] She had no illusions about achieving this goal. "It is of course the deliberate policy of the South Africans to include Namibia in its major international projects," she wrote in 1974, "thus establishing vested interests in Western countries in the preservation of the occupation régime there."[13] In other words, Western industries felt that South African colonial occupation promoted the political stability they needed for commercial success in Namibia.

In 1975 Rogers wrote a startling report that exposed a series of relationships that Rössing and its international partners would have preferred to keep secret. She identified the company's major shareholders, investors, and sales contracts, paying special attention to the history of its two contracts with British Nuclear Fuels. South Africa's uranium industry and its nuclear program, Rogers argued, had benefited from scientific, technological, and financial collaboration with the West from their inceptions. South Africa was probably selling uranium free of safeguards to France, the UK, Iran, and Israel. Evidence suggested that South Africa had launched a nuclear weapons project destined to use uranium from its new enrichment plant, which Rogers speculated would be fed by Rössing product.[14]

The report argued against the legitimacy of uranium prices. The British government had insisted that Rössing uranium would be significantly cheaper than yellowcake obtained elsewhere. Rogers knew full well that the contracts didn't specify a fixed price but were keyed to the so-called "world market price," a nebulous figure derived during yearly negotiations and loosely linked to the Nuexco spot price. "The estimated price of Namibia uranium to the UKAEA, possibly about $5.70 a pound in the original contract, is currently far below prevailing prices," Rogers wrote. "Given the expense of mining such low-grade deposits in Namibia, it

seems highly improbable that this is in fact the price that the British
Government will be paying." Even if Rössing uranium were cheaper, this
was only because it was, in essence, stolen: "[B]efore uranium mining could
be initiated, multi-million dollar settlements were recently made with the
Aborigines in Australia and the Eskimos in Canada, for use of their traditional rights in the wilderness areas involved. In Namibia, the minerals
are plundered with no attempt at an agreed compensation settlement with
the Africans."[15]

British officials had also invoked the need to protect their uranium
supply against whimsical policy changes by foreign governments. But the
notion that the South African government didn't control uranium exports,
Rogers wrote, was "so far from the truth as to be ludicrous." In addition
to having final say over all contracts, the South African government had
"a clearly stated policy that exports should be in the most highly processed
form possible, and explicitly reserves the right to ban all exports of
uranium oxide in order to channel its production into the proposed
enrichment plant."[16] Although some of her speculations proved wrong,
Rogers was, as we saw in chapter 3, correct on this and on other key
points.

The fight against Namibian uranium contracts that would unfold in the
next 15 years rested on the foundation laid by Barbara Rogers's 1975
report: No matter how much the global players tried to bury politics in
economics, the uranium trade was driven by political considerations.
Rössing was plundering Namibia, which meant that Britain was buying
stolen goods. Western Europe, North America, Japan, and Iran, by virtue
of their investments in Rössing and in South Africa, were deeply implicated in South Africa's nuclear program and hence in apartheid itself.
Finally, as imperative as Namibian decolonization was in its own right, the
likelihood of an apartheid atomic bomb made it even more urgent.

In 1977, Rogers cofounded the Campaign Against the Namibian
Uranium Contracts (CANUC) under the auspices of the Namibia Support
Committee. Constituted in 1974, the NSC distinguished itself from Britain's larger anti-apartheid movement by focusing only on Namibia. Believ-
ing that Namibian independence could precede and perhaps precipitate
the end of the apartheid regime, NSC members conducted separate political campaigns.[17] Of these, CANUC probably boasted the highest profile.
"Not an anti-nuclear campaign as such," its ambitions were both "more

limited (to stop the illegal importation of Namibian uranium—at present half of Britain's supply) and much broader (to promote political support for SWAPO and the National Union of Namibian Workers and to support the sanctions campaign against South Africa.)."[18]

CANUC quickly learned the value of cooperation, forming coalitions with anti-nuclear organizations and groups opposing RTZ. The points Rogers had outlined in her unpublished reports were ideally suited for forging these alliances. But she hesitated to publish these conclusions under her name for fear of violating the Official Secrets Act she'd signed while employed at the Foreign Office. The claims were therefore reprised and updated by her CANUC colleague Alun Roberts in his 1980 booklet *The Rössing File: The Inside Story of Britain's Secret Contract for Namibian Uranium.*[19] A documentary titled *Follow the Yellowcake Road* and covering much of the same territory was aired on British television that March. Meanwhile, CANUC forged alliances with activists in France, Japan, and the Netherlands. These coalitions met in July 1980 when the United Nations Council for Namibia convened a week-long hearing on Namibian uranium.

NUCLEAR PEDAGOGIES

Aimed at a global public, the UNCN hearings made the case for the centrality of Rössing's yellowcake in capitalist nuclear systems and established uranium as an avatar for the Namibian liberation struggle. In this section, I argue that the hearings offered a global forum for *nuclearizing* Namibian uranium, grounding that nuclearity in Rössing's technopolitical attachments.

SWAPO delegate Theo-Ben Gurirab—embodying the UN's recognition of SWAPO as the "sole authentic representative of the Namibian people"[20]—delivered a fiery opening statement in the rich language of Third World nationalisms:

[O]ur country still languishes under an illegal, colonial occupation, the African masses there suffer daily under political repression and military aggression and the working people of Namibia are held in chains as labour hostages.

In the meantime, the racist Boers and the capitalist blood-suckers reap super-profits by the ruthless exploitation of the human and natural resources of our country.

The repugnant *apartheid* colonialism of racist South Africa and the plundering menace of the transnational corporations are the twin evil forces which prevent Namibia's independence.

Mercenaries (the filthiest scums of this world, the so-called volunteers) and covert activities are used by the racists, zionists and imperialists in our part of the world to destabilize independent governments and to sabotage and undermine the armed struggle.[21]

To quash any doubts concerning Rössing's complicity with the "racist Boers," Gurirab presented a secret company memo that had "fallen in SWAPO's hands" and described a "security scheme" intended to "maintain a state of preparedness for Civil or Labour or Terrorist attack against the mine."[22] The document listed the arms held at the mine's security offices, including semi-automatic shotguns, automatic rifles, pistols, tear gas, grenades, and plenty of ammunition. It also specified possible security actions. In the Alert stage, for instance, Rössing would advise the South African police, close all bars, and remove the alcohol. In an Attack stage, "all armed units [should] defend personnel and mine property as the situation dictates" with assistance from the police and the army. "SWAPO Central Committee," Gurirab concluded,

has decided that our people will, through the government of an independent state of Namibia, present a bill of indictment and of damages and demand compensation from any person, entity or corporation, for example, Rio Tinto Zinc, which during the pre-independence period is involved in illegal acts and in plunder of our uranium and other resources in collaboration with the illegal occupational regime.[23]

Unsurprisingly, Gurirab gave no indication of the confidential assurances, which we saw in chapter 3, in which Peter Katjavivi stated that "a SWAPO Government would not disturb RTZ's position in Namibia."[24] The "security scheme" memo and the failure of independence talks in 1978 might well have changed the outlook of SWAPO's leadership. Whatever the case, the UN hearings demonstrated the political value of focusing on Rössing and its uranium.

As the hearings unfolded, Namibian nationalism gave way to nuclear pedagogy. While questioning the experts they'd called to testify, UNCN members refused to respect the structural and ontological divisions instantiated in the IAEA, the NPT, and their political instruments. For example, they rejected the suggestion that uranium ore wasn't nuclear enough to fall under the safeguards and inspection regime. IAEA official Guy Ferri had explained that, because South Africa hadn't signed the NPT, the IAEA didn't inspect its facilities. The chairman of the proceedings asked him whether "the IAEA normally inspect[ed] uranium mines in countries with full-scope safeguards—Canada, Australia—i.e., is it unusual not to inspect Namibian mines, or is that left to the countries themselves?" Ferri replied:

Nuclear material as defined in article XX of the Agency's Statute does not include uranium ore or ore residue. Therefore, IAEA safeguards do not apply to uranium mines, nor does IAEA inspect uranium mines. Moreover, Information Circular 153 . . . provides that NPT safeguards shall not apply to "material in mining or ore processing activities."

Ferri could not have found a less sympathetic reception for this denuclearization settlement. Emphasizing uranium's nuclearity had, after all, enabled SWAPO and its allies to generate tremendous political momentum. For them, uranium was—had to be—a fully *nuclear* thing.

Council members and their allies rejected the notion that the IAEA had cleansed itself when it ejected South Africa from its Board of Governors in 1977. Ferri insisted that the apartheid state hadn't benefited from the IAEA's technical assistance programs, which focused on "less developed countries." But Barbara Rogers, who'd been called as a witness and who continued to participate throughout the proceedings, issued a sharp rebuke:

I find it hard to believe that the Agency is in fact as passive, as slow moving and as stupid as some of the answers would indicate. For example, we are assured that the Agency renders no technical or any other assistance to South Africa. How can that be reconciled with the fact that two South African scientists were reported to be sponsored by the IAEA in the last few months and visiting the Argonne National Laboratory in the US? That was a visit that was protested by workers at the Laboratory, who objected to the IAEA sending scientists from South Africa. Is that not in fact a case of the Agency rendering technical assistance?[25]

Equally skeptical, Abdul Minty cited South African involvement in an IAEA-sponsored conference on uranium extraction technology. Expulsion from the Board of Governors wasn't enough, he insisted. South Africa should be expelled from the IAEA altogether, and the UNCN given membership in its place.[26] Namibian nationalism would thereby become nuclear.

All week long the hearings emphasized Rössing's imbrication in matters nuclear. Apartheid leaders counted on Namibian ore for their enrichment and weapons programs. The "security scheme" proved the mine's collusion with the South African police and its support of colonial occupation. As a partner in the illicit South African nuclear state, Rössing transferred its complicity onto its customers. In imparting these lessons, the hearings kept returning to the problem of safeguards. The inability to verify end-use made the Namibian uranium trade illicit in both nuclear and postcolonial terms. Activists also insisted that the lack of end-use restrictions enabled Rössing to charge a premium for its uranium—an interpretation that, as we'll see, wasn't quite right.

The hearings sought to establish that the entire structure of the uranium market depended on illicit southern African ore. Apartheid was built on *international* technopolitical networks, and only full-scale sanctions could bring it down. Economist Stephen Ritterbush testified that Japan and Europe had "two major desires when it comes to uranium: a desire for assured access to . . . existing uranium producers and a desire to develop new uranium reserves in 'politically stable' countries that will enable them to circumvent unilaterally imposed supplier restrictions on international uranium sales." The restrictions imposed by the US, Canada, and Australia had pushed Europe and Japan to sign large contracts with African producers.

Ritterbush also maintained that market devices (my term, not his) produced data that overlooked the "market concentration of net exporters." The US and France would soon be net uranium importers. Without them, he insisted, "five countries—South Africa, Namibia, Canada, Niger, and Gabon—account for 95 per cent of the world production that is now available for export." After instituting end-use restrictions in response to India's 1974 "peaceful nuclear explosion" (which benefited from Canadian technology), Canada did not have uranium freely available for export. South Africa and Namibia might therefore account for up to half of *net* exports of uranium.

RTZ, in Ritterbush's analysis, accounted for roughly 25 percent of "world" production. The apartheid state's influence on Rössing's operations meant that "Pretoria is able to control a sizable proportion of the world's uranium production and reserves available for export. . . . [T]hat control gives Pretoria a degree of political and economic leverage not only as

regards the supply and price of uranium but also as regards the formulation of foreign policy towards South Africa itself and . . . [its] present
position in Namibia."[27]

Making a similar point, Japanese activist Yoko Kitazawa testified to the
gap between the Japanese government's official endorsement of UNCN
Decree No. 1 and its actual business practices. Capitalism even trumped
racism, allowing Japanese businessmen to enjoy "honorary white" status in
South Africa. Japan's contracts for Rössing uranium exemplified the quasi-
sacred status of free enterprise there.[28] In sum, capitalist nuclear systems
depended on southern Africa. The lesson? Block the circulation of its
uranium.

We'll return to efforts to blockade Rössing uranium in due course. First
let's travel south to the mine itself to see how its management sought to
define "the market" in the face of these technopolitical pressures.

MAKING THE MINE'S MARKET

In the mid 1970s, as the UNCN geared up for its assault on Rössing's
legitimacy, company managers in Namibia tried desperately to get the
mine up and running. Digging the gigantic open pit proceeded smoothly
enough, but the processing plants experienced serious technical problems.
The unusually abrasive ore took a toll on machinery, causing frequent
breakdowns. It became clear that a complete design overhaul with sturdier
equipment was needed. Then, in May 1978, a fire destroyed part of the
solvent extraction section, preventing Rössing from honoring its initial
sales contracts on time. Managers decided to speed up initial deliveries by
flying the first batches of yellowcake to Europe.[29] Paying for these setbacks
meant drastically reducing operating costs and finding ways to improve
cash flow.

By the late 1970s, Rössing's customers included utilities and atomic
energy agencies in Britain, Iran, Spain, Japan, and West Germany. The mine
had solved its technical problems, streamlined operations, and even increased
plant capacity. Prices were peaking and demand seemed limitless. At least
ten large, long-term contracts priced yellowcake from just under $20 a
pound for Britain to a whopping $41 a pound for Japanese utilities. In
December 1979, the managing director estimated the year's profits at
nearly 60 million rand, over 20 million higher than anticipated. "Overall,"

he reported happily, "the mine has developed into an efficient and effective production unit, the morale of all personnel is high, [and] the competence of the labour force is increasing steadily."[30] (We'll hear personnel speak for themselves in part II.)

This commercial success was enabled in large part by a front company set up in Zug, Switzerland under the name RTZ Mineral Services. Minserve, as it was known, was expressly "formed because of the unwillingness of certain customers to deal direct with a SWA company."[31] (RTZ used the designation "South West Africa" for over a decade after the UN changed the territory's name to Namibia.) Switzerland, which did not belong to the United Nations, had no intention of respecting UNCN strictures and did not consider Minserve's activities violations of Swiss law.[32] The Zug office thus had complete freedom to serve as a middleman. Minserve customers signed contracts that didn't specifically designate the supplier while Minserve signed mirror "back-to-back" contracts with the Rössing mine.[33]

As international pressure against Namibian uranium intensified, RTZ offered customers additional layers of obfuscation. In May 1980, for example, Minserve and Rössing replaced customer names with numbers in all internal company documentation. BNFL became 6.13, the Atomic Energy Organization of Iran became 6.19, and so on.[34] The code wasn't effective at keeping customer identities secret. CANUC, for one, had a good idea of who was buying Rössing's yellowcake and didn't hesitate to publish what it knew. Nevertheless, the secrecy reassured customers, partly obscured the paper trail, and—perhaps most important—legally shielded the investors who gathered at board meetings.

Minserve also provided cover for activists' concerns about the potential military uses of Rössing's uranium. Contracts with customers in non-nuclear weapons states specified end-use restrictions in a three-line clause that enabled the South African state and RTZ to claim compliance with global non-proliferation norms.[35] Because the IAEA did not impose safeguards on yellowcake, however, no one could track the material. In any case, most Rössing customers were electric utilities that genuinely wanted the uranium as fuel. No evidence suggests that parties granted exception to the non-proliferation clause—France, Britain, and perhaps others (the archives are, as always, incomplete)—paid a financial premium for restriction-free uranium. Yet the record suggests that the absence of restrictions and the loose nature of the end-use clause served as a *political* premium. That is, these parties found reason to buy Rössing uranium *despite* the mounting international controversy.

The problem of proliferation dispatched through contractual wording, Minserve and mine management worked hard to portray Rössing as an ordinary mining company—a depiction that helped to justify their trademark contract structure, which indexed prices to a "world market price." Insistent on renegotiating price yearly with each customer, RTZ persuaded buyers not to use the Nuexco spot price as a reference, because it did not reflect long-term commitments. Rather, each buyer's price would reflect prices in other long-term contracts, which were themselves confidential.

The outcome of the yearly price negotiations varied greatly among Minserve's customers, depending in part on how buyers valued Rössing itself. Enticing customers to sign long-term commitments required a track record of reliability. After its initial troubles, Rössing shone on this front. Customers "were prepared to acknowledge our superior performance and to reward the Company with prices which were significantly above the spot market, and perhaps significantly above what we might otherwise be able to obtain."[36] As a result, Rössing's uranium often sold for twice Nuexco's spot price.[37]

Backed by huge profits, management expressed little concern when the spot price began to drop in 1980. Conceding that the reasons for the decline were "difficult to understand," the chairman expressed confidence that prices would recover by 1983 because of "the lack of any other available commodity for storing energy so easily and with such dependability as uranium."[38] The rapid spread of nuclear power seemed a foregone conclusion. Demand for uranium was bound to pick up. Yet Nuexco's indicator continued to decline.

Management began getting nervous. If the political heat caused buyers to become skittish about long-term commitments, they would begin to insist on the lower prices indicated by Nuexco. International reports on uranium supply and demand proliferated as brokers, trade organizations, and international agencies recalculated supplies using revised predictions for reactor construction and capacity. How did financing schemes and government regulation affect price? Could such factors be quantified and predicted? How would breeder reactors and fuel reprocessing—technologies designed to get more energy out of the same quantity of uranium—affect demand? As competing theories about "market structure" and "market stability" abounded, Rössing investors sought reassurance that management understood the complex factors governing supply and demand. Did Rössing have a strategy for acting in "the market?"

At board of directors meetings in 1981, Minserve began presenting reports that increasingly narrated "the market" as an autonomous entity with laws to be *discovered* rather than *made*. At first, Minserve representatives barely gestured at the Nuexco indicator, insisting that it didn't apply to Rössing's contracts. Soon, though, they began to give the spot price more attention and more agency. Descriptions of its derivation and effects grew longer and more complex. By 1983, Minserve explained that, even though Rössing didn't deal in spot *transactions*, movements in spot *prices* had begun to affect its yearly price negotiations.[39] These examples show how the power of the Nuexco indicator was shifting. From a barely relevant number, it was becoming a predictor of Rössing's future.

Investors demanded an explanation for the steady decline in the spot price. Minserve dutifully monitored international nuclear power development, concluding that delays in reactor construction helped to explain the downturn in uranium demand. Utilities with more uranium than they needed faced heavy penalties for defaulting on supply contracts. This led them to stockpile and sometimes even sell enriched uranium, increasing global supply just as demand decreased.[40]

The Minserve analyses painted the "anti-nuclear lobby" as an irresponsible impediment to progress that made unnecessary fusses over events such as the 1979 Three Mile Island accident in Pennsylvania. Sometimes, however, opposition politics benefited Rössing. Minserve gleefully reported that the Australian Labor government's policy to limit uranium mining made some customers nervous enough to come rushing to Rössing.[41] With sufficient commercial cunning, Rössing could thus turn political events to its advantage.

Minserve's portrayal of anti-nuclear politics and government policies as "constraints" on trade was standard free-market ideology. As we saw in chapter 3, such formulations had special resonance in southern Africa, where industrial and political leaders often crafted "the market" as "neutral" territory unencumbered with disagreements over apartheid and colonialism. Minserve reports enacted this perspective, offering detailed analyses of American and Australian politics while avoiding discussion of the anti-apartheid movement. Until 1985, the rare mention of southern African politics arose only as a problem of public relations or site security. Management would then reassure the board that it had the situation well in hand, and the conversation would move on.

The separation of southern African politics from market discussions implied that apartheid and the colonial occupation had no *functional* impact on Rössing's ability to participate in the transnational uranium trade. It encouraged investors to think of Rössing as a uranium mine like any other, rather than as a potentially high-risk producer. Following one report on the vagaries of supply and demand, for example, Rössing's chairman commented:

Whatever the future demand position was the Company had to live with two weaknesses in the market place; firstly we could not compete against prices offered by producers of uranium as a by-product, and secondly Rössing, as a low grade mine, was at a considerable disadvantage in meeting the competition from the high grade mines in Canada and Australia.[42]

Virulent opposition to apartheid evidently did not count as a "market weakness."

Minserve thus provided a double buffer. For customers, it legitimated transactions by separating the product from its provenance on paper. For investors, Minserve black-boxed the political specificities of Rössing's market life. Its reports reified "the market" as a field of play whose rules could be entirely apprehended using supply, demand, and the politics of nuclear things. This portrayal remained in place until late 1985, when international furor over the state of emergency in South Africa increased pressure for sanctions. At that point, as we'll see, the *narrative* separation between "the market" and South African politics became too implausible to maintain.[43]

Before we examine how Rössing and RTZ deployed more powerful technopolitical strategies grounded in material transformations, we must briefly return to the activists.

BLOCKADE

While company officials cast anti-apartheid politics as irrelevant to market narratives, European activists began to uncover the movements of Rössing yellowcake—in unlabelled containers—through European planes, ships, docks, and roads. The secrecy meant that European transport workers had unknowingly handled barrels of uranium oxide, a situation that directly contravened international norms for the transportation of radioactive substances, "nuclear" or not.[44]

In 1981, SWAPO and the Namibia Support Committee organized a seminar for West European trade unions to apprise them of the secret shipments and to appeal to the loyalties of international socialism by discussing living and working conditions at Rössing and describing the mine's paramilitary security forces. After the seminar, former Rössing worker and labor organizer Arthur Pickering toured Britain under the auspices of CANUC, seeking trade-union support for a blockade that would stop the circulation of Rössing's product. Transport unions assented readily enough. But shop stewards at BNFL's Springfields plant, which converted and enriched Rössing uranium, worried more about job losses. They successfully thwarted a national trade-union blockade.[45]

CANUC concluded that its activists had to "campaign harder over the wider political issues linking Namibia and Britain, in particular the questions of nuclear weapons and nuclear power." They redoubled their efforts to court the peace movement, especially the Campaign for Nuclear Disarmament, which had undergone a remarkable resurgence following new missile deployments in Europe.[46] But here too they encountered hurdles. Peace activists were stubbornly focused on Britain, Europe, and the superpower rivalry. "On this insular and nationalistic scale of values," CANUC complained, "the other parts of the nuclear chain from uranium mining on Aboriginal lands to weapons testing in the Pacific, have occupied the bottom rung."[47] Only repeated invocation of military nuclear exceptionalism, such as Rössing uranium in British weapons or South Africa's probable weapons program, generated enthusiasm among anti-nuclear campaigners.

In 1984, CANUC finally found strong allies within the British women's peace movement. That year a group of women took up the cause and specifically protested the enrichment of Namibian uranium at BNFL's Capenhurst plant. Arrested and fined for causing "criminal damage" to the plant, several of the women gave the penalty money to SWAPO instead. Their hopes for a blockade rekindled, CANUC and the NSC redoubled their efforts. Assisted by counterparts in the French anti-apartheid movement, they attempted to block Rössing uranium at the Paris airport, only to be thwarted by French authorities.

Undeterred, activists kept hammering at the issue in meetings and talks designed to recruit more trade-union allies. In 1986 CANUC updated *The Rössing File* in a booklet provocatively titled *Namibia: A Contract to Kill: The Story of Stolen Uranium and the British Nuclear Programme*. The publication acknowledged tensions separating the peace movement from organized (nuclear) labor, and asserted the tactical importance of not treating BNFL workers as culprits: "[T]hey do not necessarily approve of the weapons connection in their work, or of handling uranium mined under apartheid." Long-term strategy was a different matter, however, and on this CANUC aligned itself more closely with anti-nuclear activists: "It is necessary both within Britain and internationally for trade unions and the peace and anti-nuclear movements to start developing alternatives to the nuclear industry."[48]

Building on the support CANUC had received from transport workers, the publication flatly declared: "We cannot accept that [BNFL workers] should have a veto over action by other trade unionists in solidarity with the struggles taking place in Namibia and in Southern Africa in general. These struggles are of historic importance. Supporting them is in the interest of every worker and every peace activist."[49] The bet paid off. In 1987, Britain's Transport and General Workers Union officially spoke against importing Namibian uranium. The following year, Liverpool dockworkers refused to handle the stuff.

Campaigners against Namibian uranium may not have succeeded in shutting down the mine, but they had a bigger impact on operations than they may have realized at the time. To see how, let's return to Rössing.

FLAG SWAPS AND THE TECHNOPOLITICS OF PROVENANCE

Combing through company archives makes clear that the campaign against Rössing pushed Namibian uranium further into the shadows of the market. Starting in 1983, requests to defer yellowcake deliveries grew more frequent. Minserve sometimes had trouble discerning which requests were related to construction delays and similar woes and which stemmed from "political difficulties" (Minserve's delicate phrase for international opposition to its product). The distinction mattered because different disruptive

An illustration of the violence feared by Rössing's security and the South African colonial state: SWAPO soldiers firing mortar rounds against South African forces operating in Namibia across the Angolan border. (UN photo by Tadeusz Zagosdzinski)

causes invoked different contractual clauses. For common problems, standard fees would apply and deliveries would be rescheduled. But if "political difficulties" fell into the realm of outright government interdiction, customers could appeal to the *force majeure* clause and back out of the contracts altogether.

Minserve worried especially about Japan, Rössing's biggest source of profit. Concerned that their government might cave to international pressure, Japanese utilities had begun asking Minserve to substitute material of non-Namibian origin. Securing new contracts, meanwhile, required increasingly arcane arrangements.[50] Activists had made Minserve's role as a front organization an open secret. As customers sought to distance themselves further from RTZ, spot transactions became more attractive because of their technopolitical flexibility. Doubtless this was another reason why Nuexco's indicator grew in significance.

In the face of mounting international pressure, Rössing and Minserve began to fulfill contracts using "flag swaps," a practice that began as a practical means of reducing the cost of transporting uranium but turned into a method for eliding the product's origins. In one type of swap, Comurhex in France and BNFL in Britain proved particularly cooperative. After converting Rössing's yellowcake into hexafluoride, they labeled its origin as French or British on the customs forms accompanying the material to enrichment plants.[51]

In a second type of swap, a contract written for Rössing yellowcake would get filled with uranium from another RTZ mine while the contract signed by that mine would get filled with Rössing uranium. This arrangement depended on the willingness of the second RTZ customer to accept Namibian uranium. Fortunately for Rössing, "neutral" Swiss atomic power companies happily participated in such swaps.[52] Yet a third type of arrangement involved two conversion plants shuffling titles to uranium oxide and hexafluoride. All told, the quantity of flag-swapped Namibian material rose from a few hundred tons in 1982 to several thousand by 1985–86, with swap fees starting at $0.70 a pound.[53]

At first, the pressures that made Rössing uranium increasingly illicit also made it more profitable. Sales contracts were denominated in US dollars, but most costs were incurred in South African rand. As opposition to apartheid drove down the value of the rand, profits mounted. In 1985 Rössing showed its highest profit to date—more than 190 million rand.[54] This was remarkable given "the continued weakness in the world uranium market."[55] Still, a favorable exchange rate wouldn't help much if Rössing lost its buyers. As talks on Namibian independence stalled and the apartheid state's violence intensified, mandatory sanctions against South Africa looked increasingly likely. How would such sanctions affect Rössing's ability to sell uranium and buy necessary equipment?

The mine's public relations director looked into the question in January 1986. Drawing on research conducted by the South African Institute of International Affairs, he ranked nations according to their sensitivity to "domestic South African unrest." Switzerland handily topped the list as the country least likely to withhold matériel. A Swiss company had even supplied parts for the Valindaba enrichment plant.[56] Japan was next, thanks to its "seikei bunri" philosophy to "keep politics out of trade," a formulation developed to deal with the cleavage between the People's Republic of China and Taiwan. West Germany, Italy, France, and the UK followed, with Australia and the Scandinavian countries ranked as least "reliable."[57]

Evading supply sanctions would require front companies, purchasing agents, and offshore accounts.[58] Computers were particularly vulnerable to sanctions because of their "significance for nuclear explosive purposes." Although Rössing didn't make bombs, the head of purchasing worried that it remained more vulnerable than other southern African mines "due to the nuclear nature of our product."[59] Apparently, anti-apartheid activism had succeeded in making Namibian yellowcake nuclear.

Prohibitions on importing Namibian uranium worried corporate officials even more than supply sanctions. According to Minserve's London solicitor, the exact framing of an embargo affected the application of the *force majeure* clause.[60] For example, if the Japanese government specifically prohibited the use of Namibian uranium, utilities could invoke *force majeure* and renege on their contracts without penalty. But if the government issued "general policy guidelines," Rössing could claim that a utility had backed out for reasons other than "government intervention."

Each contract had a slightly different *force majeure* provision, however, and the solicitor suggested a range of precautions.[61] At best, such measures might give Rössing breathing space. "We have to accept that in any coordinated imposition of sanctions uranium is the easiest material for the authorities to trace and block," noted one Minserve executive. Uranium's resurgent nuclearity posed a clear threat to Namibian yellowcake. "Without the assistance of the converter or the falsification of origin records," he continued, "it is inevitable that the sales of Rössing material will be severely curtailed. . . . Any study on the counter effects of sanctions on Rössing has therefore to be one of damage limitation."[62]

One form of damage limitation worked the finer points of anti-apartheid policies. In January 1986, for example, the European Economic Community implemented restrictions on new collaboration in the nuclear sector. "It is unclear as to what collaboration means," mused Minserve, "but it is regarded at the moment by Euratom as *not* referring to the supply of uranium."[63] Perhaps, then, the long-standing ambiguity over the nuclearity of uranium could still serve Rössing's interests? Unfortunately for the company, the IAEA had finally climbed on the anti-apartheid bandwagon. In September 1985 it called on member states to halt all nuclear cooperation with South Africa, including "purchases of uranium from South Africa. It demands also that South Africa stop immediately all illegal mining, utilisation, exploitation and sale of Namibian uranium."[64] Equally damning, the IAEA requested that Red Book estimates exclude Namibian uranium reserves.

Menacing as UN actions were, US sanctions posed a larger threat. Accordingly, Minserve focused on anti-apartheid legislation coming up for a vote in Congress in September 1986. Assisted by the British Foreign Office, Minserve arranged a meeting with US Secretary for African Affairs Chester Crocker and Rössing's new chairman, Dr. Zed Ngavirue, a Herero academic whose skin color alone argued that Rössing didn't adhere to apartheid practices. Crocker, who had crafted US President Ronald Reagan's infamous "constructive engagement" approach to South Africa, sympathized with Rössing's arguments but offered little hope. Ngavirue and his colleagues persevered, visiting Congressional staff members, the Department of State, and "other influential people."[65] Despite these efforts, Congress overrode a presidential veto and passed the Comprehensive Anti-Apartheid Act (CAAA) in October 1986.

By proscribing US imports of southern African material, the CAAA echoed aspects of the IAEA's trigger lists, which we encountered in chapter 1. One difference was that the CAAA focused on provenance rather than on nuclearity. A considerable portion of Rössing's product went directly to plants in the US for conversion to hexafluoride. In addition, much of its yellowcake converted elsewhere went to the US for enrichment. Stopping the flow of Namibian-origin uranium oxide or hexafluoride through US plants could shut down Rössing's business entirely.

In a risky move geared to working around the CAAA, Minserve secretly hired Wrightmon USA, a self-described "consulting and information services" firm, on a monthly retainer of US$15,000. Because US law required lobbyists to reveal their employers, disclosing the arrangement would alert the UN Council for Namibia, which could nullify the effort.[66] Furthermore, the South African government—desperate to retain control over the situation—had specifically asked Rössing to "do nothing in Washington to attempt to influence the situation."[67] Minserve therefore kept its link to Wrightmon discreet, referring to it as a "consulting firm" and telling Rössing's board that they had decided against a "formal lobbying approach."[68]

Playing provenance against nuclearity, Wrightmon president, Diane Harmon, employed a double strategy to maximize the amount of Namibian uranium imported into the US. She used her contacts in the American nuclear industry to form an alliance with US conversion plants, which stood to lose significant money if Rössing's business disappeared. Next, she exploited a loophole in the CAAA that went against the interests of the US converters. Rössing yellowcake that entered the US directly counted as Namibian. But if that yellowcake got converted and relabeled as British UF_6, hadn't its nationality changed? Surely it could then enter the US as British enrichment feed.

If Rössing transferred all its conversion contracts to European plants, its customers could maintain their US enrichment contracts. Glossing over the fact that US *conversion* plants would lose business in this scenario, Harmon instead emphasized that US *enrichment* plants would suffer if they lost southern African feed. Combined with other import restrictions, this might even force one of the plants to close and cut jobs.[69] Harmon met with Congressional staffers and scurried between various government departments—Energy, State, Labor, Treasury—to make her case.[70]

At first, events tilted in Rössing's favor. Quoting Harmon verbatim but never using her name, Minserve reported to Rössing's board in June 1987 that it had used US contacts to "help persuade the Department of Treasury to change the interim ruling [which allowed UF_6 imports] to a final ruling." The Department of Energy had also "strongly supported" this outcome.[71] By the end of 1987, the Department of Treasury, the Department of Commerce, and the Nuclear Regulatory Commission had all ruled that "South African-origin uranium ore and uranium oxide that is substantially transformed into another form of uranium in a country other than South Africa is *not* to be treated as South African uranium ore or uranium oxide and is therefore *not* barred."[72] Pleased with this outcome, Minserve asked Sir Alistair Frame, RTZ's well-connected chairman, to "have a word" with BNFL and the British Foreign Office to make "absolutely sure that there is no way in which the UK will change its present policy of labeling all the material it converts as UK origin."[73]

In the Netherlands, similar practices helped stall a lawsuit in which the UNCN charged Urenco and the Dutch government with violating Decree No. 1 by allowing Namibian uranium into Urenco's Almelo enrichment plant. Because the Netherlands recognized the UNCN's sovereignty, the suit argued that the court should prohibit Urenco from carrying out enrichment of Namibian uranium and require Urenco to obtain certificates proving that incoming material was not Namibian. In response, Urenco claimed that British and French conversion plants routinely mixed Rössing yellowcake with material from other nations, making it impossible to determine which bits came from Namibia.[74] Activists could neither confirm nor counter this claim without auditing plants, but they remained undeterred, commenting that "by placing emphasis on the *physical* qualities of the uranium, the government was able to avoid dealing with the chain of *contracts* under which the uranium is presented to Urenco from Namibia."[75]

Meanwhile, flag swaps were becoming increasingly expensive for Minserve. That turned them into lucrative opportunities for other industry players.[76] One particularly complex deal, which involved nearly 1,000 tons of yellowcake, cost Minserve over $1.8 million in swap fees, transportation, and storage. In February 1988, Nuexco offered a swap that would cost Minserve $2 a pound in swap fees alone, over twice the previous year's going rate. Expenses aside, Minserve noted ruefully that swapping was "becoming increasingly difficult."[77] Even Euratom, which had previously participated in swaps, began to express reluctance.[78] So did Japanese customers who demanded, along with a "more favourable price mechanism," evidence that showed which Canadian mine would substitute its uranium for Rössing's. They threatened to declare *force majeure* and withdraw from their contracts if adequate documentation were not produced. [79]

In 1988, as anti-apartheid measures continued to gather momentum, congressional Democrats and the Department of State's Office of Non-proliferation and Export Policy began to close the UF_6 loophole. They declared the generic practice of flag swapping legitimate in cases where buyers and sellers of uranium simply wanted to "avoid transportation costs associated with physical transfers." Relabeling Namibian or South African uranium, however, did not fall into this category: "It is not possible to avoid the provisions of the Comprehensive Anti-Apartheid Act by swapping flags or obligations on natural uranium physically of South African origin before it enters the USA."[80]

Technopolitical manipulations of provenance had limits, but they succeeded in delaying the implementation of restrictions that could have put
the Rössing mine out of business. In the end, delay sufficed. SWAPO, the
South African state, and other negotiating parties signed an independence
accord in December 1988. Under the circumstances, the UNCN decided
not to pursue its lawsuit against Urenco. Namibia's transition to independence began in April 1989. Elections were held in November. In March
1990, Namibia formally became independent.

Even before the accords were signed, Minserve used independence as
a marketing tool, invoking it in contracts and using it as a trump card in
price negotiations. When one Japanese utility—having already demanded
delivery of hexafluoride rather than oxide "for reasons of political
camouflage"—pressed Rössing to delete the floor price from its contract,
Minserve refused "on the grounds that the integrity of the contract needed
to be kept for the incoming Government."[81] Executives planned presenta-
tions in London, Brussels, and Washington on "the importance of Rössing
to an independent Namibia and the serious effect that a major reduction
or closure would have on the local economy."[82] If Rössing lost its market
share, they told current and future customers, the postcolonial government
would lose a primary source of income.

Of course, an assured alliance with the postcolonial government was
crucial to the effectiveness of independence as a marketing tool. In mid
1988, RTZ had initiated an "informal approach" to SWAPO leaders by
inviting them to a briefing in Zimbabwe. As soon as the independence
accord was signed, RTZ began lobbying to remove Namibia from the
provisions of the CAAA. Perhaps most significant, it used independence
as a tool for luring new customers with the argument that buying Rössing
uranium was a moral imperative, an investment in the future of Namibia.
Sweden, which had a strong nuclear power program and had served as a
haven for Namibian exiles (including Ngavirue himself), proved particularly attentive to these overtures.

CONCLUSION

Uranium's effectiveness as a global symbol of the Namibian liberation
struggle depended on the perception of it as a nuclear thing. To achieve
results, activists had to build a case for the nuclearity of Rössing's product.

An apartheid bomb, a specter both nuclear and illegitimate, constituted their most important resource. But did an apartheid bomb truly exist? If it did, was it really fueled by Namibian uranium? Secrecy made proof impossible. And so activists found additional sources for nuclearity in yellowcake—not in its physical properties, but in its movements through transnational circuits of French conversion, Dutch enrichment, Japanese atomic power, and British bombs. Namibian uranium grew increasingly nuclear as it traversed these technological systems, whose nuclearity was not in doubt.

Drawing attention to the travels of ore through nuclear systems enabled Namibian liberation activists to profit from the moral economy invoked by anti-nuclear movements just as these movements achieved peak recognition. Provenance mattered just as much. As an avatar for the liberation struggle, the campaign against Namibian uranium also relied on the broader technopolitical case against apartheid. Rössing's role in Western nuclear systems reflected (and stood in for) the interdependence of apartheid and Cold War capitalism. Hence the significance of "the uranium market" in struggles over the legitimacy and movement of Rössing's yellowcake. Resources such as copper and diamonds were also targeted by the UNCN's Decree No. 1, but none commanded as much global attention.

Confronting challenges to their legitimacy, Rössing and RTZ initially relied on the depoliticizing discourse of market-talk, along with the most straightforward of market devices: contracts. Location mattered here too. Switzerland didn't belong to the UN and could easily justify noncompliance with UNCN edicts, giving Minserve's back-to-back contracts indisputable legality *in Switzerland*. Furthermore, Swiss companies had a track record of supplying technology for apartheid industry. When back-to-back contracts faltered in their fight against politics, RTZ turned to more complex technopolitical strategies. Once again Swiss companies came to the rescue by participating in flag swaps.

Elsewhere in Europe and the US, conversion and enrichment plants also performed technopolitical maneuvers to keep Rössing's product in circulation. The farther uranium traveled down the processing chain, the murkier its origin became. This technopolitical camouflage worked well enough to keep Rössing in business through Namibian independence. After that, postcolonial promise became the mine's best marketing tool. During this journey, RTZ and its allies sought to turn uranium into a common commodity by making each batch of yellowcake indistinguishable from any other.

Yet RTZ's attempt at banalization, keyed to erasing uranium's provenance rather than its nuclearity, carried a price in real monetary terms. Political challenges foreshortened the time horizon of Rössing's contracts. When customers balked at long-term commitments, Nuexco's short-term indicator acquired more relevance. In return for *not* invoking political opposition as a *force majeure* allowing default, customers demanded and received the more favorable prices promised by Nuexco's falling indicator.

This increased the spot price's power as a market device. By the late twentieth century, it had become a widely used tool in uranium market-making. The early years of the twenty-first century saw the creation of a uranium futures market, with new formulations of "price" and new investors eager to outflank the old market makers. As we'll see in chapter 10, this has sharpened the contradictions between the market pressures that drive claims for uranium's banality and the proliferation anxieties that militate for its nuclearity.

At what point does licit trade become black (or gray) marketeering? Illicitness is often in the eye of the beholder. Many commodities—from oil and diamonds to coca and cigarettes—pass in and out of legality as they ricochet through their chains of production and consumption. For regulators, consumers, brokers, industries, states, informal economies, regional accords, and global treaties, some goods may be legitimate under some criteria but not others, licit in one system of exchange yet not in another. "Legal," "licit," and "legitimate" aren't necessarily co-extensive. The first term refers to compliance with laws. The second and third make room for ethical conundrums, especially those concerning human rights. Spread over time and across space, trade of all sorts can combine acts both legal and illegal, licit and illicit.[1]

The Treaty on the Non-Proliferation of Nuclear Weapons and the International Atomic Energy Agency aimed to resolve or at least manage the boundaries of legitimacy and legality for the development, use, and circulation of nuclear technologies. From the beginning, the settlements enacted by the treaty and the IAEA were entangled with Cold War nuclearity and postcolonial geographies. Uranium ore's exclusion from the ontological state of nuclearity made it easy to treat African uranium mines and the countries where they resided as non-nuclear places. Yet despite their universalizing aspirations, ontological edicts remained fragile and insufficient. Since its creation, the IAEA has issued over 300 "circulars" pertaining to safeguards, a testimony to the ever-changing challenges of regulating the circulation of *nuclear* things to particular *places*.

Recent events highlight how idealized accommodations between nuclearity and geography failed to manage the messiness of actual practices in real places. In 2004, Pakistani metallurgist A. Q. Khan confessed to having sold atomic weapons technologies to Libya, Iran, and North Korea through a network of legal and illegal suppliers. In 1998, India and Pakistan, neither one an NPT signatory, both tested the atomic bombs they had been developing for decades. After the 1991 Gulf War, the IAEA dismantled a clandestine nuclear weapons program in Iraq. As I've shown in part I, the boundaries between licit and illicit practices proved porous well before these indubitably nuclear events demonstrated the fragility of the NPT world order. The uranium cartel's price-fixing efforts, Niger's yellowcake sales to Libya, Gabon's promises to the Shah, apartheid South Africa's uranium enrichment plants, and flag swaps for Namibian uranium all occupied the borderlands of il/licitness, legitimate on some meridians, downright illegal on others.

Market devices such as the Red Books and the Nuexco indicator were portrayed by promoters and users as mechanisms for *dis*entangling politics and economics. Aided by IAEA definitions that de-nuclearized uranium and made it safe for trade, market devices set the terms of its commodification. The banalization of uranium, I argued in chapter 2, grew more robust through distribution into reserve mapping, demand forecasting, "world market price" contracts, and cartel price fixing. Uranium became an economic object, whose terms of economization eschewed "politics." Industry leaders construed government policies, anti-apartheid actions, and environmental opposition as blemishes on the purity of "supply and demand," inappropriate intrusions into commodity pricing. Disentanglement by way of market devices, though, could never be more than a provisional achievement.

Throughout part I, we saw that attempts to exclude politics from trade required constant work. Conflicts over Namibian uranium in the 1970s and 1980s offer a clear example of the multiple and persistent entanglements between politics and economics, nuclearity and provenance, licit and illicit exchanges. Even as they engaged in camouflage, neither Rössing, nor Minserve, nor the plants and utilities that conducted the swaps imagined themselves part of a "black market." They saw themselves as legal businesses engaged in legitimate profit making. Neatly bracketed from political context by market devices, their practices involved standard tools of transnational trade.

For anti-apartheid activists, however, the trade in Namibian yellowcake was born illicit. As Rössing uranium became more contested, technologies of transportation, conversion, and enrichment increasingly served to conceal its origins and movements. The *techno*politics of the capitalist uranium market worked hard to make yellowcake from southern Africa sufficiently generic to enable its circulation, while activists strove to reveal these efforts as techno*politics*.

Thanks to its dense involvement in European and American technopolitical networks, white-ruled South Africa played a decisive role in shaping the uranium trade for decades. Because of its partners in Europe and South Africa, Nuexco could claim that its data incorporated knowledge of all African uranium transactions. Red Book data for Gabon, Niger, and other Francophone African countries came from the French. Gabonese and Nigérien uranium was sold by French broker Uranex both independently and within the cartel, and was subsequently represented by French companies at the Uranium Institute.

The design and distribution of these supposedly global market devices excluded black Africans from knowledge production and decision making. Absent from these sites of power, black Africans had great difficulty exercising sovereignty over their countries' uranium or even claiming that it belonged to them in the face of "market" mechanisms that assumed it didn't. El Hadj Omar Bongo, we saw in chapter 4, found it easier to rely on longstanding Françafrique networks to assert limited influence on the COMUF than to use uranium as a means of expressing significant political autonomy vis-à-vis France. His efforts to deploy uranium a tool of international diplomacy meeting with limited success, Bongo settled on using the mine domestically as a signifier of national modernity and state power (and, no doubt, a source of personal wealth).

In part II we will see that the difficulties uranium engendered for postcolonial state power in Africa were by no means limited to yellowcake sales. Occupational health and environmental regulation, treated as particularly inconvenient forms of politics by the makers and users of market devices, posed even more significant challenges. African states, whether white-ruled whites or black-ruled, proved largely unable (and often unwilling) to meet these challenges. The problem did not merely stem from the cozy relationships that mining industries nurtured with state officials, though these certainly didn't help. More profoundly, it arose from the *regimes of perceptibility* that structured the production of knowledge about occupational exposures and environmental contamination.

Throughout the African mining industry, assemblages of social and technical things made certain people, hazards, and health effects visible while rendering others invisible.[2] These assemblages included instruments, laboratories, clinics, labor relations, scientific conferences, expert and lay knowledge, and people of all sorts. They spanned continents, though their distribution was uneven. The resulting holes in data and knowledge production mattered not just for the people left out, but also for the conflicts and consensus generated in their absence.

Deeply contentious in many places, the nuclearity of uranium mine work was especially difficult to produce and sustain in African mines. Achieved in one place, it did not automatically transfer to another. Once achieved, furthermore, nuclearity could not necessarily sustain itself. Though *radiation* could persist well after industrial activities ceased, *nuclearity* could have a very short half-life. This, in turn, had deep consequences for health, environments, governance, and citizenship. To understand these consequences, we need to change perspective and shift the scale of analysis.

II NUCLEAR WORK

In the rest of the book, I aim to make visible the radiation exposures of African uranium miners by highlighting the moments and mechanisms that rendered them invisible. Let's start, though, with a quick visit to the Colorado Plateau. The history of uranium mining on the Plateau has been central to anti-colonial and environmental activism, popped up in Congressional testimony and courtrooms, and made up a chapter in the huge *mea culpa* investigation conducted by the US Department of Energy during the Clinton administration. It has been filmed, put on CD-ROM, and posted on websites.[1]

The story goes like this. Early in the Cold War, US atomic weapons requirements and the huge monetary rewards offered by the US Atomic Energy Commission spurred a domestic uranium boom reminiscent of the mid-nineteenth-century gold rush. Armed with Geiger counters from the Sears catalog, prospecting handbooks from the AEC, and inspirational pamphlets (with titles like "You Can Find Uranium!"), prospectors combed the deserts and mountains of the Colorado Plateau, often undergoing tremendous hardship. Boomtowns sprouted. Speculators started stock markets. Some people got rich. Most didn't.

Navajos who helped to locate uranium deposits on their land were typically cheated out of royalties by avaricious speculators. Those who worked in uranium mines elsewhere on the plateau were paid less than white workers. Their families often camped by mine entrances and drank mine water. Their children played in mine galleries. As with Anglo, Hispano, and Mormon mineworkers, who died in significant numbers, no one told the Navajo about the dangers. Many succumbed to lung cancer before they could figure out why. Navajo activists refer to this as "radioactive colonialism."

The plateau's environment suffered too. Mine tailings poured into the Durango River. A third of the buildings in Grand Junction, Colorado used concrete mixed with tailings from uranium mines and mills. In the 1970s, more than 600 of these structures underwent remediation to mitigate the hazards posed by radon emissions. Perhaps *no one* really understood the health hazards of radon exposure during those first few decades of uranium mining? That's what the US uranium industry liked to claim.

But as Wilhelm Hueper, director of the Environmental Cancer Section of the National Cancer Institute (NCI), wrote in a 1942 textbook, lung cancer "resulting from the exposure to radioactive ores" was "one of the oldest occupational tumors known."[2] In the sixteenth century, Paracelsus and Agricola had described lung disease in miners in the Schneeberg district of Saxony, attributing the ailments to the inhalation of metallic vapors. Similar symptoms were found in pitchblende miners in Jáchymov, Czechoslovakia. After the symptoms were linked to malignant tumors in the late nineteenth century, "Schneeberg lung cancer" was listed as an occupational disease.

In the 1920s, researchers began to suspect inhaled radon as the trigger. One study published in an American journal in 1937 reported that lung cancer had killed some 30 percent of autopsied uranium miners.[3] In 1940, the new Nazi administration in Karlsbad issued regulations governing radiation exposure in the uranium industry. Although these rules were controversial and ultimately ineffective, their precedent (not to mention the publications) belied claims that no one knew about radon hazards in mines.[4]

In 1948 two scientists from the US Atomic Energy Commission's Health and Safety Lab and Medical Division found high levels of radon and dust during a preliminary survey of Colorado Plateau mines. They recommended that the AEC adopt a policy similar to that for beryllium sales by including a clause on health and safety standards in uranium purchasing agreements. The scientists were warned off, as was the NCI's Dr. Hueper when he began calling attention to the hazards of uranium mining. In 1949 the Office of Indian Affairs, the US Public Health Service, and the Colorado Department of Health expressed concerns about health risks. A PHS survey of radon and dust in Navajo mines expanded to the rest of the region, but only on condition of secrecy. Private mining companies let PHS project director Duncan Holaday and his colleagues take readings in the mines only if they kept their true purpose hidden from the miners.

Preliminary results reported in 1952 and 1957 showed high levels of radon along with abysmal ventilation. The PHS scientists urged the adoption of a provisional "working level" for radon. But who could establish or enforce it? No one in a state agency, the federal Department of Health, or the Atomic Energy Commission wanted to accept responsibility. Only a few mines acted on their own to improve ventilation systems and provide workers with respirators. Already by 1963, the PHS study showed excess lung cancer among miners who knew nothing of this occupational danger. Still, a standard had not been set.

More and more fingers pointed at the AEC, which insisted that it didn't have jurisdiction over uranium until the ore entered the mills. Overlook the fact that the AEC offered a monetary prize for uranium strikes and was the sole legal purchaser and consumer of uranium. Ignore that by paying a bit more for uranium ore (as it did with beryllium) it could have helped finance ventilation improvements. Forget that experts in matters radioactive populated its divisions and were often eager to help. Mining just wasn't in the regulatory portfolio. As this book has argued concerning other actors on the international stage, the AEC didn't find uranium mining *nuclear* enough to bother with.

Eventually the situation did change. Widows went to Washington. Articles appeared in newspapers. The Departments of Labor and Interior came under fire for not getting involved sooner. (Bureau of Mines inspectors did not receive authorization to enter uranium mines until 1966.[5]) Both government departments found themselves in agreement with the Public Health Service. In 1967, the Department of Labor set a maximum acceptable threshold for radon and threatened to close any mines that didn't meet the standard within 18 months. In Congressional hearings, the Secretary of Labor testified that "the best available evidence is that over two-thirds of the approximately 2,500 underground uranium miners are working under conditions which at least triple their prospects of dying from lung cancer if they continue this work and these conditions remain unchanged."[6]

Everybody hedged. The AEC and the mining industry complained that the new standard couldn't be met. Mines would shut down. Workers would lose their jobs. The US would lose access to its Cold War fuel. The growth of nuclear power would be inhibited. More research was commissioned. Agencies were reorganized. More miners died. Widows from Colorado received compensation; those from Utah did not.

At last, regulatory resolution: As of July 1971, the threshold was firm, the standard established. Yet a gap remained between standard and practice. Mines were given six months to comply. In reality, it took much longer. Though working conditions began to improve in some mines, for many miners it was too late. More got sick. More died. In 1979, Stewart Udall, a former Secretary of the Interior, filed lawsuits on behalf of Navajo miners against the Department of Energy (the new incarnation of the AEC). The case went nowhere. The courts denied jurisdiction, citing cold war national security imperatives.

Eventually the matter went back to Congress. In 1990 the Radiation Exposure Compensation Act (RECA) became law. It covered residents exposed to fallout from atomic weapons testing in Nevada ("downwinders") and uranium miners who had worked before 1971. RECA presumed that uranium mining after 1971 was no longer intended primarily for military applications and therefore was subject to federal health regulation. Miners who toiled *after* momentum started building to treat uranium as an ordinary commodity thus were not eligible for compensation. RECA offered reconciliation for the sins of the Cold War, not capitalism.

Technically, RECA does not provide "compensation," but rather "compassionate payments" in recognition of miners' contributions to national security. The process for obtaining payment is not simple, however—especially for Navajo miners, some of whom didn't learn of their eligibility for 10 years or more after the law was passed. To receive payment, miners have to undergo complex medical tests to prove that they've acquired a radiation-related illness. The Navajo Nation is huge, and many homes are far from the clinics. Documenting employment in a long-defunct uranium mine can also be difficult for elderly Navajos who don't have birth certificates. One of my collaborators, who practiced medicine at the Northern Navajo Medical Center in Shiprock, New Mexico, spent two or three days a week screening former mineworkers for RECA eligibility, using his off hours to help them navigate the paperwork and bureaucracy.

For the first two and a half decades of production, as we've just seen, American uranium mines were not treated like nuclear workplaces. Miners didn't know that their occupation entailed radon exposure and certainly didn't know how much radiation they'd received, either because measurements weren't taken or because workers were deliberately kept in the dark. Even if they had known about their exposures, most miners wouldn't have known how to interpret the information. From the 1940s through the 1960s, the monstrous connotations of radiation were associated with acute, high-level exposure, not with the slowly accumulating effects of radon exposure. Assessing the dangers of radon exposure in a politically meaningful way required complex sociotechnical assemblages that generated plenty of controversy and disagreement.

Our frame of reference now established, we can shift our perspective to African mines. I do not claim that the experience and exposure of African uranium miners were *just like* those of American workers. There were significant differences. I simply want to make sure that we don't reduce those differences to "development" or "modernity" or "Africanness" or even "colonialism." If we did, we'd miss the central dynamic at work in shaping the nuclearity of work in uranium mines: distribution.

The link between radon exposure and cancer remained deeply contested for decades as scientists, industry officials, and regulators remained unpersuaded by the evidence that Nazi bureaucrats and Dr. Wilhelm Hueper of the National Cancer Institute had found so convincing. US mine operators, for instance, held that poverty, malnutrition, and the arsenic content of the ore explained the Jáchymov miners' high mortality and morbidity rates. In the early 1940s, one Belgian scientist insisted that pitchblende miners at Shinkolobwe in the Congo's Katanga region hadn't exhibited the slightest sign of lung cancer despite "the inhalation of massive amounts of dust." In countering such claims, Hueper usually highlighted the long latency period between exposure and illness.[7] By the time the irrefutable proof of cancer appeared, he warned, thousands more workers would be exposed.

Few scientists wondered about the health effects of uranium mining on the Katanga workers, Congolese men living under colonial rule. Those who did, such as PHS epidemiologist Duncan Holaday, could not garner enough data in the mid 1950s to reach meaningful conclusions. Tantalizing hints suggest that the outcome could have been different. Although the company that operated Shinkolobwe claimed in annual reports to have monitored and controlled radiation exposure, it provided little data to Holaday and none to university researchers from Léopoldville (now Kinshasa). After the mine closed (in April 1960, just before Congolese independence and the Katangan secession), there was no way to check if practice matched description. No one followed up to see if the monitoring took place or was effective in preventing cancer.

on facing page: Underground gold miners in South Africa in the 1950s. (Margaret Bourke-White/Time & Life/Getty Images, used with permission)

By the mid 1960s, the Shinkolobwe workers who mined the high-grade ore that wound up in the Hiroshima bomb had dropped out of scientific conversations on radiation exposure in mining. Not that they'd been terribly visible in the first place. The invisibility of their exposure cannot be corrected through archival diligence, because the records—assuming they were even kept—do not appear in the inventory of the company's Brussels archives. But the absence of facts like these gives us analytic insight, a way to imagine writing the history of irradiated bodies. How was radiation exposure to anyone, anywhere, made visible, and to whom? Did uranium workers know they'd been exposed? How did they understand it? What instruments, laboratories, political structures, and medical practices were required not just to make exposure visible, but to link illness to occupation? Addressing such questions is the goal of part II.

We must begin, though, with a history of how the invisibility of African uranium miners affected data and standards for radon exposure. This chapter argues that this absence was the fallout of complex and often contradictory processes of producing knowledge. As we'll see, experts put a great deal of work into framing questions about quantities and measurability. As a scientific object, "radon exposure" became visible only through instruments, roads, laboratories, data sets, regulations, and ventilation systems. How these were distributed geographically, politically, and technologically made some contaminated bodies visible while keeping others invisible. Causality was a slippery notion that varied by discipline. Epidemiologists couldn't answer questions about direct causality; their data established correlations and associations. Experimentalists looked for causality in animals rather than humans. Each discipline could question the other's conclusions, and both were vulnerable to challenges from industrial leaders and policy makers.

This chapter pays particular attention to the encounters among American, French, and South African experts. Their differences of opinion stemmed in part from the frequently shifting nuclear status of uranium mines, which affected how different disciplinary and professional groups took the lead in monitoring radon and its effects. The specific national and international contexts for research mattered a great deal, and we'll get into the South African story in detail in chapter 8. Experts, as we'll see, had a range of reasons for reaching out to colleagues in different countries. Sometimes they were seeking data, other times customers. International conferences and manuals weren't just spaces of knowledge exchange; they served as stages on which to display mastery.

RADON AND ITS DAUGHTERS

By the 1950s, a host of international institutions applied themselves to turning data about radiation exposure into "global" knowledge. They were motivated by the need to set working limits on occupational exposures to radiation. Such limits were crucial for nuclear power to overcome the fears generated by atomic bombs. Public attention focused on external exposure to gamma radiation—the stuff that produced the frightening keloids, hair loss, leukemias, and other sicknesses that ravaged survivors of Hiroshima and Nagasaki. In mines, however, the greatest danger came from internal exposure by alpha particles inhaled along with radon gas.

Dangers from radiation exposure had been widely known since the 1920s and 1930s, when radiation sickness felled Marie Curie and a number of New Jersey women who had been employed painting radium on the dials on watches. These cases prompted early efforts to limit workplace exposure and sparked extensive debates on what constituted a "safe" dose. In the late 1940s, the US Atomic Bomb Casualty Commission (ABCC) began to study survivors of Hiroshima and Nagasaki, initiating what would become the longest study of the biological effects of radiation on humans. The American experts proved more interested in the genetic effects of radiation than in the somatic effects that caused disease.[8] For decades, this fascination with chromosomal mutations would shape studies of radiation exposure. (In chapter 9 we will see it again in Namibia at the end of the twentieth century.)

In the 1950s, mounting anxiety about atmospheric testing led to the establishment of the United Nations Scientific Committee on the Effects of Atomic Radiation, which was tasked with collating and analyzing all available data on the biological and environmental effects of ionizing radiation. Presenting these as global data, UNCSEAR framed the problem of dose limitation as one that required an international solution. But who should take responsibility for setting internationally authoritative standards? The International Commission for Radiological Protection (ICRP), founded in 1928 by physicists and radiologists seeking to define limits for their own occupational exposures, was an obvious candidate. After the war, ICRP's membership grew, its aims broadened, and it began issuing recommendations on permissible radiation doses in all manner of occupations.[9]

Other international institutions had interests in radiation protection, notably the International Labor Organization (ILO), the International Atomic Energy Agency (IAEA), the European Atomic Energy Community (Euratom), and UNSCEAR itself. As these institutions competed for primacy, the pool of scientists available to serve them remained small well into the 1970s, with experts often working on several committees at once. In principle, the division of labor had UNSCEAR collating scientific data, the ICRP articulating the "fundamental philosophy" of radiation protection in the form of quantitative and qualitative recommendations, and the IAEA and Euratom developing "codes of practice." All lacked enforcement powers, however. Only national authorities could translate recommendations and codes into legal limits and regulatory structures. Far from achieving global coherence, therefore, radiation regulation remained fractured as institutions jockeyed for jurisdiction and legitimacy.[10]

When international bodies turned to regulating radon in uranium mines in the 1960s, institutional and disciplinary fissures gaped wide. Epidemiologists, radiation biologists, and health physicists strongly disagreed about how to isolate persuasive data on the effects of radiation hazards amid the confounding data from contaminants such as dust and tobacco. Their disputes intersected with national differences over whether (and how) to treat uranium mines as nuclear workplaces. Finally, many studies succumbed all too readily to corporate interests or military agendas.

At the heart of these debates lay the challenge of making radiation exposure visible. In the 1960s and 1970s, only *external* exposures could be detected directly by worker dosimeters such as film badges that captured gamma rays able to penetrate the skin. Because gamma levels varied directly with the grade of ore, external exposures were relatively easy to predict, and low-grade mines rarely required workers to wear dosimeters. *Internal* exposures, however, were more complex.

Uranium decays into radon, which further decays into radioactive products known as *radon daughters*, whose concentrations depend not on ore grade, but on a range of less predictable factors. As radon daughters decay, they emit alpha particles that cannot penetrate skin but may lodge in the lungs and damage bronchial tissue. This internal exposure proved extremely

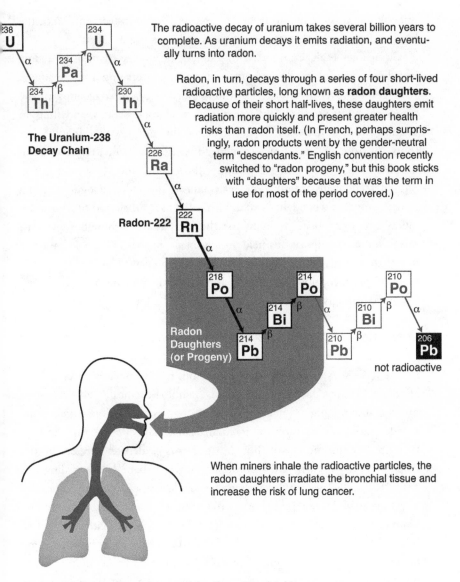

The radioactive decay of uranium takes several billion years to complete. As uranium decays it emits radiation, and eventually turns into radon.

Radon, in turn, decays through a series of four short-lived radioactive particles, long known as **radon daughters**. Because of their short half-lives, these daughters emit radiation more quickly and present greater health risks than radon itself. (In French, perhaps surprisingly, radon products went by the gender-neutral term "descendants." English convention recently switched to "radon progeny," but this book sticks with "daughters" because that was the term in use for most of the period covered.)

The Uranium-238 Decay Chain

Radon-222

Radon Daughters (or Progeny)

not radioactive

When miners inhale the radioactive particles, the radon daughters irradiate the bronchial tissue and increase the risk of lung cancer.

Radon inhalation. (figure prepared by Evan Hensleigh)

difficult to measure, requiring heavier, more delicate instruments—not the kind of equipment that miners could readily carry around.

Companies typically resorted to ambient dosimetry to measure radon, installing instruments throughout a mine and averaging their readings. Averages, however, were engines of invisibility. They didn't account for the surges in radon produced by activities such as rock blasting and shifts in ventilation. Nor did they gauge the exposure of men working in "hot spots" far from air intakes, where reduced ventilation caused elevated radon daughter levels. This posed a dilemma throughout the 1950s and 1960s. Should technicians measure the real culprits, the radon daughters, using instruments that were delicate and difficult to use? Or should they measure radon and extrapolate a figure for the daughters? Although subject to uncertainties, extrapolations held an advantage in that radon was easier to detect than its daughter products, enabling technicians to take more readings and produce more meaningful averages. American and French experts disagreed strongly about which approach was better.

HOW NUCLEARITY SHAPED DATA COLLECTION

Before the IAEA officially demoted uranium to non-nuclear status, a 1963 conference looked at the problem of "radiological health and safety in mining and milling of nuclear materials."[11] Such occasions often transcended Cold War divisions by including participants from across the Iron Curtain. For the most part, though, the terms of debate were set by the Americans, the French, and the conflict between them. Undergirding this conflict, I argue, were differences in the nuclearity of uranium mines.

As we saw in the prologue to this chapter, the US Atomic Energy Commission refused to treat uranium mining as sufficiently nuclear to warrant oversight. In the 1950s, the Public Health Service proved willing to measure radiation in mines. The PHS was built on epidemiological and medical (not nuclear) expertise, however, which profoundly shaped how (and why) it tackled the problem. In contrast, the French Commissariat à l'Énergie Atomique, which portrayed itself as the motor of national energy independence, took the nuclearity of uranium mines for granted. Radiation helped to justify keeping uranium mines under CEA purview, and the agency offered the mines (along with reactors and other facilities) to French health physicists and radiation specialists as venues for developing and displaying their unique expertise. This gave a different shape to how (and why) radiation was measured.

From the beginning, PHS project leader Duncan Holaday focused on radon daughters rather than radon. As an epidemiologist, he wanted to isolate the carcinogens directly. The PHS study began with two closely related goals. The first was to investigate correlations between cancer and radon daughter exposure among US uranium miners. The second was to develop a federal standard for maximum permissible levels (MPLs) of radon daughters in the mines.

Holaday dismissed early ICRP recommendations for *radon* concentration as "not usable" because they referenced the wrong substance.[12] Nor was he satisfied when, in 1959, the ICRP proposed changes that expressed MPLs as an equilibrium between radon and its "uncombined" daughters, which were the radioactive elements not attached to dust particles and therefore most harmful when inhaled. Holaday found that measuring the fraction of uncombined daughters was "very laborious."[13] His team had developed a "field method" to measure alpha particles, but the method couldn't distinguish between attached and unattached daughters.[14] Measurement capacity and standards had to match. There was no point in setting MPLs for a substance that couldn't be readily measured.

The PHS field method yielded a crude map of alpha radiation in US mines. But limitations in the system often meant that actual radiation levels had to be estimated from "measured levels in other years, in nearby mines, [or] mines with similar ore bodies."[15] Translating daughter concentrations into worker exposure, furthermore, required guesswork about how much time miners spent in which spots. Last but by no means least, weak infrastructures limited the scale of the dosimetric system and the ability to produce data promptly. Mines on the sparsely populated Colorado Plateau lay far apart, prompting PHS scientists to complain that "the road system serving the mines was inadequate. . . . Water had to be transported to the mines, and their remote locations often delayed or prevented their being serviced with adequate electric power. These factors mitigated against the use of modern environmental controls generally found in the metal mines of the United States of America."[16, 17] Ironically, then, the mines serving the most "modern" industry in the most "modern" nation had the least "modern" working conditions.

Confronting the AEC's recalcitrance, Holaday repeatedly insisted that uranium mining *was* nuclear work. As an epidemiologist with no earlier experience in matters atomic, Holaday framed nuclearity in terms of hazard: "Among workers in the nuclear energy industry, uranium miners constitute a unique group, in that the effects of exposure to excessive amounts of radon and its daughters were observed and studied long before the fission of uranium was discovered." Miners, he affirmed, were "exposed to higher amounts of internal radiation than . . . workers in any other segment of the nuclear energy industry."[18]

On a mission to provide an epidemiological demonstration of "the reality of the health hazards associated with uranium mining," Holaday sought to correlate his admittedly crude exposure estimates with miner morbidity. Because lung cancer typically took 10 to 30 years to set in, he needed more time and better data sets to make an effective study. He appealed to foreign colleagues "to collect and correlate data from the various mining populations throughout the world."[19] Only then could persuasive, enforceable standards be established in the face of a reluctant AEC.

In stark contrast to the AEC, the French CEA measured radiation levels in its uranium mines almost from the get-go, deploying the same scientific infrastructure used to monitor reactors and other "nuclear" workplaces. Spearheaded by a young mining engineer named Jacques Pradel, the CEA's protection program set maximum permissible levels, monitored radon and gamma radiation, and tracked exposures for each worker. If levels exceeded the limit, experts affirmed, "all activity is momentarily suspended at the workings, the mining engineer is informed, and the measures deemed necessary are taken without delay."[20]

Pradel and colleagues liked to cite the high-grade Margnac mine as the exemplar of their approach to minimizing exposure. Operators isolated inactive shafts where lack of ventilation could lead to "prohibitive" radon levels. They carefully channeled radon-charged water away from work areas. Margnac's gamma levels, meanwhile, were so high that operators had to rotate jobs to avoid high doses, the quintessential workplace expression of nuclearity. "Protection against radiation," experts explained, required "limiting working time in the mine to 2–8 hrs a week." At conferences and in publications, CEA experts proudly displayed how mine nuclearity was mapped, sampled, logged, processed, and controlled. "Despite difficult workings [*sic*] like the Margnac mine," they declared in 1958, "there has not been one instance of over-exposure." To prove it, they generated tables quantifying the radon inhaled by personnel in each of the "mining divisions in Metropolitan France."[21]

Let's pause to notice two aspects of these assertions. First, the claims made by French experts concerned dosimetry, not illness. Second, their data concerned metropolitan France. The CEA mines that had been operating in Madagascar since the early 1950s were utterly invisible—experts did not mention them at all, not even to note that data had not been collected there. We'll revisit these points in chapter 7.

Displaying their dosimetric mastery, CEA experts took more readings than experts anywhere else. In 1962, for example, the CEA amassed 35,000 radon samples compared to the PHS's 6,000 daughter readings.[22] Whereas PHS scientists took measurements under the sufferance of mine operators and had no power to remove irradiated workers, Pradel and his colleagues emphasized the "exceptional policing role" that in principle allowed them to outrank mine superintendents whenever exposures exceeded permissible levels. In France, dosimetry conferred social power on a new group of experts and turned uranium mine shafts into nuclear workplaces.

The French also measured different things than the Americans. Although agreeing that radon daughters were the true culprits, French experts measured radon, not alpha particles. The PHS approach—despite Holaday's claims—was *not* practical for mine operators, because it entailed either analyzing captured air samples above ground before radioactive decay tarnished the readings or using a delicate portable device that could easily get damaged underground.[23] The inefficiencies of these solutions severely limited the number and frequency of readings. Furthermore, the PHS field method did not account for wide-ranging temporal variations. The CEA's radon sampling was more practical for regular, operational hazard control. To overcome its limitations, CEA experts gathered data on equilibrium levels between radon and daughters, and established a "correction factor" that correlated radon levels with alpha activity from the daughters.[24]

In contrast to US mines, the nuclearity of French mines was *distributed* in a dense network of experts, laboratories, paved roads, underground shafts, rocks, dosimeters, and state power. Nuclearity came from the concentration of the entire "fuel cycle" in a single state-run institution that (for better *and* worse, as we'll see) regulated itself. Nuclearity also had to do with the short distances that air samples, dosimeters, and film badges had to travel between mines and processing labs, and with the fact that those labs already processed badges from other nuclear workplaces. None of these conditions obtained in the dispersed, privately run, diffidently regulated US mines.

Ultimately, US and French experts were interested in different questions. Duncan Holaday focused on epidemiological correlation: What was the relationship between alpha exposures and lung cancer? PHS scientists had neither mandate nor equipment to limit operational workplace exposure. They wanted to measure radon daughter levels and correlate them with morbidity and mortality, and to use that connection to set maximum permissible levels. After that, monitoring and enforcement would be up to someone else. Jacques Pradel, in contrast, focused on control: How to record and limit worker exposure in actual mining practice? CEA mine directors later lauded Pradel for his ability to speak "the language of the miners" and to "make them understand what atomic scientists and biologists wanted to impose on them."[25] The CEA had already defined its MPLs. Pradel wanted to measure radon so that he could control its presence on a regular basis.

The contrast had other implications. PHS experts saw no need to measure dust or gamma radiation, because radon daughters were the culprits for lung cancer. In addition to radon, French experts measured dust and gamma rays, which in some mines were significant enough to be hazardous. Gamma exposure too could be controlled through individual dosimeters, which were the norm throughout installations like reactors and plutonium separation plants under CEA purview. The same labs could process film badges for reactor workers and uranium miners alike.[26] But was it enough to establish nuclearity and record radiation? The French did not seek to correlate exposure data with medical records; the two data sets remained separate. So they didn't check whether their limits—even if rigorously applied—sufficed to keep miners healthy.

SOUTH AFRICA'S ELUSIVE RADON

To make a persuasive case for strict, enforceable standards for radon daughter exposures in mines, Duncan Holaday looked for more data than the PHS study provided. The obvious sources were the Belgian Congo and

South Africa. Mines in both places had produced uranium for the US and thus gave American scientists connections they could draw upon. The Shinkolobwe and Witwatersrand mines, seemingly poised on opposite ends of the radiation spectrum, could together provide a broad range of data. In 1956, Holaday traveled to Africa in the company of two US AEC scientists who had an interest in the problem of radiation in mines even though their agency had long refused to treat American uranium mines as nuclear workplaces.[27]

The American team learned that the first load of Shinkolobwe ore—the stuff that had gone to the Manhattan Project—had been "very high grade and hand picked." The film badge program hadn't started until 1947, so gamma measurements for that load were unknown. But levels were estimated to have reached 50 millirems per hour. According to later ICRP recommendations, this meant that Congolese workers could have been exposed to a year's worth of radiation in about two weeks of hand picking. Though not as high grade, the ore mined in the 1950s still emitted significant gamma levels. Radon levels seemed comparable to those of US mines—high enough to cause concern. Out of a total workforce of 1,155, Shinkolobwe's medical unit reported one case of lung cancer among 64 men who had worked six or more years at the mine. Six of the 64 had liver cancer, which the Belgians claimed accounted for "at least 50 per cent of all cancers in the African native." In any case, Holaday was interested in lung cancer.

The medical team at Shinkolobwe promised to keep the Americans "in touch with the progress for study, which is really only getting under way now."[28] If the Belgians and the Americans stayed in touch, however, the exchange left no trace. In my search of archives in nine countries and four continents, the AEC report on the Shinkolobwe visit was the only document I found that gave any numbers for Congolese radiation exposures.

Holaday and his colleagues do not appear to have pressed the Belgians for more figures. But clearly they nurtured high hopes for South African data. Roy Albert, one of the AEC experts who accompanied Holaday, had briefly sampled radon levels in some of the Witwatersrand shafts and had found much lower levels than in the US. Self-satisfied South African scientists attributed these results to "high ventilation standards" and concluded that "probably as the result of the stringent safety precautions the radioactivity in South African mines does not represent a health

hazard." Albert had also analyzed Johannesburg hospital autopsy data
without finding "any evidence of increased incidence of lung cancer in
miners."

The South Africans dismissed Albert's recommendation to conduct a
more detailed follow-up, insisting that their data featured only "the usual
defects common to hospitals all over the world." [29] That these data covered
only white patients presented no difficulty.[30] When Albert raised the possibility of including "natives" in a follow-up study, the South Africans
persuaded him that short employment contracts and high mobility made
"the native population unsuitable for the radon study."[31]

South African scientists may have deemed further research unnecessary,
but scientists elsewhere found the Witwatersrand shafts intriguing. Their
seemingly low radon levels meant that they could illuminate ongoing
debates over the biological effects of exposure to low doses of radiation.
Was there a threshold below which radiation exposure had no deleterious
health effects? Or did health effects remain proportional to exposure no
matter how small the dose? Canadian scientists at the ICRP joined Holaday
in urging the South Africans to conduct more extensive research to help
them settle these questions.[32]

In 1960 South African scientists agreed to a two-year study to "assess
the nature and extent of the radon problem in South African mines."
Funded in part by the mining industry, the project used Holaday's field
method to sample radon daughters in seven mines. It elided the racial
organization of labor, describing the workforce as 10 percent "highly
trained employees" and 90 percent "unskilled" workers. South Africans
reported radon daughter levels well below the (still unenforceable) US
limit, although they admitted to occasional variation and excess. Drawing
on the same hospital data Albert had seen, they insisted that the "exposure
of many thousands of workers over many years . . . has not apparently
resulted in any unusual incidence of lung cancer." Their study could thus
help determine "the levels of radon and radon daughters which can be
tolerated without harm over many years' exposure."

Holaday expressed caution. Though the South African data seemed
promising, only "a detailed analysis of [miners'] mortality experience"
could provide proof. "I should like to encourage the South African group
to conduct such a study, as the information obtainable would be invaluable
to the uranium mining industry."[33] Yet the South Africans remained reluc-

tant to engage in epidemiological investigations, let alone institute radiation monitoring in the mines.

In the late 1960s, however, the South African scientific community began to feel the effects of growing international isolation. With opportunities for research exchanges shutting down, acceding to foreign pleas would offer an occasion to influence international standards and to legitimate South African practices. Carefully avoiding the term "epidemiological," J. K. Basson of the South African Atomic Energy Board agreed to run a pilot "biostatistical study."[34] In 1971, he concluded that "the death rate from lung cancer among White South African miners has not been increased by radon exposure" and that "although this investigation was undertaken as a pilot study, it appears that no improved results would be obtained by increasing the sample size."[35] This was a common South African refrain: no problems detected, no further study needed.

Like its predecessors, Basson's study involved only white miners. But the assertion that "the Non-White group" need not be studied because those workers came "from rural areas and work[ed] for intermittent periods varying from a few months to 1½ years before returning to their homelands" didn't raise any foreign eyebrows. In any case, for US experts accustomed to race-based epidemiology, the omission would have paled next to Basson's conclusion, which explicitly addressed *American* debates about lowering permissible levels in mines: "Although the induction of lung cancer by high concentrations of radon and radon daughters cannot be questioned, this study has produced no evidence for any effect at the cumulative exposures encountered in South African mines. . . . Consequently there is no support for the proposed decrease of the permissible radon daughter levels . . . as envisaged in the USA."

To make matters worse, the report landed at the US Public Health Service *after* Basson had sent it directly to Union Carbide, which in turn had forwarded it to other mining corporations and the US Atomic Energy Commission.[36] Frantic that the study would serve "as ammunition to repudiate the PHS data and conclusions," PHS experts responded harshly to the betrayal, accusing the South Africans of "gross under-reporting" of lung cancer and urging "that a competent epidemiologist, above suspicion of any possible conflicts of interest . . . be employed to pursue the problem in a technically competent manner."[37] The report's scientific (de)merits aside, American epidemiologists feared that it would jeopardize their laborious efforts to set maximum permissible radon levels. Reading a classic capitalist conflict between corporate interests and state regulation into Basson's analysis, PHS scientists did not ask whether racial exclusion might have also skewed the data.

On the broader international stage, the report became the standard reference on South African uranium mines and gave that country's experts sufficient credibility to participate in international meetings on radon in mines. Basson presented it without incident at the 1971 Geneva conference on the Peaceful Uses of Atomic Energy.[38] The report's focus on white miners seemed in line with other study populations. Its bland conclusions reassured those who promoted nuclear power and sought to treat uranium as an ordinary commodity. With no contradictory research on the effects of low-dose radon exposure, the study nurtured the hope that radon posed a negligible risk, soon to be managed through international standards.

In sum, Basson's study, which we will revisit in its South African context in chapter 8, made South African mines relevant to nuclear research outside the country while keeping them safely non-nuclear within.

THE COSTS OF CONTROL

When PHS officials accused the South Africans of letting "conflicts of interest" taint their research, they doubtless had in mind how American mine operators resisted investing in anything that didn't contribute directly to profits. Effective radon control often required a significant overhaul of ventilation systems and the addition of monitoring systems to check radiation levels. Costs varied hugely depending on the mine. The complex computations involved calculating figures like exposure reduction per worker per dollar, or ventilation dollars per pound of uranium oxide produced.

Americans, French, and South Africans had different approaches to calculating the costs of radon mitigation—and differing abilities to do so. They compared notes at a 1973 meeting in Washington. By that time the AEC had been dragged into playing a role in radon regulation. But, as F. E. McGinley (an official from the AEC's Grand Junction office) told his foreign colleagues, the actual costs of radiation control in the US were "not available for review," because private mining companies kept their books closed. McGinley therefore relied on studies commissioned during debates over lowering federal standards for radon in the late 1960s. The most prominent of these, conducted by consultants at Arthur D. Little, concluded that compliance with the new federal standard averaged $0.24 per pound of uranium oxide, though it could reach $.93. The mining industry claimed that the cost averaged $0.70 per pound of oxide and could reach $1.75. Further reductions in permissible radon levels, McGinley affirmed, would devastate the US uranium industry. The Canadians at the meeting reached a comparable conclusion. So despite the recent climb in the Nuexco spot price, North American uranium producers claimed that they couldn't afford a more stringent radiation protection program.[39]

Unlike their North American counterparts, the French did not see uranium mining as a discrete, non-nuclear activity with autonomous profit and loss margins. The nuclearity of their mines was embedded not only in their radiation data, but also in their cost calculations. Jacques Pradel and his colleagues could marshal detailed statistics, graphs, and tables with a confidence that the North Americans couldn't begin to match. For example, they calculated the average alpha dose absorbed by French uranium miners. In 1972 that figure fell comfortably below the new limit legislated in the US. Since French mines also monitored dust and external radiation, they could also produce a figure for cumulative exposure, a feat unmatched by other producer countries that did not legislate limits on cumulative exposure in mines.

The total cost of France's monitoring program (readily obtainable because all relevant data were produced by a single institution) amounted to roughly 10 percent of "the world price of uranium," though the exact time and pricing model (Nuexco? cartel?) were not specified. The French noted with satisfaction that their program could accommodate somewhat lower exposure limits without significant additional expenditures. In sum, they did not worry much about the impact of radiation protection costs. Energy needs would drive uranium sales, which would ultimately proceed "independently of cost and instead depend on the number of nuclear power plants in operation."[40]

The South Africans, for their part, made a virtue of vagueness. "Systematic" surveys, they asserted unblushingly, showed that average radon daughter concentration fell below the US maximum permissible level in all but three mines that exceeded maxima just slightly. Radon levels had been reduced considerably by the excellent ventilation practices achieved by "co-operation between the Department of Mines and the mines con-cerned." True, South Africa didn't regulate radiation levels in mines. But with so much good will and cooperation, surely there was no pressing need to do so? As for the effects of tighter standards on mining costs, such calculations were "a nearly impossible task" because "the economics of exploiting the uranium resources of the Witwatersrand series are . . . indis-solubly linked with those of gold production."[41] In short, precise numbers were as impossible as they were unnecessary.

Juxtaposing these three approaches illustrates the problems of cost com-mensurability. Experts were measuring different things. The ADL cost estimates for radon mitigation translated into an average of 4 percent and a maximum of 15 percent of Nuexco's exchange value for 1968–69. The industry figures translated into an average of 11 percent and a maximum of 28 percent of the value. (Refer back to chapter 2 to see how handy a single price for uranium is.) Even though the cartel and its "world market price" were in full swing, let's assume that the French referred to the Nuexco value when they invoked the "world price." This immediately raises a problem with the comparison. In 1972, when the French cost calculations were performed, Nuexco's value had dropped to $5.95 per pound. Expressing radiological control costs as a percentage of uranium price thus produced different figures as "the price" rose and fell. Yet how else to compare the figures?

Experts concluded the 1973 meeting with a collective shrug: "it is dif-ficult to compare the costs of radiation control."[42] So what exactly did they hope to get out of the effort? Pradel and his colleagues probably said it best:

One must consider that, inasmuch as the cost of an effective radiation protection program does represent a significant proportion of production costs, a mining company will only put it in place if it has assurances that its competitors are subject to the same constraints. In this regard, the implementation of regulations specifying in detail the way in which doses absorbed by personnel should be measured is indispensable.[43]

In other words, achieving a global standard with uniform measurement methods would level the competitive playing field. Unless everyone adhered to the same radon standards, companies would cut corners to gain competitive advantage.

For companies operating in newly independent African countries that had weak regulatory regimes, corners were easy to cut. Absent a dense network of institutions, infrastructures, and instruments to make radon contamination perceptible, exposures effortlessly faded from view. The French insisted that in "the mines in Africa with which the CEA is associated, the working regulations for these mines are more or less the same as those we have in France." But what did it mean to apply "more or less the same" regulations? Who would check?

Indeed, who verified compliance anywhere? It was one thing to have standards, quite another to enforce them. As one official from the US Bureau of Mines pointed out, inspectors had found that only half of American mines were in compliance with regulations in 1972. Some mines still exceeded maximum levels by a factor of 10. "So you see," he remarked, "we have reason for some skepticism with respect to the exposure records of operators."[44] The US Bureau of Mines functioned independently of mine operators, but it had the funds and personnel to inspect each mine only once a year. The CEA engaged in continuous monitoring, but it was beholden only to itself. Any form of radon regulation—whether independent or not, whether in France, Madagascar, Gabon, or Niger—necessarily relied on CEA expertise. We'll return to the theme of surveillance throughout part II.

EPIDEMIOLOGY VS. DOSIMETRY: HOW MUCH IS TOO MUCH?

By the early 1970s, many countries had legislated limits for maximum permissible occupational exposure. But research and controversy over the effects of low-dose exposure to radon and other radiation sources continued. The ICRP recommended a whole-body exposure limit of 5 rems per year for all workers in potentially radioactive conditions, not just uranium miners. What did this imply for radon exposure specifically? The answer depended on how scientists studied the relationship between exposure and cancer.

The US Public Health Service scientists who characteristically deployed an epidemiological approach to setting radon standards could not establish causality per se. Their statistical correlations—the core of epidemiological knowledge—left ample room for challenges to causal claims. In the early 1970s, for example, one researcher found that only two of 150 American uranium miners who'd died of lung cancer did not smoke cigarettes.[45] Did this mean that radon daughters weren't hazardous in the absence of cigarettes? Or did tobacco merely aggravate the carcinogenic effects of exposure? Following a long tradition of blaming disease on worker habits instead of working environments,[46] US uranium mine operators and many AEC scientists plumped for the first interpretation and commissioned new studies to explore the synergy between tobacco and radon. With exposure levels stated as estimates, and with subjects less than forthcoming about their smoking habits, the data were weak enough to continue blaming workers for their own lung cancers up to the 1980s, when non-smoking Navajo miners were shown to have developed lung cancers in excess of statistical norms.[47]

French scientists, in contrast, scrutinized the link between radon and cancer through experimentation. When rats exposed to radon daughters developed lung cancer, CEA scientists took it as conclusive proof of a single cause for cancer.[48] Only well after further experiments demonstrated the aggravating effect of tobacco did the CEA attempt to ban smoking among miners—with the predictable poor results. Though the CEA felt no need for fuzzy statistical correlations to set exposure limits, the absence of epidemiological research on French uranium miners had implications. Because the CEA did not correlate medical records with exposure records, it didn't check whether its exposure limits protected workers sufficiently. In the 1990s French epidemiologists studied French uranium miners (those in Francophone Africa were excluded) and found a statistically significant excess in lung cancer deaths—a result all the more striking because of the precision of the dosimetry.[49]

Situated between epidemiology and experimentation, a third strand of research, conducted in a variety of US and European research centers, studied the deposition of radon daughters in human lungs. Using lung models to investigate how unattached radon daughters affected different bronchial tissues, scientists aimed to ascertain the biological mechanisms by which radon daughters induced cancer. Their research evolved from broader, earlier investigations of the "radiosensitivity" of different organs, undertaken partly to develop organ-specific dose limits. Their results would eventually graft onto those from animal experimentation to produce what became known as the "dosimetric approach" to radon exposure.

Experts disagreed about the relative merits of these approaches, which measured different things using different instruments. American epidemiologists didn't think French experiments on rats said much about humans, because different species respond to toxins differently and because, to save time and money, the experiments typically exposed animals to high doses of contaminants over a short period. Even if dose *intensity* induced cancer, high-dose exposure didn't predict low-dose response.[50] French health physicists likewise critiqued the epidemiological approach for measuring averages instead of actual exposures and then extrapolating the maximum permissible exposure for each miner from those averages. This was problematic because the high end of exposure could be up to three times the average for any given mine.

DOSIMETRIC MASTERY ON DISPLAY

In September 1974, the CEA and the International Labor Office (ILO) hosted an international symposium in Bordeaux to sort through some of these "passionate discussions."[51] Debates didn't bring closure to the controversies, but they did enable French experts to display their dosimetric mastery. The timing couldn't have been better. Just a few months earlier, the national electric utility EDF had announced a massive expansion of France's nuclear power program. The CEA courted Egyptians, Iraqis, Libyans, and other attendees for its expanded training program, aimed at countries aspiring to develop atomic research and energy programs. French dosimetric mastery showed potential customers that the CEA had mastered the atom from the moment it left the ground.

By 1974, as we saw in part I, sellers of French-produced yellowcake had begun to frame uranium as a banal commodity. By displaying French-run uranium mines as *perfectly controlled* nuclear workplaces, I argue, radiation protection experts decoupled exceptionalism and nuclearity. They thus retained the power that nuclear expertise conferred upon them while turning radiation exposure into a readily manageable—and ultimately banal—occupational hazard.

Papers at the Bordeaux conference emphasized the superiority of the CEA's meticulous system of monitoring over the American approach. The CEA insisted on calculating cumulative exposures from radon, dust, and gamma rather than assuming (as was standard in the US) that only radon posed a hazard. Most striking for my purpose here was the presentation by Massan Quadjovie, an official in the Gabonese government's Direction des Mines, who described the radiation protection program at the COMUF mine in Mounana as a textbook example of the French approach. Listening to Quadjovie, delegates might well have concluded that operations in Gabon offered both a model of French dosimetric mastery and proof of its portability.

According to Quadjovie, film badges measuring gamma exposure were sent monthly to the CEA's lab in France, which reported the results back to Mounana. Underground radon was measured by ambient sampling, a method that Quadjovie spelled out in tremendous detail. If any "abnormal results" occurred in an employee's gamma or radon readings, "an investigation is immediately conducted to determine the causes of the anomaly and take the necessary measures," such as moving workers to less radioactive environments. Quadjovie dutifully admitted to some imperfections in the system: "Obviously this type of control implies that one can trust the personnel, each employee being responsible for his film badge. There are still some cases of forgetfulness, or of imaginative use of the film. Sometimes the badge is lost or put into an environment that isn't representative of the working environment. One must therefore take all these anomalies into account when compiling the results, and periodically run checks at the worksite."[52] Making Gabonese mine workers visible (however dimly) made it possible to blame them for failures. It also enabled the Gabonese state to display regulatory competence abroad even though, as we'll see in chapter 7, it did not exert any regulatory muscle at Mounana.

Performing all of this at an international conference helped CEA experts display the meticulousness and portability of their approach. But had Mounana residents spoken for themselves at the conference, what might they have said? Let's listen to one of them:

Uranium caused many deaths, but the COMUF didn't want to recognize that. . . . Nor did the state, because this was the big company of the territory, whose secrets couldn't come out . . . so as not to scare the workers. But many people were contaminated, many have already died, many are still suffering. Even those of us who stay in town, we have been affected by radiation. We weren't protected at all. The uranium was left anywhere and everywhere.[53]

We will return to this bitter appraisal in chapter 7 when we meet its author. For now, let's use it as a lens to notice Quadjovie's concern for radiation protection *principles*—not outcomes. To what extent did the well-ordered, highly routinized procedure he described correspond to daily practice? What exposures resulted despite all this monitoring? Where and how densely was the nuclearity of COMUF mines distributed? Only the mines themselves can answer these questions. Here, we let them lurk in the background, refracting our gaze as we examine the emergence of prescriptions that claimed global purview.

THE (NON)NUCLEAR VALUE OF LIFE

It wasn't just in France that radiation protection experts sought to turn exceptional nuclear hazards into banal radiation hazards. With its *raison d'être* anchored in nuclear exceptionalism, the ICRP used the unique hazards posed by radiation exposure to justify its existence as an international organization that held a distinct *and urgent* mandate. As a political tactic, however, exceptionalism wasn't reserved for experts. With environmental and peace movements gathering momentum in the 1970s, antinuclear activists increasingly deployed exceptionalism to contest the very existence of the nuclear industry. In response, the ICRP and other international organizations began to render nuclear hazards ordinary by developing global prescriptions to manage them.

Publications issued by international organizations often gain legitimacy through the anonymity of authors collectively bound in objective agreement. But texts are negotiated by humans who have stakes in producing particular forms of consensus. So it was with the 1976 *Manual on Radiological Safety in Uranium and Thorium Mines and Mills*, published by the IAEA and the ILO to teach mine operators how to implement radiation protection methods. Briefly outlining likely hazards and ICRP exposure limits, the booklet mostly described monitoring techniques, instruments, and procedures before concluding with a nod to the need for conducting medical examinations and keeping health records. The content came from a small international panel of experts, including France's Jacques Pradel and South Africa's J. K. Basson, who left clear marks on the text, particularly in the sections on lung cancer research and monitoring methods.

The manual reported increased lung cancer mortality among certain North American and European miners but emphasized their exposure to *high* radon levels. "In contrast," it stated flatly, "no increased mortality was detected in South African gold/uranium miners exposed to lower levels of radon." Basson's 1971 report—the controversy over its conclusions already buried in the archives—constituted this evidence. The manual also echoed Pradel's skepticism toward the PHS study, noting that "a major uncertainty affecting the conclusions lies in the assessment of the individual exposures." Uncertainties notwithstanding, the manual clearly stated that radon exposure presented health hazards. But the section on monitoring methods suggested that South African practice represented the state of the art. Two of its five photographs depicted a young white man measuring "radon daughters in the return air from a stope in a South African gold/uranium mine." How could any reader fail to conclude that South African mines exemplified responsible radon monitoring?

Shortly after the IAEA–ILO manual went to press, the ICRP also issued a booklet on radon.[54] Clearly, atomic agencies continued to compete for authority over matters radioactive.[55] The pool of experts remained small, however. The ICRP's task force included Duncan Holaday and Jacques Pradel, but no South African scientists. Along with overlaps between the two texts, there were palpable differences. In contrast to the IAEA–ILO manual's overt skepticism about PHS epidemiological results, the ICRP's statement seemed to reflect a compromise between Holaday and Pradel: "Substantial epidemiological information, which has been developing in recent years, will eventually constitute a very solid basis on which exposure limits will be based."

The assertion that this evidence would soon be "converted into [a new] exposure limit"[56] came to fruition a few months later when the ICRP issued Publication 26. The first comprehensive revision of ICRP recommendations in over a decade, Publication 26 addressed radiation exposure in all its forms and laid out global prescriptions for nuclear banality.[57] Its main tool was ALARA, the idea that all exposures should be kept As Low As Reasonably Achievable. ICRP members (and others) referred to ALARA and the recommendations that ensued as a "philosophy," an articulation of the ethics that should guide nuclear activities.[58] With ALARA, the ICRP could style itself the conscience of nuclear work.

The phrase had a history.[59] ICRP recommendations had called in 1955 for exposures to be reduced to "the lowest possible level." In 1959 this became "as low as practicable." The 1965 recommendation was "that all doses be kept as low as is readily achievable, economic and social considerations being taken into account." All of these locutions raised hackles. For example, the nuclear power industry fretted that the 1965 version imposed new constraints on operations, while critics thought it gave the industry too much leeway. Both sides demanded more specificity. In 1973 the ICRP's Publication 22 turned the unsavory "readily achievable" into "reasonably achievable."[60]

Publication 22 had begun to frame nuclear activities as ordinary industrial work. ALARA aimed to keep the hazards of nuclear work comparable "to those that are accepted in most other industrial or scientific occupations with a high standard of safety."[61] The ICRP stressed that balancing "detriment" and "benefits" was by no means limited to radiological protection.[62] The right calculations could attenuate the exceptionalism of nuclear risks. For example, the number of deaths resulting from nuclear work should be comparable to other "safe" industrial activities. Similarly, the cost of preventing deaths from exposure should compare to that spent per life "saved" in other industries.

Unfortunately for the ICRP, Publication 22 generated more controversy than it resolved. It included only *fatal* cancers in its risk calculations, critics noted, raising objections to its dose modeling. ALARA promised a practical way to navigate the low-dose debate by charting a course between the linear hypothesis, which held that all low exposures were detrimental to health, and the industry's desire to set a permissible threshold for exposure. The ICRP used the linear hypothesis because no credible alternative existed, but it insisted that the "social gain" from reducing exposures was far less at the low end of the exposure curve. In practice, any maximum permissible limit constituted a threshold below which a reduction in exposure wasn't worthwhile. Some found this explanation confusing; others, disingenuous.

Finally, the ICRP conceded that exposure prevention had to reach beyond the simplified model given in Publication 22 for calculating the "optimum benefit" of a given protection program.[63] Some exposures were less "necessary" than others, and deciding where to draw this line involved social and economic judgments that might vary among different nations.

The more comprehensive Publication 26 attempted clarification by presenting a tripartite framework for the ICRP's radiation protection
philosophy:

(1) Justification. All exposures had to be justified—they had to balance
the risks and benefits of the activity occasioning the exposure.
(2) Optimization. Assuming a given activity was justified, its associated
radiation protection had to be cost-effective. This meant that the money
spent on saving a life had to be "reasonable."
(3) Individual risk limitation. To ensure that undue risks weren't borne
by individuals who might not benefit from taking them, a ceiling was set
on the total exposure permissible for any given individual.

The ICRP stressed that its recommended limits were "boundary conditions for the justification and optimization of procedures *rather than* . . . *values
that should be used for purposes of planning and design*."[64] In other words,
employers shouldn't *plan* for workers to absorb maximum permissible
doses. The existing whole-body limit was the *outer* boundary, the limit of
acceptability. In organizing workforces, employers should try to keep
average exposures to a tenth of that figure.[65] Nevertheless, ordinary cost
considerations could still guide decisions. Minimizing *average* exposures had
to be balanced against expense. Hence the emphasis on "optimization."

Units and standardization

Publication 26 also mandated a change of units. The rem, the rad, and the
curie should now be replaced by the Sievert, the gray, and the Becquerel.
Thus the whole-body limit for "nuclear workers" of 5 rems per year should
now be expressed as 50 millisieverts per year. Few were pleased by this
change. Measuring instruments and information systems were designed for
the old units, so until replacement time came, conversion would involve a
lot of work—the cost of which ICRP had not apparently considered. In
what must be one of the best typos of the nuclear age, two members of
the US Nuclear Regulatory Commission commented that the "new units
are, and probably will continue to be, a source of irradiation [*sic*] to many
and will complicate communications for years." (R. E. Alexander and W. S.
Cool, "Practical Implications to the United States Nuclear Regulatory
Commission of the ICRP Recommendations (1977)," in IAEA, *Application
of the Dose Limitation System*, p. 430)

The ICRP refused to set a cost for human life. Such a figure, it insisted, could emerge only from national regulatory authorities as "a value judgement of political rather than scientific nature." One member explained that, although "this 'monetarization' of human lives has been criticised as . . . cynical and inhuman," its purpose was "to get as much detriment reduction as possible for the limited amount of money that society is willing to set aside for protection. If that money is not used as efficiently as possible, lives are unnecessarily lost. Quantitative assessments and comparisons are therefore indispensable."[66] Which parts of "society" would provide the funds and make spending decisions? The question itself was outside the ICRP's purview, since the answer varied by place.

By separating politics out from the science of their risk calculations, the ICRP sought to preserve its legitimacy as a body of disinterested experts who maintained moral integrity while holding absolutely no regulatory authority. Yet embedded in the refusal to place a monetary value on human life was the fundamentally political assumption that the value of life varied by *nation*. In this scheme, I argue, the biggest gap lay not between America and France, but between the "First World" and the "Third World." Of course the matter was rarely put so starkly. In later years, one of the ICRP's scientific secretaries preferred to express it this way:

It's a matter of national and local judgment as to how much it's worth spending to reduce risk. . . . Take the Third World, for example. It's not necessarily appropriate to spend the same amount of money on reducing radiation risks in an environment where there may be other competing risks that are clamouring for a small amount of government resources.[67]

The assumption was that expenditures for reducing radiation risks would come from government, not corporate funds. Poor nations had to choose which environmental health problems mattered most. Bodily ailments had to be separated into those laying claim to public resources and those that could not. All the while, the multinational corporations that generated the contamination made no appearance in ICRP philosophy.

If ICRP members had hoped that Publication 26 would resolve disagreements over radiation protection, they were disappointed. As antinuclear activism gained momentum and radiation exposure caught the attention of European and North American labor unions, controversy over putting their recommendations into practice intensified. Industry officials criticized the ICRP for not going far enough to de-exceptionalize uranium mining. The mining industry already spent considerable sums on occupational safety. Adding substantial radiation control costs would unreasonably burden uranium mining with higher safety standards.[68] By contrast, some trade union leaders thought that excess exposures could easily slip through the cost-benefit back door opened by the "banalization of nuclear technologies."[69] "Who bears the cost?" raged one French trade unionist. "Where do the benefits go? . . . One can easily imagine workers accepting an exposure that is significant but still under the limit if the goal is to stop the accumulation of hazards in a nuclear installation, but refusing if the exposure is aimed only at speeding up initial operations."[70]

By the late 1970s, just about everyone agreed that radon exposure had to be controlled.[71] Everyone, that is, except South African scientists and mining officials who clung to the claim that "uranium does not present a hazard during the mining and extraction processes."[72] By 1979, it was doubtful that anyone heeded this claim. Even US mining executives admitted that radon was hazardous, though they opposed labor union demands to lower permissible levels. Engineers had begun developing radon control techniques such as ventilation and backfill. Personal radon dosimeters—portable instruments that recorded alpha radiation—began to appear. The French once again leapt to the forefront with an "Integrated System of Individual Dosimetry" that cost $36.70 per month per worker to implement and maintain. Their plan to equip "all French mines" with the system in 1982 meant, of course, all mines in France.

True to its mandate, the ICRP continually evaluated new research and new measurements. In 1981, it announced a partial compromise between the epidemiological and dosimetric approaches for setting permissible radon levels.[73] Admitting that the dosimetric approach allowed higher permissible levels, the ICRP chose to err on the side of caution. For exposure limited to radon daughters, it recommended a maximum level derived from the epidemiological approach. If gamma exposure also occurred, the "additivity rule" ensured that total exposure did not exceed the whole-body occupational limit of 50 millisieverts.

This solution quelled debate for nearly a decade. But controversy resurfaced after 1990, when the ICRP revised the whole-body occupational limit from 50 millisieverts down to 20 millisieverts. Previously, the ICRP had considered only fatal cancers and two generations of hereditary effects in its calculation. Taking non-fatal cancers into account precipitated a new limit that pertained to all radiation workers (including miners) and would inevitably affect permissible radon levels. Mining industry officials from the US, Canada, South Africa, and France charged back to the battlefield with updated models of old weapons. The new limits would force some mines to shut down! They would affect non-uranium mines where radon could also lurk! Instead of banalizing uranium mines, this move could nuclearize non-uranium mines![74]

To make their case, mining officials highlighted uncertainties in the ICRP's epidemiological studies. Data from Japanese bomb survivors came under attack because that population had "not been followed to extinction," in the unfortunate words of one Canadian official. Japanese exposure had been a one-time, high-intensity affair rather than the spread-out, low doses that characterized occupational exposures.[75] (Scientific consensus eventually concluded that repeated low exposures spread over a long period were more noxious.) The upheaval went on and on. By the early 1990s, nurturing scientific uncertainty in order to defer regulation had become a well-honed tactic in many industrial sectors, so much so that scholars have dubbed participating scientists "merchants of doubt."[76]

Amidst the hand-wringing and foot-stomping, radiation risks in African mines remained almost invisible. Mining officials and their allies clung tightly to the ICRP's ever-present recommendation to allow for "economic and social considerations" to accommodate national variations in value judgment. The aforementioned Canadian official spoke for many in industry when he declared in 1991 that

rigid enforcement of the new ICRP recommendations is not desirable. Some Third World countries are very dependent upon the uranium business as a primary source of foreign exchange. It is unlikely that some of these mines could rigidly adhere to the new recommendations and continue to be economically viable. The economic and social dislocation which would be caused by the closure of one of these mines would be far worse in its impact than the radiation risk to the mine workers. . . .[77]

Such statements subsumed Africans under the "Third World" rubric even though no "Third World" countries outside of Africa produced uranium. Since no data on African exposures had entered global scientific circuits, these sentiments about the "Third World" traded on widely shared assumptions that exposures would be difficult and expensive to control, that states would adequately reflect the needs and desires of their citizens by choosing foreign exchange over regulation, and that the economic cost of closures trumped that of exposures.

No Africans were invited to the endless meetings in which the ICRP or the IAEA derived limits, set standards, and developed codes of practice. French experts continued to speak for Niger and Gabon. By the 1990s, J. K. Basson was out of the picture, but the contacts he'd cultivated helped white South Africans to remain a strong presence. Those who sought to counter the "merchants of doubt" had no way to include black African miners. In the early 1990s, for example, an international group of epidemiologists reanalyzed data from the eleven *existing* studies of radon and lung cancer risk that covered miners in Australia, Canada, China, Czechoslovakia, France, Sweden, and the US. African exposures were excluded from reanalysis because they did not exist *as data* in the first place.[78] These systemic invisibilities penetrated, in deep and lasting ways, the efforts to produce universally applicable prescriptions and place-less knowledge.

CONCLUSION

Although experts attempting to understand the dangers of radon exposure could get caught up in the epistemological thrill of disciplinary tensions, they tried hard to achieve a workable compromise. Consensus was elusive because it involved so many questions that could be addressed in different ways. How much exposure was likely to trigger cancer? What were the mechanisms? How could worker exposures be effectively measured? Who would set workplace limits on radon levels? At the ICRP and in other venues, the process of developing working guidelines involved pooling data sets, assessing their strengths and weaknesses, comparing the maximum limits suggested by different disciplinary approaches, and so on. Sociologist Steven Epstein calls this kind of work—that is, the process of getting different types of categories to match up for the purpose of producing usable knowledge—*categorical alignment*.[79] In analyzing how data produced in different places are transformed into knowledge with universal aspirations, historian Paul Edwards refers to such processes as a movement from "making global data" to "making data global."[80]

No data set could be perfect. Holes persisted (notably at the low end of the dose response curve), and plugging them was beyond the capacity of any single research group. Along the way, the process of categorical alignment produced alignment over the nuclearity of hazards at uranium mines. Whether or not nations treated uranium mines as nuclear workplaces, the ICRP, the IAEA, and even the ILO converged, in the mid to late 1970s, on an international regime that promoted the treatment of occupational radiation exposures as just another industrial hazard.

Many in the mining industry argued that radon was *less* significant than other perfectly ordinary workplace hazards. Correspondingly, less money should be spent on its mitigation. Ultimately, industry leaders were primarily concerned with *financial* alignment. They did not want to incur mitigation costs unless these were also borne by their competitors—or unless they were borne by "society," perhaps in the form of reduced taxes or other subsidies.

Still, the invisibility of African miners didn't necessarily result from deliberate strategy. As data sets circulated through international conferences, committees, and publications, they acquired heft, making exclusion progressively more difficult to notice, and inclusion progressively more difficult to achieve. Inclusion would have required a denser distribution of experts, infrastructures, instruments, and labs in African mines. The most salient instance of deliberate exclusion—South African arguments that black Africans were "not suitable for study"—disappeared from view as a commonplace standard for selecting populations to study. In this sense, the absence of African workers and data was a textbook case of what historian Robert Proctor calls *agnotology*: the "conscious, unconscious, and structural production of ignorance."[81]

Scientific ignorance affected both the conflicts and the consensus generated in the absence of data. The suggestion that "Third World countries" should spend less than industrialized nations on reducing radiation risks could remain an abstract matter of principle, the objective result of cost-benefit calculation, as long as African workers remained invisible. Agnotology also mattered for the people who were left out of data collection. But how it mattered—even how it came about in different places—could vary dramatically. Its significance cannot be addressed by gesturing toward "Third World" or "African" conditions. We must examine how the nuclearity of uranium mines and the (in)visibility of contaminants were (or weren't) made material *in each place*.

What did life in Africa mean to a Frenchman in the 1960s? Imagine the rugged male explorer, the civilizing agent, the avatar of a gloriously global nation. In the era of empire, such icons offered ready roles for metropolitan voyagers. But were exploration, civilization, and perpetuation of national glory still viable as empire waned?

The Commissariat à l'Énergie Atomique thought so. It cultivated a sense of destiny among employees preparing for "overseas" duty, albeit one tempered by postcolonial caution. As the new carrier of French global "radiance," a nuclear industry largely fueled by African uranium would restore national glory lost through war and decolonization.[1] Although a CEA employee could still see himself as a brave pioneer whose "personal mental, intellectual, and temperamental resources constitute[d] his only wealth," he had to display more enlightenment than his colonizing predecessors. Africa had changed. Adventure was still possible, but "not adventure with a capital A. Gone is the time when one could succeed after departing on impulse, as an escape, with a desire to restart a life from scratch."[2] Rupture, yes—but not rebirth.

CEA employees received a script to help them along: a handbook that recast imperial travel as overseas deployment. From this they learned that technological success—indeed, the very future of Franco-African relations—rested on the metropolitan man's ability to break from the clichés of colonial expectations and behavior. Newly independent Gabon, Madagascar, and Niger had laws that French emissaries must respect. Frenchmen had to dismiss claims that Africans mistreated mechanical objects and were "incapable of analysis, unable to conceive of both the whole and its parts, and scornful of the laws of causality." The break from the colonial past had to take place deep within the European himself. Much rode on his success: "[R]emember that the CEA and France are often judged through you."[3]

Ethnology would help supervisors treat workers in a more reasonable fashion. Behavior that might appear irresponsible often had deep cultural meaning. When African workers abruptly left for days or weeks at a time, they weren't necessarily being fickle. They were likely responding to a higher responsibility: "The call of the father or the brother living in the bush who needs the urban worker to resolve an issue for the family or the village is far more powerful that the lure of financial gain or the desire of security which might keep a man on the job."[4]

Technology would fuel postcolonial transformation, but expertise should not translate into arrogance or violence. "Your skills, which constitute your instrument of work, should contribute to the enrichment of Africa; this is your contribution to the work of civilization, not a source of superiority that entitles you to be haughty and brutal."[5] Technology transcended racial distinctions. Skill sets, not skin color, differentiated individuals. Nevertheless, the script left colonial *categories* undisturbed.

Africans at best counted as *évolués* ("evolved ones"), colonial subjects who'd been successfully civilized and uplifted through European-style education. This category continued to hold tremendous sway as the emblem of potential Modernity, the proof that Tradition could be transcended. Of course the *évolué* required an Other, and that Other remained the "tribe." The CEA's thin descriptions were thick with colonialism's obsession with skin tone and tribalism. After gesturing toward theories of human migration to Madagascar, for example, the handbook asserted that "there currently exists a light race . . . and several tribes of black race whose characteristics are mostly defined by differences in custom." Inhabitants of Gabon broke down into "multiple tribes of different dialects," all "Black Bantus." Niger's two main groups of inhabitants were those "of black race: Hausas, Djermas, etc." who lived in the southern third of the country, and those "of white race: the Touareg" who lived in the semi-desert zones.[6]

Such tribal categorization served as a mechanism of invisibility and control. Employment difficulties were explained away by citing characteristics supposedly specific to a particular ethnicity or by invoking tribalism and tradition generically. Adapting the CEA's script to work in the mines, expatriates used ethnicity, tribalism, and tradition as tools of labor management.

CEA geologists first traveled to Madagascar in the 1940s, well before the end of empire. They emulated colonial officials by treating Merina men from the central high plains as *évolués*. Educated in French-speaking schools, these men were hired in intermediary positions such as lab technicians, mechanics, chauffeurs, and clerks. Starting in 1953, the CEA transferred these intermediaries south to help operate uranium mines in the Androy desert, a region that had managed to avoid the rule of the Merina kingdom in the previous century.[7] Many Merina specialists were reluctant to live in a spiny desert whose inhabitants had a harsh, forbidding reputation. To entice them, the CEA promised high wages and separate housing—better than locally hired workers received, if not as good as the French enjoyed.[8]

Merina men made no secret of their discomfort in the Androy. It was hardship, *not* adventure.[9] For some, though, the experience led to a global career. When the CEA left Madagascar in 1968, it lured a dozen Merina specialists to other uranium sites in Africa, treating them (almost) as full-fledged expatriates who enjoyed social privileges but were not allowed to supervise Europeans. Some traveled even farther afield. One man I interviewed ran uranium prospecting missions in Indonesia for over a decade before moving to France to conclude his career.

Employment rosters for Androy mines listed most manual laborers as Tandroy or Betsileo. In 1998, however, the French Madagascar "veterans" I interviewed lumped all such laborers together as Tandroy. French memories focused on relations not between French and Tandroy, but between Tandroy and Merina employees. One man told me that Tandroy workers viewed the French as their liberators from Merina oppression, a statement thoroughly in keeping with colonial sentiments even though the Merina kingdom had never ruled the Androy.[10]

Relations with Merina technicians did not frame the memories of Malagasy miners who had worked in the Androy, however. The workers I spoke with insisted that they had no trouble with the other ethnic groups on the island. The real problem was with the CEA *vazahas* (foreigners), who didn't understand the first thing about zebus, the humpbacked cattle that were a linchpin of Tandroy society and the primary investment for workers' wages. "I tell my children to take good care of the zebus we have, because I worked very hard [in the mine] to buy them, and I'm not sure they would be able to work hard enough to buy as many."[11] *Vazahas* did not respect local practices.[12] They partitioned land with fences and shot stray zebus, returning carcasses with a stern warning instead of an apology and compensation. This was deeply offensive to Malagasies. A zebu that died outside of ritual sacrifice had no value.

The uranothorianite deposits in Madagascar were the first substantial source of uranium that CEA geologists had found outside metropolitan France. For that reason alone, they were worth mining. But they did not contain enough uranium to justify building an extensive infrastructure. The CEA's buildings, roads, and uranium mills in the Androy all had a makeshift quality.

Gabon, however, was another story. The first discovery in 1957 made clear that eastern Gabon had much more uranium than southern Madagascar—enough to justify substantial investments in housing, roads, plants, and running water. With the end of empire on the horizon, the CEA planned to stay in Gabon for the long haul.

As we saw in part I, the CEA joined forces with private mining company Mokta to form the Compagnie des Mines d'Uranium de Franceville. "Development" became the order of the day in Gabon. At first, the COMUF imported most of its mid-level technicians from the Congo. These men, trained by the CEA during prospecting missions, were desirable not just for their skills but also for their "modern" lifestyles, their abandonment of "traditional" ways. By the mid 1960s, the COMUF had trained enough Gabonese *évolués* to replace the Congolese. It contrasted these skilled men to a common workforce that was "unsuited to real transformation, continuing to follow their ancient customs, and polygamous as soon as they have the opportunity. Racial rivalries subsist: Bendjabis and Batékés are especially ready to come to blows."[13]

Feeling little need to place ordinary workers in modern housing, the COMUF director proposed grouping them "in villages, according to race." The company would provide basic building materials, and each worker would get a fixed number of days to complete his dwelling. On their own time and at their own expense, polygamists could build huts for their additional wives behind their main cabins. Monogamists were welcome to move into concrete houses in the company town. Those who followed modernity's script well enough to deserve promotion would also be "rewarded by the installation of water and electricity."[14]

Of course, Gabonese had different ideas about how to live their lives. Beginning in the late 1960s, Bongo's postcolonial state pressured the COMUF to build concrete housing for all workers. As worker *cités* grew, households took in their extended families, sometimes as many as twenty people in a two-room building. People modified their dwellings, kept livestock, grew crops. They used communal water taps for washing as well as drinking. All these actions violated rules, occasioning fines and sometimes expulsion.[15] Gabonese who couldn't stand the regimentation left of their own accord. A few created a new community of houses fashioned from discarded materials, nestled in thick vegetation, surrounded by chickens and goats. They ironically dubbed it "Cité du Silence."[16]

Most workers, though, preferred steady salaries, running water, electricity, and concrete houses that kept out torrential rains. At its peak, the company town of Mounana housed up to 7,000 people in reassuringly solid structures made from the plentiful waste rock that was easily transformed into concrete. In the early 1970s, as residents of Grand Junction, Colorado began to demand remediation for thousands of houses built with uranium mill tailings, Gabonese families started to inhabit their new homes. Another three decades would pass before they learned that their gleaming white emblems of modernity were bursting with radon.

It was easy to lose sight of nuclear things in Madagascar and Gabon.

Making nuclear things visible in the first place—at least some things, in some ways, to some people—looked straightforward. Geologists combed the southern Malagasy desert or bushwhacked the eastern Gabonese rainforest, listening for the tantalizing clicks that signaled radioactivity on their Geiger counters. After that, though, practices quickly became more pedestrian: sinking bore holes, conducting chemical tests, painstakingly mapping geological indices. For worthy deposits, operations focused on ordinary aspects of rocks, pits, and underground shafts. The only distinctions that mattered in digging a mine had to do with the rock. How much rock had to be removed to get to the ore? Which chemical process would pry the uranium from the matrix that bound it? Of what grade would the ore be?

By the time geologists, mining engineers, and metallurgists solved all their problems, the uranium extraction operation looked a lot like any other mining endeavor. Besides, the desert of Madagascar's Androy region and the forest-savannah mosaic of Gabon's Haut-Ogooué province weren't exactly replete with images that made Europeans and Americans speak—whether in rhapsodic or in horrified tones—of the "nuclear age." Electricity was in short supply, certainly not "too cheap to meter." Images of Hiroshima and Nagasaki had not penetrated the colonial press in a lasting way. Ghanaians and Senegalese may have expressed outrage when the French tested atomic bombs in Algeria in 1960, but the news barely penetrated Francophone territories further south.[17] Radiation was thus doubly

on facing page: COMUF company housing built with mine tailings in the 1970s. (photo courtesy of Cogéma) invisible: undetectable by physiological senses and absent from cultural and political imaginings.

This chapter shows how the nuclearity of uranium *ore* didn't automatically make uranium *mining* a nuclear activity, even in mines run by the atomic-minded French. The fault lines loomed large in the open-cast quarries of the Androy from 1953 to 1967. Far from CEA headquarters in France, and even from the capital city of Antananarivo and the CEA's laboratory in nearby Antsirabé, these unprofitable quarries had no experts dedicated solely to radiation protection. In Madagascar, assemblages of instruments and expertise like those that made exposures visible in France were fragile in the extreme—and so was the nuclearity of uranium mines. Some mineworkers in the Androy had inklings of the potential danger, as we'll see, but they lacked access to scientific and cultural networks that would make sense of those impressions. Their labor failed to acquire a politically or medically meaningful nuclearity.

The COMUF mine in Gabon had denser, more extensive links with metropolitan experts and surrounding territories. Size mattered, as did timing and location. The Haut-Ogooué had enough uranium to keep operations going from 1958 to 1999. The COMUF's first director, a veteran of uranium mining on French soil, was committed to the nuclearity of mines, treating radiation as a distinct hazard that required separate monitoring. Yet this regime made exposures visible only to upper management, producing technological data, not health data. Interpretation was subject to managerial whim, as became clear after 1968 when a new site director reframed the data to legitimate higher exposures. The COMUF's vaunted medical service didn't treat exposure data as relevant to clinical work. Separating radiation monitoring from mine medicine made it possible to maintain that workers didn't suffer from occupational disease. As we will see, however, workers did not accept such claims passively.

Fault lines also divided the mine's nuclearity from that of its surroundings. No one had expressed concern about mine tailings in Madagascar. By the time such problems attracted significant scientific interest anywhere, the CEA had left the island. In Gabon, however, mines and mills generated waste for decades—plenty of time for the topic to undergo study, for residents to take notice, for the rivers to turn strange colors. The French proved reluctant to ascribe nuclearity to the tailings. Nature had made the rocks radioactive. Mining removed the highly radioactive portions and

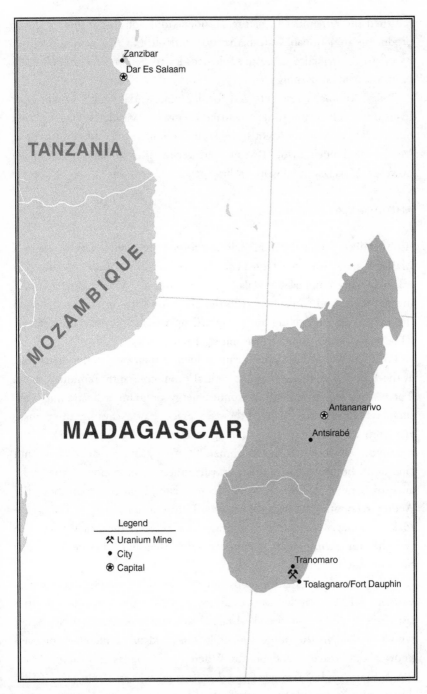

Locations of uranium mines and cities in Madagascar and in Tanzania. (map prepared by Evan Hensleigh)

returned the remainder to nature. Residents did not accept this logic, but producing an actionable alternative was difficult. No matter how obvious it was to the naked eye, waste took decades to become technopolitically visible as a nuclear thing.

Today, Mounana residents and former mineworkers are learning how to invoke nuclearity in order to make themselves visible beyond Gabon's borders. They claim the right to be seen, accounted for, and compensated. Movies have been made, lawsuits threatened. In the ongoing struggles, however, Malagasy mineworkers have remained invisible.

AMBATOMIKA, MADAGASCAR

Ambatomika. Some say a Frenchman forged the name. Others claim a Malagasy man from the high plains was responsible. Either way, the pun was peculiarly colonial. In Malagasy, "amb" means "where there is," "bat" means "rock," and "mika" refers to the glittering slabs of mica that littered the nearby desert. Adding an "o" spiced up the mix for the nuclear age: "The place where there are atomic and mica rocks."

For French expatriates, nuclearity resided in their link to the metropole, as the French flag flying over the central camp constantly reminded them. Their work fed their nation's atomic energy program, a heartening reassurance because little separated the daily practice of excavating these quarries from work done at other mining concessions. Yearly trips back to the metropole also comforted Frenchmen,[18] offering images of reactors (and later atom bombs) that let them visualize their contribution to the "radiance of France."[19] Both then and decades later, French stories about the Androy recounted the distance between European nuclearity and Malagasy rocks. Engineers reveled in recounting local uses for the heavy *vatovy* ore (weights for fishing lines, ammunition for slingshots) before its "true" purpose was discovered.

Visions of reactors and bombs did not, however, transfix local mineworkers. When I traveled to the Androy in 1998—three decades after the last uranothorianite pit closed—I talked to some twenty men and women who'd worked in the mines and mills. They did not remember (or even know about) reactors and bombs. When I explained, many laughed and shook their heads. "You crazy *vazahas* [foreigners]," said one man. "Why do you want this stuff?"[20] Another—thinking of the region's new sapphire mines, where my translator sometimes worked—shrewdly asked what sapphires were used for.[21] In their eyes, I'd simply joined a long line of foreigners interested in rocks.

I visited some of these former mineworkers in their villages, which I found with help from their family members or former colleagues. Others tracked me down. When they heard that a *vazaha* staying in the village of Tranomaro wanted to talk about *vatovy*, several men walked 10 kilometers or more to find me on market days. Their motives for doing so differed, but some suspected that the CEA wanted to reopen the mines and had sent me as a scout. They hoped that I might provide them with a pension, or jobs for their sons.

The time of *vatovy* was indeed exceptional for the Androy residents who'd lived through it. But that exceptionalism had little to do with radiation or the things that their French supervisors considered nuclear. It had a lot to do with value, especially wages, and the investments that employment made possible. Fanahia worked in the mines for 13 years. "I bought 50 zebu," he said, "and a bicycle . . . and a cart, and a radio, and a watch that I ordered from France. . . . I did some trading in watches, I would order them from Besançon and resell them to other men who worked with the *vatovy*."[22] Above all, *vatovy* exceptionalism had to do with the backbreaking tasks of attacking rocks with jackhammers and loading the broken bits into wooden carts. Tales of rockslides and lost body parts abounded. Mahata worked in the quarries with his father and two brothers until his father fell on a pneumatic drill and lost a leg. "We tell our children, you must guard the zebu carefully, because the work we did to get them was painful. We broke our legs and our feet doing that. So the zebu that are there must be well-guarded. Because you, you aren't able to do that hard work. . . . Better to guard the zebu than to work there."[23]

Radiation wasn't totally absent from memory in the Androy, but it appeared indirectly, nested in needles, displaced into dosimeters, yoked to discipline, and merged with medical monitoring. Some workers, for example, used Geiger counters daily to sort rocks into "good and bad piles."[24] The needle on the counter told the whole story: "When there's *vatovy*, the needle goes to 500 or higher."[25] The presence of *vatovy* made the needle jump. For French managers, radiation connected Geiger counters used for radiometric rock sorting with dosimeters worn by employees.[26] Both instruments recorded radioactive specificities.[27] For workers,

Open-pit mining in the Androy, 1950s or 1960s. (undated photo, courtesy of Cogéma)

though, dosimeters seemed disconnected from Geiger counters, less instruments of work than objects of discipline. "It was the boss who put them
on us, he fixed them on our clothes," remembered a woman who had
worked in one of the mills.

Dosimeters directly diagnosed illness and enabled employment.[28] "If
you didn't wear them, you were out. They kept track of that," said Joseph
Ramiha.[29] He continued:

Every week the doctor looked at dosimeters and said "this is good, this is not
good." It was the dosimeter that told him what kind of illness. Some had tuberculosis; it was the dosimeter that told [the doctor] if there was tuberculosis. If
there was tuberculosis you were forbidden from working. . . . Workers wore them
in the mills and the quarries, but mechanics didn't. They were for the smell of
the *vatovy*. . . . They dried the *vatovy* and that's what made the smell. . . . I saw
two men in the quarries get fired because the dosimeter said they were sick. . . .

The doctor explained that the dosimeter detected illness. . . . I didn't understand
too well. . . . The sickness came from the smell.[30]

Thus the dangers of radiation exposure acquired substance. The invisibility
of radiation remained, recast as odor. The illness-inducing potential of
exposure became tangible, transmuted into tuberculosis. Everyone knew
you couldn't work if you were diagnosed with tuberculosis, even if you
were asymptomatic. "That's normal, of course," said my translator. Ramiha,
who understood French even though he'd lost his speaking fluency,
agreed.[31] No surprise, then, that dosimeter readings and TB tests merged
in local logics.

Those who remembered dosimeters often linked them to illness and
doctors. Some stories resembled radiation rumors from elsewhere, complete with fears of sterility. I asked a group of women who had sorted ore
in one of the mills about the dosimeters. One of them replied:

Yes, we asked why they were putting them on and the boss said there was sickness
inside, there was gas. . . . Yes, he said what kind of sickness but we didn't understand anything about that. . . . Yes, we were worried. . . . [The boss] said that
maybe there was sickness in there. There were others who said that you couldn't
have children with the sickness from *vatovy*. We were afraid at first, but then there
was nothing.[32]

If women remained fertile, perhaps there was no danger after all?

In the Androy, no formal guidelines governed the handling of radioactive materials. The CEA's prescriptions for radiation monitoring were applied unevenly depending on the individuals involved. The lumpiness started with the transmission of information. Some supervisors tried to explain radiation hazards to their employees. Others didn't bother. One report portrayed Malagasy manual workers as so irredeemably primitive that it wasn't worthwhile developing a job training program for them.[33] Why even try to induct them as nuclear subjects?

The lumpiness continued with dosimeter distribution. Many workers didn't get one. A handful of mines and mills operated by private colonial contractors sold the uranothorianite ore they produced to the CEA. [34] Fanahia moved from a CEA mill to a privately operated one because "it paid better, there were bonuses for tonnages." He remembered clearly that "at the CEA they had dosimeters, but not elsewhere."[35] Reporting on one private contractor in 1964, a CEA manager explained that production "is done by piece work over a rich vein where the thorianite is visible. The women grind the rocks with mortars and then wash them in a basin. The thorianite they recover is purchased by the kilo."[36] There was no mention of the dust produced by grinding.

Labor conditions in the private mills were even more difficult than at Ambatomika. Robert Bodu, a CEA metallurgist sent to inspect all the uranothorianite mills in 1960, remarked of the private operations:

One gets the sense that all these installations have something provisional about them, something very rustic. The chutes are usually made from recycled metal sheets taken from old drums, the inclines are poorly thought out, the distribution of pulp throughout the machinery is very random, etc.[37]

Private operators, Bodu continued, wanted to "get the most out of equipment that's already been amortized, and engage in minimal expense." Feeling that this "rustic" approach impeded productivity, Bodu offered a long list of technical suggestions. He conceded, however, that implementing change in the private mills would require a "minimally competent" foreman. "This is often not the case and the mill is confided to a European who, despite his good will, is sorely lacking in technical experience."[38] Being European did not guarantee technological knowledge. Private operations tended to be managed by former French *colons* who had no science or engineering background. Given how many corners private contractors already cut, the absence of radiation monitoring was hardly surprising.

Yet making a sharp dosimetric distinction between the CEA and its private contractors would be too facile. Some of the operations at Ambatomika also struck Bodu as "rustic." He especially bemoaned the rudimentary methods used to dry the concentrates in the mills:

Concentrates are spread out in the sun on big sheets of corrugated metal and turned over periodically by a worker. . . . This procedure is clearly archaic, long, and above all dangerous because the worker is exposed to dust and radiation.[39]

Bodu's mandate did not, however, include measuring specific worker exposures—let alone mitigating them.

The production managers who ran CEA operations in the Androy filled the pages of their activity reports with tales of inclement weather and technical woes rather than discussions of radiation levels. They clearly knew that jobs such as packing uranothorianite concentrates for shipment to France presented significant exposure hazards.[40] But they did not report on the distribution of film badges or provide specific dosimeter results. If Ramiha displaced dosimeter readings into tuberculosis tests, the operations directors buried them in productivity tables, invoking them only in footnotes to reports on person-days worked. Here's one note from July 1959: "Five workers who exceeded the maximum admissible radiation dose were transferred to 'pasture' on the BETI site."[41] Perhaps Ramiha's friends weren't laid off for high dosimeter readings, but rotated to another, less radioactive workplace? It's impossible to know.

The silences in activity reports speak to the looseness and fragility of Ambatomika's ties to metropolitan infrastructures of nuclearity. Of the three hazards tracked by the CEA back in France—radon, dust, and gamma rays—Ambatomika managers tracked only external gamma exposure. By far the simplest to measure, this hazard was also the most salient in open-pit mining. But international experts had found that radon could accumulate rapidly around stored piles of untreated uranium ore. Furthermore, Androy ore contained thorium, which decayed into equally noxious thoron gas. Practices that converted uranium mining into nuclear work in France—such as measuring gas levels and estimating exposures by cross-referencing time sheets—were well beyond the technical capacity of Ambatomika managers. Dosimeters were distributed and gamma doses recorded just long enough to determine whether job rotation was required. There the official activity ended.

By 1967, around the time that the CEA and other Western uranium producers began serious efforts to create "the uranium market," the people, equipment, and quarries in Madagascar were exhausted. The CEA packed its bags and went home.

Even for CEA experts, the nuclearity of the Androy mines was brittle and intermittent. Threads of geological and metallurgical nuclearity ran through consultants like Bodu who visited occasionally to advise managers about prospecting or treating ore. These experts noticed high radiation levels, but not in ways that shaped design choices as they had in French mines. Radiation monitoring did not empower a distinct class of experts in Madagascar. I found no evidence that anyone had compiled exposure figures to produce scientifically legible or portable data sets.

Here's one thing I do know. A keyword search on "Madagascar" in the archives of France's Institut de Radioprotection et de Sûreté Nucléaire (a recent CEA spin-off) returns results, but the archivist will not reveal the quantity or contents of the files. "For the keyword 'Madagascar'," she wrote me, "the documents cannot be communicated because they contain names and medical data."[42]

Do the files pertain to French or Malagasy employees? Are they generic or do they contain dosimetric data? Television documentaries criticizing France's uranium endeavors at home and abroad, along with increasing pressure to account for and compensate exposures, make it unlikely that the files will be released anytime soon. Laws restricting archival access are remarkably useful in that way.

Here's another thing we know: Uranothorianite ores could attain grades up to 20 percent uranium and 50 percent thorium. The exceptionally high thorium content produced high levels of gamma radiation.[43] Eight years after the CEA left Madagascar, this was unusual enough to warrant special mention in the IAEA's 1976 *Manual on Radiological Safety in Uranium and Thorium Mines and Mills*, which we encountered in chapter 6. In a passage describing the potential for high external exposures in uranium mills, the manual singled out Malagasy ore as emitting two and a half times as much gamma radiation as pitchblende, the highest grade of uranium ore.[44]

Mind you, these levels hadn't been measured in Madagascar; they were recorded in France at the plant that extracted uranium and thorium from the ore. Did the manual's authors consider the exposures of the Malagasy men and women who had mixed those uranothorianite concentrates? Did the manual's users wonder about those workers? If so, their musings failed to make it into the international scientific literature. Uranium from Madagascar remained disembodied, an object of geological interest but not a site for public health measures or epidemiological research.

The nuclearity of Malagasy mines thus remained lumpy, as refractory as the ore itself. The workplaces were "nuclear" for the CEA in some aspects, but they were not fully woven into its network of nuclearity. Many loose ends remained. Dosimetric films—and a few willing supervisors—suggested to some workers the presence of invisible dangers. But knowledge about radiation disappeared without a sustained infrastructure. Ore concentrates got the full nuclear treatment only upon their arrival in France as hazardous material. Even as high radioactivity wrote Malagasy uranothorianite into instructions on radiological protection, the conditions of its production vanished.

Even these absences can teach us about the power of nuclear things. We notice not just that this power is distributed unevenly, but also how that unevenness matters. We glimpse how that power resides in the countless images that usher "the nuclear age" into some parts of the world but not others. We observe that nuclearity in a single place can be fractured, expressed in some practices but not others. We see the purpose of laying claim to nuclearity through some practices, and the outcome of not doing so in others. Following the French CEA to Madagascar teaches us that even within a single, centralized, state-run agency, nuclearity was not a full-fledged, immutable property of a technological system. How it was concentrated in some places and fizzled out in others had to do with race, geography, and colonial power.

Yet we cannot gesture grandly at "colonial power" to explain the unevenness of nuclearity. The imperceptibility of exposures and the long-term invisibility of Malagasies as radiation workers resulted from *geographically and temporally specific* colonial and postcolonial circuits of power. Place mattered tremendously to what was rendered perceptible, to whom, and with what physiological and political results. Historical rhythms mattered too. Nuclearity had uneven temporalities as well as uneven spatial distributions. Let's pursue these themes by traveling to the French uranium mines in Gabon.

MOUNANA, GABON

The first open pit at Mounana lay at the bottom of a hill where thick rainforest met patches of savannah. Another mine had preceded it in the early 1950s, a French-owned manganese mine that had broken ground deeper in the rainforest some 25 kilometers south of Mounana. That mine built a cableway to transport ore through the rainforest to Congo-Brazzaville, where the manganese boarded a railway to the port of Pointe-Noire. In 1961, the Compagnie Minière d'Uranium de Franceville used this shipping route to export its first batch of uranium ore, suspended in cable cars chugging through the forest canopy.

Xavier des Ligneris, the COMUF's first director, had previously run CEA uranium mines in France. In Gabon he tried to follow the CEA's prescriptions directly, treating radiation as separate from other health and safety issues in the workplace.[45] Although he left the development of all other health and safety guidelines to his managers, he personally wrote and signed those pertaining to gamma rays, radon, and dust. With so many other things to worry about, the newly independent government quickly issued a stamp of approval.[46]

Getting workers to wear the dosimetric film badges according to prescription was less straightforward. At first they didn't wear the films regularly enough. Operations managers repeatedly issued warnings that films were "absolutely obligatory" or sanctions would ensue.[47] Then workers apparently wore the films too much, so directives began warning employees not to take their films outside the workplace. Speculating that recurring spikes in external exposures resulted from workers putting their films directly on the ore, managers kept trying to "tighten surveillance."[48] Supervisors were to approve time cards only if workers returned dosimetric films correctly.[49]

Des Ligneris expected that enforcing correct procedures would automatically control exposures, but it didn't. Up to eight weeks could elapse before exposure data returned from the CEA lab that processed the films in France. Coupled with the inherent unpredictability of the ore body, this meant that spikes in external exposures continued.[50] Once underground mining began, in 1966, radon added to managers' anxieties. The average concentrations of radon in the shafts regularly exceeded maximum permissible levels—sometimes by a factor of 12.[51] Many employees consistently exceeded their annual exposure limits, often in less than eight months. All the surveillance in the world couldn't stop inexorable radioactive decay in the rock.

Nor could high exposures be easily attributed to African incompetence. For one thing, ambient sampling meant that good results didn't depend on individuals' wearing instruments correctly. For another, European employees also charted high readings. When reports indicated the difference in African and European exposures, the amount of overexposure between the groups was comparable.[52] Of course, many more African workers got overexposed, because many more Africans worked in the mines.

Radon turned labor management into a calculus of exposure. Employees worked in high-level shafts until they'd reached (or even exceeded) their annual limit, at which point they were moved to workplaces that had lower levels. Before ALARA, working "to the limit" before rotating jobs was a standard way to manage exposures in nuclear workplaces. Des Ligneris would have become well acquainted with the practice in France, where one mine's high gamma levels limited individuals to four-hour shifts every two weeks. With high external gamma exposures and a proximity to labs that could turn around the dosimetric results quickly, rotation could be planned in advance. The less predictable internal exposures to radon were tougher to control. Plus, the low grade of the ore meant that Mounana ran on a tighter budget. By 1967, production had fallen well behind schedule.[53]

All this worried Xavier des Ligneris. In late 1967 he commissioned a study of the doses absorbed by 166 workers over a one-month period. Thirty-eight men had exceeded the monthly limit and had to be moved to less radioactive environments. The *average* dose absorbed by the remaining men came disturbingly close to the maximum. This left little wiggle room. Des Ligneris predicted that in the sites with the highest radon levels workers would absorb their annual maximum dose in just four months. "If we exceed this limit," he noted, "we will quickly get to the point where we can't even staff the sites with [lower] radon levels."[54]

As the mine got deeper and radiation levels increased, management feared a shortage of skilled workers. Continually hiring new personnel offered one solution, since new hires were assumed to be radiation virgins. But training new workers took time, canceling out the benefits from labor turnover.[55] Finally, des Ligneris decided to upgrade the ventilation system, a solution that worked temporarily. From March to May 1968, radon levels decreased significantly.

In the meantime, corporate headquarters called for leadership change at Mounana. As a CEA man whose career had focused entirely on uranium ore, Xavier des Ligneris had played an important role in finding and mapping the deposit, drafting the initial mining plan, and building a strong prospecting team. His expertise also ensured that Mounana's production schedule corresponded closely to the CEA's fuel requirements. But Mokta, the CEA's corporate partner in the COMUF, wanted someone better attuned to budget constraints and less concerned with the nuclear dimensions of his work. The ventilation upgrades had been expensive, and although the CEA predicted rises in uranium demand and prices, the immediate prospects for "the uranium market" looked grim.

In mid 1968, Mokta sent Christian Guizol, a former manager of a manganese mine in Ivory Coast, to replace des Ligneris and shake things up.[56] Gabonese employees remembered Guizol as a hard, uncompromising man.[57] It soon became clear that Guizol didn't view Mounana's disciplinary measures as sufficiently strict in the workplace or the living quarters. When gamma exposures climbed back up in late 1968, Guizol deemed his predecessor soft on Africans and blamed the workers for not wearing their films correctly. He tightened disciplinary and surveillance measures around film use. But the test dosimeters he placed in the shafts disappointed him when they corroborated the gamma exposures indicated by worker badges. Radon levels also climbed back up, with 78 workers registering overexposure in November 1969.[58]

Guizol responded by reconfiguring the calculus of exposure. Rather than intensifying job rotation as des Ligneris had done, he raised the maximum for exposure levels. He'd noticed that the 1968 radon guidelines issued by the ILO used a formula that (under Mounana conditions) yielded a higher permissible level than the French guidelines used by the COMUF.[59] A few numerical gymnastics enabled Guizol to increase the radon MPLs threefold in 1970. Insisting that the new levels matched ILO guidelines better, he remarked frankly that they were also "more advantageous" to the company. The effect was immediate. As of March 1970, not a single worker registered overexposure.[60] No wonder that, as we saw in chapter 6, Gabonese state officials could report complete success with the COMUF's radioprotection program at the 1974 Bordeaux conference.

Who would object to this change in regulation? The Gabonese state, which had supported the COMUF in quelling strikes and recruiting labor, had no expertise in the matter. I found nothing in the (admittedly chaotic) COMUF archives to suggest that Gabonese state officials ever inspected radiation, radon, or dust in the mines.[61] The fact that MPLs had been realigned to accommodate high radon exposures apparently just disappeared into a closet.

The state may not have questioned the company's good faith, but at least one COMUF worker did. Not long before Guizol raised the MPLs, Marcel Lekonaguia began to express doubts about his working conditions. Born and raised in the village of Massango (which was later expropriated for the mine), Lekonaguia worked for the first wave of CEA geologists before later joining the COMUF. In the mid 1960s, he became a shift boss in charge of blasting in the underground shaft. Company guidelines specified waiting 15 minutes after a blast before returning to the workplace.[62] Lekonaguia probably didn't know that French radiation guidelines set a waiting period of at least 30 minutes to allow dust to settle *and* give the ventilation system time to evacuate radon released by the blast.

What Lekonaguia did know, all too well, was that "after the blast, there's a lot of dust. . . . It's the dust that wasted us. . . . You swallow it, you breathe it." Protective gear that dissolved in water didn't help: "Those little masks, they didn't hold up well. They're made of paper. . . . If it gets a little wet—paf!" That, he insisted, was how he developed the cough and other assorted ailments that would plague him for the rest of his life.

Lekonaguia's doubts represented mistrust not just of the company, but also of the labor union launched at Mounana in 1965. One of the union's first acts was to negotiate a schedule of hardship bonuses. Underground workers got the highest premium of 15 percent over base pay, a response to the danger of rock falls, the threat of flooding, and the reluctance of Gabonese workers to go underground because of evil spirits. Radiation appeared nowhere in the new labor contract.[63] "Insalubrious work" was defined as any task that required a dust mask (the kind that went "paf!"). Tasks that required dosimetric film badges did not fall under this definition. Indeed, film badges did not appear anywhere in the new contract.

Underground miners and plant operator, 1980s. (undated photos, courtesy of COMUF)

Did radiation seem too intangible to union leaders, its effects too distant to be worth haggling over? Or did workers not know enough about its hazards to insist on its inclusion? Whatever the case, masks served to signal routinely compensable "insalubrity." Dosimetric film badges did not.

The union's lack of concern didn't stop Lekonaguia from wondering about the films—especially given the tight discipline that they incarnated. "They said this film here, you must always keep it. At the end of the month, they check them, they send them to see if the men reached [the limit]. The results, they don't give them to people."[64] All he ever learned was whether he'd reached some threshold that prompted job rotation. He never found out what the numbers were, how close to the limit he'd come, how much he'd accumulated, or even what the limit meant. What, he wondered, was all the secrecy about?

Lekonaguia's brother, Dominique Oyingha, became convinced that the company and its doctor, Jean-Claude Andrault, were hiding something. Like Lekonaguia, Oyingha had joined the mine when it opened but had never cared for the COMUF. He seized an opportunity to get better schooling in Libreville, returning to the Haut-Ogooué after a few years to launch his own transport business and eventually enter regional politics. When Lekonaguia got sick, Oyingha immediately suspected workplace conditions and complained to the company. Nothing changed. His brother got worse. "Uranium caused many deaths, but the COMUF didn't want to recognize that," he told me. "Nor did the state, because this was the big company of the territory, whose secrets couldn't come out . . . so as not to scare the workers."[65] Because only independent, external expertise could be trusted, Oyingha took his brother to hospitals in the Congo for tests. "The COMUF," he recalled,

had representatives in the Congo, because the uranium went on the Congo-Ocean railroad. So this secret extended all the way there. So I played clever, I said "we are not Gabonese." But the doctor . . . he said "What do you mean, you aren't Gabonese?" I said, "No, we're not Gabonese." He said "But has your brother ever worked in a mine?" I said "Well, working in a mine, maybe that's saying too much, but yes . . . we may have been in a place where uranium is mined." So then the doctor said that my brother's disease was due to uranium. As of that moment that I said to myself, okay, it's because of uranium that my brother is suffering.[66]

Why, I wondered, did they choose the Congo in the first place?

Because there was a uranium mine in Zaire, in what's now Congo-Kinshasa. And the people who came from there brought us the news: "this product that you are mining is dangerous." Because the COMUF had also brought workers from there. People who knew things, who came to teach the Gabonese how to mine. So it was these people who, secretly, shared the news with their friends that this product that we are mining is a toxic product.[67]

Hadn't the COMUF realized that its employees were talking about this? "No," Oyingha replied. Had the Congolese mentioned disease openly, they would have been fired. Besides, all Congolese workers in Gabon had returned home in 1962, expelled after violence erupted over an acrimonious football match.

Several years had elapsed before the two brothers, faced with Lekonaguia's declining health, had remembered the rumors and gone to the Congo in search of answers. Upon their return, Oyingha confronted Xavier des Ligneris, who by then must have been getting ready to depart for France:

I said, "Monsieur des Ligneris, my brother suffers from uranium." Des Ligneris says to me "But what, how did you come to this conclusion? Who told you that uranium makes people suffer?" I said, "Monsieur des Ligneris, I know. I will give you the list of diseases that are caused by uranium." He said, "WHAT?!?"[68]

Des Ligneris agreed to have the doctor examine Lekonaguia again. But Oyingha knew that Andrault wouldn't side with them. When they went to see him, Andrault scoffed: "Are you crazy? . . . Who told you that uranium made people sick?" Oyingha laughed as he remembered this response. He respected—even loved—this doctor, who offered free medical care to everyone in the region, who'd taken the time to learn about local customs, who allowed traditional healers into his clinic. That was precious beyond measure; as another Mounana resident put it, "Andrault is a god around here."[69] But everyone had limits, and Oyingha didn't expect the doctor to acknowledge the possibility of occupational disease. He threatened Andrault:

I said, "My friend, you are my friend, we have known each other for a good bit of time, but let me tell you that the sickness that my brother suffers from, it comes from uranium. And if you don't want the news to spread . . . [so that] your workers don't become afraid, take proper care of my big brother. If he dies, I'm coming after you."[70]

The COMUF granted Lekonaguia sick leave but refused his request for permanent leave and compensation, insisting that he return underground if he wanted to draw his paycheck. In 1970, the two brothers filed a complaint with the state social security office in Libreville. But "it was a one-party [state], and in a one-party [state] you can't say anything, you can't oppose the state."[71] Their complaint produced a perfunctory inquiry. After this the company agreed to move Lekonaguia to the open pit.[72]

Undeterred, Lekonaguia asked for his file. Andrault refused, citing medical confidentiality. No surprise there, the brothers said—"The doctor, he's just a lawyer for the COMUF." At this point in the telling, Lekonaguia's nephew chimed in: "If they admit to occupational disease, a lot of people will come after them." The more the COMUF resisted, the more Lekonaguia and his family became convinced that his illnesses were work-related. Through the 1970s and 1980s, more and more young people from Mounana (including Lekonaguia's nephew) went to study in France, often on scholarships provided by the COMUF. They got their first taste of independent expertise by reading the press and occasionally witnessing anti-nuclear protests. When they returned home, some confirmed what their parents had first learned from the Congolese technicians: "This product that we're mining, it's a toxic product."[73]

Finally, Lekonaguia decided that if COMUF managers kept rejecting his demands, he would rebuff theirs. He began refusing to return his film badges for monthly readings. He suspected that his diagnosis—along with the chain of causality that linked work to illness—could be read directly from the films. One day, his nephew explained to me as Lekonaguia displayed one of the films, they would find someone else to read the results.

Among the two dozen Mounana employees whom I interviewed at length, Lekonaguia and Oyingha seemed exceptional in their outspokenness. I had been introduced to the brothers by a young computer technician, a second-generation COMUF employee whose father had been one of the first Gabonese men promoted to management. It had taken a couple of weeks for that introduction to happen. I suspected that the young man first wanted to make certain that I wasn't working on behalf of the company. Even then he distanced himself from their narrative, implying with a vague hand gesture that Lekonaguia in particular was a little crazy. Nevertheless, archival documentation—albeit scant—supported their story.[74]

Lekonaguia probably wasn't alone in hoping that keeping his film badges would pay off in the long run, though it's difficult to tell from the available evidence. In the mid 1980s, COMUF quarterly radiation protection reports listed the numbers of non-returned films, a statistic that had not appeared in reports from the late 1960s and early 1970s. Some months, over 25 percent of films didn't get returned.[75] Archival disorder made it impossible for me to systematically document dust and radiation exposures over a sustained period. I couldn't even determine whether full records still existed. If they did, they were buried deep in the termite-ridden files.

Intermittent reports suggested that radiation levels continued to fluctuate around the limit permanently realigned by Guizol. Keeping equipment in working order was a struggle, and by the mid 1970s the COMUF's dosimetric equipment was outdated and poorly calibrated.[76] A series of fatal accidents triggered an investigation in 1977 that found poor maintenance, inadequate warning notices, careless handling of explosives and flammable materials, absent supervisors, and sloppy housekeeping throughout the mines. The investigator deplored the lack of protective clothing and equipment for plant workers, who typically wore only "shorts, a shirt (sometimes made of nylon) and bare feet or light (unreinforced) boots."[77] Where were the overalls, helmets, gloves, safety glasses, and dust masks found in French plants? He also expressed dismay at ventilation systems that provided fresh air at half the needed flow or less. In some galleries, the ventilation recycled rather than evacuated the air. The resulting heat, humidity, and dust produced "extremely difficult working conditions."[78] Doubtless because he hailed from the "conventional" mining sector, the investigator said nothing about radon. But ventilation that couldn't control temperature, humidity, and dust wouldn't have evacuated radon effectively either. Recycling air also recycled radon, producing even higher concentrations of its carcinogenic daughters.

In 1989 the COMUF finally sought to upgrade its radiation monitoring equipment.[79] Six years had elapsed since French mines had introduced personal dosimeters that gave each worker an individual reading of exposure to dust, gamma radiation, and radon. Better equipment, however, did not necessarily produce lower exposures. In April 1992, all three of the regulated contaminants exceeded operational guidelines throughout the site. In the yellowcake packing zone, they were nearly double the permissible levels.[80]

In July of 1998, the year before the COMUF closed shop, one of the yellowcake plant operators spent an afternoon showing me the facility. Makeshift planks patched up holes in the flooring, labels and buttons were missing from several panels in the control room, and I didn't see a single sign warning workers about radiation hazards. No one in the area wore protective gear. In one section, sulfuric acid vapors escaped from holes in the tanks, permeating the air and causing my eyes to water, my nose to run, and my throat to close up. My guide seemed unfazed, remarking ruefully: "Ah well, working conditions aren't like in France, eh?"[81]

As for Marcel Lekonaguia, he periodically received awards for long service to the company, medals conferred at ceremonies presided over by state representatives, sometimes even El Hadj Omar Bongo himself, as we saw in chapter 4. At the end of one of our conversations, Lekonaguia changed into a pristine blue uniform adorned with these medals so that I could take a photograph (which unfortunately didn't turn out well enough to be printed here). He felt these decorations to be cheap rewards, inadequate recompense for the hardship he'd endured. A pickup truck, he suggested, would have been more appropriate compensation, trailing off: "If I were in America. . . ."[82]

NATURAL RADIATION, OR RADIATED NATURE?

Even before the COMUF arrived, Dominique Oyingha told me during one conversation, local people had health problems—"especially those of us in Massango, because the whole village washed themselves at Okeloleni, drank at Okeloleni. That place where we drank and washed, there was an atomic reactor." Oyingha was referring to the Oklo fossilized reactor mentioned in chapter 4, the one that caused such big financial problems for the company in the mid 1970s. Many people died, he continued, and "we knew it wasn't normal. . . . At the time, people accused each other of witchcraft. . . . Only when the COMUF did its mining did we see that we really were in danger-danger." When the COMUF expropriated Massango's land and moved the village a short distance away, however, it didn't have radiation hazards in mind. The danger did not become nuclear for quite some time.

In the 1960s and 1970s, the first and most obvious hazard was to the area's food supply, because many people abandoned their plantations to work at the mine. "The soil was really fertile. It gave bananas, tobacco, even rice," Oyingha reported. "I myself ate that rice when I was young. They also planted coffee. When the COMUF arrived . . . there wasn't much coffee anymore because coffee paid less well than the COMUF. That's how people neglected agriculture." Even worse, he said, "Our environment was destroyed. . . . The COMUF exploited the bush for its mining operation. It didn't have enough advanced technology to support the tunnels, the galleries, the shafts and all that, they just did it all with wood. There was a sawmill. . . . In exploiting the timber, I tell you, our environment was devastated."[83] The mine's appetite for wood thus contradicted Oyingha's "expectations of modernity,"[84] symbolizing both the limits and excesses of its otherwise "advanced" technological systems.

The yellowcake plant that opened in 1982 severely aggravated environmental problems by evacuating its waste effluents into the waterways. The COMUF distributed drinking water to its company towns but not to outside villages that relied on the rivers for water. Troubled by this new assault, area residents expressed their concerns about water pollution to the provincial governor. When dead fish began floating through the backwaters in larger and larger numbers, alarmed residents worried that their "lives [were] in danger, because they often get serious intestinal and other diseases." They also complained that the medical care they received had "little value." The governor's office instructed the company to redirect mine water away from residential zones. "Almost a thousand people are struck by radioactivity," the memo concluded sternly, adding that the "local dispensary should reinforce its monitoring" and deliver "more effective care."[85]

The company accepted responsibility for some of the problems. It admitted that the chemical effluents increased water acidity, but it insisted that the state's recent public works projects had also damaged the riverine system. It categorically denied that radioactivity contributed to the problems. Radiation levels had been under "extreme surveillance since 1962, following international norms. All the tests are done in France. There has never been any irradiation of COMUF workers, let alone village inhabitants." The company also defended its medical services vigorously, noting that no epidemic could be attributed to pollution. Half of the COMUF's hospital patients did not work for the company, its clinic conducted many exams and tests each day, and the medical service was committed to the "perpetual improvement" of its methods. Area residents were therefore "very medicalized."

The COMUF insisted that the complaints were aimed at obtaining potable water for the nearby villages not controlled by the mine. "If an effort is made on this point, we will certainly circumvent many of the complaints."[86] The company bet that "development" would palliate local criticism. Yet some in top management knew that there was cause for concern. A few months after residents filed their complaint, the IAEA held a seminar in Libreville on "Radiation Protection in the Exploration, Mining and Milling of Radioactive Ores for Developing Countries in Africa." Featuring "international experts" from Europe and North America, the seminar offered both "practical demonstrations" and lectures. The audience included representatives from seven African countries in addition to Gabon.[87] The COMUF was expected to host participants during a weekend visit to the mine.

"America," that distant promise of the good life for Lekonaguia, was a source of anxiety for COMUF management. Company officials feared that the North Americans speaking in Libreville would criticize the COMUF's approach to radioactive waste disposal and occupational exposures. "We at COMUF," the head of the Paris office wrote to CEA experts, "are counting on you, not only to limit the risks of this seminar but also to ensure that it has beneficial consequences for COMUF's future."[88] Happy to oblige, the CEA contacted one of the American experts to ask that she present "dilution and dispersion" as a viable means for long-term management of uranium mill tailings. If Africans learned that North American regulations mandated that tailings be "confined and contained" rather than dispersed, they might insist on implementing more costly waste-management strategies.

Dilution and dispersion, the CEA insisted, was the only "practical" solution "under conditions that [were], in the long term, acceptable economically and credible geotechnically for the two African countries that we know well and that we wouldn't want to see 'contaminated' by the mentality that predominates in North American administrations!"[89] In Niger, CEA experts explained, "wind erosion always wins. Recent data indicates that the dispersion of dust from a hot desert like the Sahara is global." As for Gabon,

In the long term I don't see how we can guarantee the geotechnical stability of a surface or sub-surface confinement structure: landslides will end up transporting everything to the nearest river. Currently at Mounana a large quantity of residues have passed into the Mitembé. Fortunately <u>the flow rate is high</u>, and I think that <u>the fairly powerful rivers [into which the Mitembé flows] would have the capacity, in the long term, to transport</u> all these sediments to the Atlantic via the Ogooué. <u>A sedimentology study remains to be done, of course.</u>[90]

French-run uranium operations in both Niger and Gabon, therefore, should continue releasing mill tailings directly into the environment. Experts hadn't actually *studied* dispersion or dilution patterns. Rather, they *assumed* that the winds were reliably strong and the rivers sufficiently powerful.

The Libreville meeting didn't live up to the COMUF's fears, but it did put corporate management on alert. So far, French experts had provided all information about radiation exposure, mining pollution, and related matters. If non-French experts got involved, it would be only a matter of time before even the most favorably disposed state authorities aligned Gabonese regulatory provisions with European ones. The time had come for a more technopolitically robust plan for waste disposal.

In 1984, COMUF headquarters sent an engineer named Victor Jug to Mounana to develop a plan. His biggest puzzle concerned the disposal of "sterile" rock (ore with insufficient uranium content) and liquid effluents. The idea was to dump or pump these into the depleted quarries, allowing liquid waste to empty out into two rivers, the Mitembé and the Lekedi. Because the standard international practice of treating waste before dumping it was costly, Jug suggested proceeding in stages, "depending on the pressure exerted on COMUF."[91] Best begin with "very limited ambitions" as justified by the depressed uranium market.[92] Dumping minimally treated waste directly into the river could be acceptable in the short term. In the long run, though, to "conform to the regulations of developed countries, water from the quarry could not be rejected directly and should undergo treatment."[93]

By 1986, Mounana did treat some liquid effluents, but the process "was not very effective by normal criteria" and methods for analyzing radium content didn't work properly. This "could be embarrassing if precise questions arose concerning the results." Worse, Jug had found that "in transitioning through the quarry, water became recharged with radium." A conventional tailings dam could provide a remedy, but it would be "very expensive and awkward to build given the local topography and rainfall." It would also saddle COMUF with "heavy maintenance and surveillance obligations if Gabonese regulations eventually came close to those being implemented in Europe."[94] The company eventually built a dam in 1990, which I visited eight years later. Both the French associate director of the mine and the Gabonese operator who toured me around the plant assured me that children "just knew" to stay away from the tailings pond.

Jug also commissioned a study of the "radiological state" of the site and its environs. Conducted in 1985–86 by Nicole Fourcade of the CEA, the study measured the area's radon levels over a one-year period. Relative to radon levels near uranium mines in France, levels around the COMUF were "in general well below the limit for the public of 3 [picocuries/liter] of additional radioactivity."[95] The key to this optimistic assessment was the word "additional." French regulations (Gabon had none at this stage) limited public exposure to no more than 3 picocuries *over background levels in a particular region*. This meant that if the "natural level" of area radon were 3 picocuries at one region and 5 picocuries at another region, total radon near mining activities shouldn't exceed 6 picocuries and 8 picocuries, respectively. The model assumed that people could easily tolerate whatever radon levels they'd been living with. Regulation should merely limit *additional* exposure.[96]

What if the "natural level" of radon was already high? As Dominique Oyingha and others told me, local residents had long suffered strange diseases and early deaths in the Mounana region. Even if Fourcade had heard these stories, however, she probably would have dismissed them as mere anecdotal accounts. Acknowledging that buried uranium ore "must have been a significant source of radon relative to the region's natural sources," she took radon readings on forest trails as a baseline for "natural levels."[97] Although these could climb as high as 10 picocuries, the report used as reference the average of 3.3 picocuries. Fourcade focused on "excess" radon on COMUF property, using the company's office complex as ground zero. Overall, levels stayed below those recorded around defunct uranium mines in France.[98]

As we have seen, however, radon per se wasn't the problem. Determining the level of hazard caused by the radon daughters required measuring alpha particles directly, which by the mid 1980s was possible using well-developed instrumentation. The alpha readings for 1985 were striking. In the COMUF offices and the *cité cadres* (management housing), levels were relatively low. In Massango village and the *cité* for specialized technicians, levels were 3 times those at the offices. In *cités* for manual workers, they were 4.5 to 13 times the office levels. Nevertheless, Fourcade concluded that her research results were reassuring. The mill tailings, which emitted low levels of radioactivity, were "of natural origin" and no more harmful than buried ore. As long as the mine avoided using tailings in "inappropriate" ways—"like for example . . . [in] the construction of dwellings without taking particular precautions"[99]—everything would be fine.

In 2007—two decades after that warning and eight years after the COMUF closed up shop—subsequent investigations admitted what everyone living in Mounana already knew. The housing developments, the dispensary, and many other buildings—all touted by the company and the state as motors of modernization—had been built using "sterile" rock from the mine tailings. A fatal flaw of Fourcade's research suddenly became clear. Her study had measured levels on the *streets* of the *cités*. *Inside* the houses, radiation and radon levels could reach eight times the internationally sanctioned maximum.

DISTRIBUTED CLAIMS AND THE FAILURES OF CITIZENSHIP

After the Mounana mine shut down in 1999, COMUF workers and area residents grew increasingly suspicious about their health. The steady salaries that palliated concerns about pollution now gone, they turned their focus to the debris that mining had left on their land and in their bodies. In 2001—perhaps thanks to a former COMUF employee subsequently elected to Parliament—Gabon created a state agency to monitor radiation exposure.[100] Another three years elapsed before the agency came into existence, however, and it's still not clear whether it has much efficacy. In any case, former employees and area residents continued to mistrust the state. Complaining that remediation work focused only on containing loose ore left behind by the mining activities, they sought a *medical* nuclearity—and compensation—for their work.

In 2005, inspired by Aghirin'man, a non-government organization concerned with occupational and environmental illness in active Nigérien uranium mines, a group of Mounana residents formed the Collectif des anciens travailleurs miniers de Comuf (CATRAM). CATRAM demanded from Areva (a French corporation that inherited responsibility for the COMUF's legacy in 2001) a health and environmental monitoring program and a fund to disburse medical compensation claims.[101] Gabonese retained a powerful sense that their plight wasn't "like in France." Invoking "ethics, equity, and social justice," one CATRAM member noted that Gabonese workers "did not, during their entire careers at the Mounana uranium mine, benefit from the attentive medical surveillance reserved for their expatriate colleagues. During their leaves in France, the latter systematically underwent hematology examinations and cancer screening."[102] Sick Mounana residents, in contrast, often did not know whether they had cancer or some other disease.

As the Gabonese soon learned, screening had not immunized French expatriates from cancer itself. The same year CATRAM was formed in Gabon, Jacqueline Gaudet formed the Association Mounana in France. Gaudet had lived in Mounana for 15 years, first with her parents, then with her husband. After her father died of lung cancer in 2000, she approached Areva for compensation. Because he'd worked abroad, however, her father was covered not by France's national social security system but by a separate fund dedicated to French expatriates. That fund covered everything *except* occupational illnesses, which allegedly were covered by Gabon's social security fund even for French citizens. But the Gabon fund required employees to file a claim within 48 hours of their diagnosis. It was far too late for Gaudet's father and others like him. Furious, Gaudet vowed to make Areva pay.

CATRAM, the Association Mounana, and Aghirin'man joined forces with three other French NGOs to send a small team to Mounana in June 2006. Group membership included representatives from Sherpa, an organization of high-profile legal experts formed in 2001 to investigate global human rights and environmental justice violations perpetrated by French companies. Also involved was CRIIRAD, an independent laboratory created after the 1986 Chernobyl accident to develop nuclear expertise unbeholden to the French state, and Médecins du Monde (MdM), a humanitarian organization focused on bringing medical care to vulnerable populations. The Mounana team took independent environmental readings and surveyed nearly 500 former COMUF employees about their health and work experience, publishing their results In 2007.[103]

Survey responses echoed narratives I heard from Lekonaguia, Oyingha, and other Gabonese in 1998. The vast majority of workers reported no formal training on radiation or radon-related risks. At best, they learned about risks by word of mouth from other workers. Employees were not required to wear protective gear, and all work clothing was washed at home. "We were so unaware of the risks that we smoked and ate at the workplace, and since we never wore protective gloves, we ate and inhaled whatever was on our hands and in the air, [including] after maintenance operation[s] that left yellowcake powder suspended in the air."[104] Employees did not receive reports of their radiation exposures. All agreed that the Gabonese state had done nothing to monitor working conditions or occupational health. One former medical doctor testified that company clinicians had no training in uranium-related occupational health, and that the company's radiation protection division consistently refused to provide dosimetric readings to the medical division.

COMUF employees, 1950s. (undated photos, courtesy of COMUF)

Absent the ability to conduct full medical examinations, the NGO team's assessment of health outcomes couldn't be conclusive. One clear pattern emerged, however. Half of the Gabonese workers surveyed reported pulmonary distress, which also appeared (at least anecdotally) to affect their families disproportionately. Workers did testify to satisfactory medical attention during employment—even though, the report added parenthetically, "the doctor didn't have nuclear expertise and the nature of the examinations wouldn't have detected internal contamination." After the site closed, Gabonese "felt abandoned" at the very moment that their health problems had become more severe.[105] Back in France, eleven of the seventeen expatriates surveyed suffered from cancer.

The 2007 report also addressed environmental contamination, estimating that the COMUF had generated about 7.5 million tons of waste. The company had dumped waste directly into the river, buried some in the old Mounana pit, and scattered some around the site under a light layer of dirt. After 1990, it stored the remainder in its tailings pond. Most alarming of all the findings, the maternity ward, the Massango school, and other public buildings exhibited high levels of radioactivity. As residents had long known, the structures had been built with mine tailings.

Gabon had received 35 million euros from the European Union to rehabilitate the site. "One wonders about the quality of these efforts," the report noted with dismay, given that remediation plans consisted only of covering contaminated zones with vegetation. CRIIRAD deemed this a short-term solution at best; only outright removal of contaminated soil would produce "definitive rehabilitation."[106] A newly created Gabonese regulatory agency tasked with monitoring radiation took samples regularly to ensure compliance with the remediation plans and international norms. But lacking its own laboratory, it sent the samples to France for analysis.[107]

In Paris in April 2007, the NGOs released their reports on Mounana and Areva's mining operations near Arlit and Akokan in Niger (more on this in chapter 10). Areva responded by promising to install "health observatories" in Mounana and Arlit.[108] Yet "observatories" were a far cry from remediation and compensation. Discussions between Areva and the NGOs dragged on. Any hopes Areva may have nurtured that delay would dilute public outrage were dashed by a pair of documentaries shown on French television in 2009. The first, which aired on a high-profile current-affairs program, documented radioactive contamination produced by uranium mining in France itself. Among its revelations: roads, football fields, and parking lots had been built with radioactive waste rock, and the Limousin's water supply had been contaminated by mine run-off. The second film, airing on a less popular channel, essentially presented a visual accounting of the problems that NGOs had identified in Gabon and Niger.

Between the two documentary airings in June 2009, Sherpa and MdM announced that they'd reached an "unprecedented" agreement with Areva to form a "pluralist group" of experts to oversee the health observatories set up by Areva in Gabon, Niger, and elsewhere. The committee's tasks included defining protocols for data collection, analyzing the results obtained by all the health observatories, and making proposals to improve occupational and environmental health at the sites. For Areva, the accord marked "an important stage in the necessary dialogue between a responsible mine operator and civil society." For Sherpa, it displayed "a new maturity" for all parties: "Areva's openness to dialogue testifies to its willingness to respond to the worries of citizens who are ever-better informed and advised. By allowing itself to engage in discussions with Areva, civil society, for its part, shows that an accord can have a constructive impact on local populations."[109]

Others were less sanguine. Expressing skepticism about the prospect of prompt and effective remediation, CRIIRAD declined to participate in a process that might legitimate Areva as a responsible corporate actor without producing real change. Though the agreement laid out procedures for ensuring balance between Areva and the NGOs in selecting committee members and charting goals, inclusion of residents and workers was glaringly absent. One clause, for example, specified that an MdM doctor would participate in each local observatory unless the national government objected, in which case a "disinterested" replacement would be found. No one sought to include local representatives on the pluralist committee or in the health observatories— striking oversights, especially given how little COMUF workers trusted the state. The "unprecedented" agreement, in sum, did little to bring irradiated Africans into transnational knowledge production.

CONCLUSION

Uranium mines in Madagascar and Gabon were not born nuclear. Nuclearity, when it emerged at all, did so slowly, unevenly, according to different

historical rhythms. Radiation did not, by itself, turn uranium mining into
nuclear work. It had to be made perceptible and allied to human agency.
Where such perceptibility and such alliances marshaled nuclear exceptionalism effectively, radiation could serve as a mechanism for forming, maintaining, or disrupting power relationships.

As we saw in chapter 6, dosimetric mastery had empowered French
radiation protection specialists in French mines and in dominant circuits
of global knowledge production. But in Madagascar dosimetry was filtered
through other experts. It served management only as a short-term tool
for making labor decisions. Malagasy mineworkers, for their part, saw
dosimetry as a tool for exerting power over colonial subjects (that is,
themselves). Radiation was first a mysterious residue, then a distant memory.
Their work *never* became nuclear. Their exposures *never* served as a resource
for postcolonial claims making. They did not—could not—use it as a tool
in making citizenship claims.

In contrast, Gabonese and French employees of the Mounana mine
eventually found ways to claim nuclear exceptionalism for themselves.
They learned how to represent their exposures as the distinctive conse-
quence of globally known hazards, as corporate malfeasance, as (post)
colonial injustice, as politically accountable. This process, spanning decades,
was far from satisfying for those involved. A huge gap separated accountability and compensation. Demanding compensation brought French expatriates up against the limits of their citizenship and their own expectations
of modernity. Gabonese workers, long accustomed to limited citizenship
and crushed expectations, nevertheless hoped that transnational political
alliances would offer more than their state and their former employer did.
For some, "America" shimmered in the distance as an emblem of prosperity and justice.

Juxtaposing these histories exposes not only the uneven spatial distribution of nuclearity but also its uneven temporalities. There was no moment
when the nuclearity of uranium mines and mills became settled and
forever mandated. Radon could be ignored in Madagascar and carefully
measured in France. Houses could be built with radioactive rocks in Gabon
even as their counterparts in the United States were being demolished.
Differences among places had to do with time as well as space, with *temporal* frictions between mine closures, transnational activism, global knowledge production, capital flows, postcolonial politics, and more.

These juxtapositions, in turn, bring into focus the double edge of governmentality. Dosimeters established forms of legibility whose first and sometimes only effect for workers was discipline. But records also carried the potential to incriminate mine operators. As a form of power distributed in things and inscribed in bodies, nuclearity could—under the right circumstances—serve workers. Clearly understanding this, COMUF managers kept their records tightly guarded.

While the French took pride in their mastery of measurement on their own soil, their African operations were threatened by North American approaches that called attention to alternatives for measurement and management of wastes. Further down the African continent, South African experts and industrialists fully understood the technopolitical perils of measurement. Obsessive record keepers in some respects, they also knew that some records were best not produced at all. To understand why, let's head south.

Mounana residents had an adage to express their frustration over the disappointments of independence: "It's like South Africa at Mounana: blacks at the bottom and whites on top." But what would these Gabonese have seen had they actually traveled to apartheid South Africa?

The history of modern South Africa was profoundly shaped by mining. Diamond mining took off in the 1870s. At the end of the century, the Witwatersrand gold mines pulled migrant labor from all over southern Africa. The rapidly expanding gold industry soon consolidated into a few large mining houses. These created the powerful Transvaal Chamber of Mines to represent their interests, formulate industry policies, and conduct technological and scientific research that benefited its members.

The mining houses pooled their recruitment efforts through the Native Recruiting Corporation and the Witwatersrand Native Labor Association (WNLA). By 1920, these two agencies were transporting more than 200,000 African men to the mines every year.[1] Needing to sustain rural homesteads and pay punitive colonial taxes, many of these workers signed on for six to nine months, often returning for repeat contracts.

Geology, technology, and politics combined to make cheap black labor the keystone of profitability. Gold was plentiful on the Rand, but the low-grade ore had to be hauled up from the deep veins through long, narrow stopes. Greater depth meant higher rock pressures and temperatures, which in turn required more timbering and stronger ventilation. The narrowness of the stopes made it difficult to introduce large-scale machines. Ore extraction was extremely labor intensive, particularly compared to mines in the US and Australia. Because the price of gold remained externally fixed until 1971, mine operators controlled their production costs by keeping wages down.

The system relied on legalized racial discrimination. Unionized white workers advocated a "color bar" to protect skilled jobs. Though mine owners chafed at the constraints that a color bar placed on cost reduction, racism won. Following strikes, revolts, elections, and negotiations, the government passed legislation in 1926 that prohibited black Africans from holding "skilled" jobs in the mines. This is not to say that white workers had it easy. White and black mineworkers alike suffered from occupational disease. The white Mine Workers Union made silicosis among white miners visible in the early twentieth century, leading to treatment, compensation, and some early reduction in dust levels. Still, getting mine owners to recognize and ameliorate health problems was an ongoing struggle.

Although South Africa's racial order originated in settler colonialism, the mining industry pioneered the mechanisms of spatial and racial control

South Africans reading a newspaper account of a clash between police and black miners at the Western Deep Levels Gold Mine, September 1973. (Corbis Images, used with permission)

that underpinned formal apartheid after the National Party's 1948 electoral victory.[2] The disciplinary mechanisms of the mines—most notoriously the passbooks that recorded workers' labor histories—offered templates for building a "new" society whose flows would be regulated by the racialized demarcation of space.[3]

Conceived as a technopolitical project and a sweeping modernist system, grand apartheid built on the mines' technologies of surveillance. As historian Keith Breckenridge has shown, ideology and practice were deeply bound to personal identity documents coupled to an increasingly automated data-processing system.[4] Biometric identifiers (fingerprints), the baroque taxonomy of racial identity, and the elaborate system of architectural zones and geographical and temporal borders formed the pillars of apartheid. These mechanisms sustained the technopolitical illusion of complete knowledge and control over Africans.

Hostel at Robinson Deep Gold Mines, 1950. (Margaret Bourke-White/Time & Life Pictures/Getty Images, used with permission)

Living and working conditions belied apartheid's claim to offer black Africans "separate but equal development." Though the conditions found throughout the mining industry had improved substantially in the early decades of the twentieth century, they worsened again under apartheid. Hostel life could be excruciating, the quarters cramped, and the sanitation rudimentary at best. Food was dreadful and often insufficient. When I asked workers at the NUFCOR plant about their memories of the 1970s and 1980s, they shuddered as they recalled the food:

The meat was rotten. We used to eat some—it was bad, very bad. . . . The mine used to give us so little that we couldn't do anything to buy our own food. [But if you say that] the meat is rotten, you got fired. . . . Ya, you can even find the worms inside, we used to eat that. Then the tea. . . . They started cooking [the tea] at four o'clock afternoon until six o'clock in the morning—you would drink it [at] six o'clock in the morning. . . . You were not allowed to take a cup, put some water there, put some sugar in yourself. No.[5]

The color bar remained:

They didn't want a black man who's educated at NUFCOR—only the white guy must know how to write and read. If you just show them that you [were] reading the newspaper you [were] fired for that. And another thing is at that time when a black man [was] employed at NUFCOR . . . he couldn't have a . . . pension fund. . . . A white guy once he's employed, first pay[check] he's got pension fund, medical everything. For a black man there's nothing like that. . . . When you fall you just go like that, nothing, no money, nothing, you just go.[6]

Although deplorable conditions constrained black mineworkers' options, Africans found ways to escape management control and maintain their personal and social integrity, as sociologist Dunbar Moodie and other scholars have shown.[7] Black workers could not legally organize until the late 1970s, but in the early 1940s some of them overcame formidable obstacles to form the African Mineworkers Union. In 1946 over 287,000 black miners went on strike to demand higher wages, prompting the Chamber of Mines to summon the full force of its allies. Using army equipment, the police and the Native Affairs Department put a violent end to the demonstrations and the AMWU. By the 1970s, black mineworkers had begun once again to organize and protest, forming the National Union of Mineworkers in 1982.

Maintaining personal integrity was particularly important in light of the dangerous and sometimes degrading nature of work that was often enforced with brutal authority by white supervisors. In principle, workplace temperatures had to remain under 27°C (80°F), but in practice they could rise much higher. High humidity made work particularly onerous. Underground water could flood and drown miners. Sudden rock bursts could maim or kill. Diesel fumes and methane gas made for nauseating, sometimes explosive air. Everywhere, always, there was dust.

Mine officials claimed high standards of ventilation, but in 1995 a government-appointed commission on mine health and safety found that dust levels had not changed in the preceding half-century.[8] The racialized division of labor meant that all these hazards disproportionately threatened black workers, who couldn't move to cooler or better-ventilated areas even if they experienced heat stroke or difficulty breathing.

Even though mine owners had greatly expanded medical services for black workers in the first half of the twentieth century, the changes often stemmed from concerns about the impact of worker mortality and morbidity on mine productivity. Still, the mines devoted more resources to medical services than many other industries in South Africa. Some mine doctors and researchers advocated for continued reforms, yet black workers' health worsened under apartheid. As historian Jock McCulloch has shown in his study of the asbestos industry, research results were suppressed thanks to the powerful alliance between apartheid state regulators and the industry, promulgating the fiction that black mineworkers didn't contract occupational lung diseases in large numbers.[9] But the medical services didn't examine black workers as carefully as they did whites. For example, smaller x-ray plates for black workers made disease harder to spot.

Perhaps the biggest problem stemmed from inadequate medical services that made detecting disease nearly impossible after migrants returned home to rural South Africa or to Malawi, Mozambique, and Lesotho, countries that supplied the majority of the labor force before the mid 1970s. Mechanisms for rural hospitals to report disease rates to a centralized database did not exist. The lack of adequate compensation schemes meant that retired workers did not report illness or debility to former employers. And it left sick workers and their families to carry the burden of treatment and

An expert panel examines miners making injury claims, 1943. (Hart Preston/Time & Life Pictures/Getty Images, used with permission)

care.[10] Then there was the industry's stance that developing silicosis required at least four years of continuous exposure to dust. This allowed mine officials to claim that black mineworkers did not readily succumb to the disease thanks to short labor contracts interspersed with sojourns in clean, rural air.[11] Safely buried in foreign soil or "ancestral homelands," the lungs of migrant miners were unlikely to prove otherwise.

Which brings us back to radon.

Nuclearity, as we've seen throughout this book, requires work. It comes in different technopolitical registers: geological, metallurgical, technological, managerial, medical. Nuclearity in one register does not automatically transpose into another. South African diplomats abroad claimed that uranium production made their country sufficiently nuclear to qualify for a seat in IAEA governance. But that didn't mean that uranium extraction counted as nuclear labor in South Africa itself.

South African radon researchers long insisted that black miners were "unsuitable" research subjects because they were short-term, migrant workers. This justified an exclusive focus on white miners and promoted the apparently self-evident conclusion that if white workers didn't contract lung cancer from radon exposures, black workers wouldn't either. As we saw in chapter 6, many overseas experts interested in South African research results accepted this logic, probably without realizing that it had long underpinned industry claims about the low prevalence of other occupational diseases among migrant workers. Nor, perhaps, did they realize the sheer scale of the migrant workforce. In 1963, the year of the first international conference on radiological hazards in mining, over 350,000 black men migrated to work in South African mines.

Before World War II, the uranium ore removed from South African gold mines was discarded as useless. With the advent of atomic weapons, the gold industry learned that profit could be made from the active shafts that still contained uranium and from the enormous tailings dumps that surrounded the mines. By the late 1950s, the industry was extracting uranium

on facing page: Men getting their chests x-rayed during a pre-employment medical examination, 1955. (Evans/Hulton Archive/Getty Images, used with permission) from 27 mine shafts and removing uranium from gold tailings in 17 extraction plants. Slurry containing uranium traveled from these plants to the Calcined Products facility (CalProds, which became NUFCOR in 1967), where it underwent further processing, got packed into drums, and was exported. Estimates suggest that uranium production accounted for 20–30 percent of the gold industry's revenues in the late 1950s, enough to keep some marginal mines in business and to finance the expansion of others.[12]

The co-production of gold and uranium meant that underground workers could spend their entire laboring lives extracting uranium without realizing it. Those who operated the treatment plants, however, must have known that they were processing ore for its uranium content. The historical record contains a few tantalizing hints that this may have raised early concerns about radiation exposure among some white workers. I found no evidence that such concerns—if they persisted—afforded workers significant political traction.

I originally intended to explore the memories and experiences that miners in South Africa had with uranium extraction, much as I had in Madagascar, Gabon, and Namibia, but I eventually concluded that this approach wouldn't yield sufficiently fruitful results in South Africa. My interviews with retired workers demonstrated the success of structural invisibilities (those I spoke with didn't know they'd been producing uranium). They also made clear that I would be replicating the work of other scholars of South African mining.[13] Meanwhile, I discovered that the nuclearity of South African uranium mining had been subject to contestation for decades.

This chapter, therefore, focuses on the mechanisms that made possible the invisibility of nuclear things. It also demonstrates the tremendous work required to build the regulatory capacity that must underpin even the most rudimentary forms of technopolitical governance.

With plenty of help from the state, the South African mining industry made some lung diseases hard to see. For decades, such systemic agnotology facilitated the invisibility of radiation exposures. Powerful though it was, however, "the system" was neither homogeneous nor monolithic. Invisibility in one domain didn't *automatically* transpose into others. Just as some scientists had investigated asbestosis in the 1960s (only to have their research suppressed), radiation did draw the attention of a few mining industry researchers. The intense interest that some international radon experts had in South African mines, furthermore, meant that questions about radon levels and exposures kept resurging even after industry officials thought they'd been settled. Some scientists worked to maintain the invisibility of radon exposures. Others developed deep suspicions and repeatedly sought to improve radon measurement.

The invisibility of radiological hazards in South African uranium extraction thus had to be made and remade. With each passing decade, industry experts and officials found new ways to stave off radiation monitoring and regulation by proclaiming uranium production non-nuclear work. In the 1950s and 1960s, summary dismissal of radiological risks and selective interpretation of international science sufficed. Focusing on the chemical toxicity of uranium ingestion at the expense of its radiological hazards, some industry experts apparently fooled themselves into thinking that they had created an effective monitoring system. As so often happened under apartheid, copious data created an illusion of control that masked inadequate data analysis.

The unspoken subtext was that radiation risks threatened only the white technicians and experts who worked in research, or reactors, or medicine—the domains that symbolized South African "modernity." Mineworkers—whether black or white—were not nuclear subjects. Under apartheid, South African nuclearity was national, middle-class, and white.

Compartmentalizing nuclearity grew harder over time. The international networks that sought South African data and conferred authority on South African experts threatened the comfortable consensus on the non-nuclearity of mines. Tentatively in the 1970s, then with greater authority in the 1980s, a small group of foreign-born scientists employed by the AEB's licensing branch struggled to establish a credible dosimetric regime in the mines. They saw existing radon surveys as *apartheid* science, at odds with the norms and findings of global practices. Along the way, they succeeded in transforming the AEB's licensing branch into an independent regulatory body.

Unsurprisingly, the Chamber of Mines fought hard against external regulation of radiation in mines. In the 1990s, the battle blasted past national borders. Both sides found allies in international debates over the implications of newly restrictive International Commission for Radiological Protection regulations. And once again, international bodies accepted South Africa as suitably representative of the African continent.

Yet none of this was pre-ordained. At its inception, uranium extraction in South Africa seemed poised to become a fully nuclear activity.

MARGINALIZING NUCLEARITY UNDERGROUND

In the late 1940s and early 1950s, some of the first research projects carried out under the auspices of South Africa's newly created National Physics Laboratory (NPL) involved measuring radiation and radon levels in a few mines.[14] The results prompted the NPL to distribute film badges to workers at a small thorium mine that had exceptionally high gamma readings. The badges revealed that 25 percent of black manual laborers and 14 percent of white supervisors absorbed the ICRP's annual maximum dose in just four months. Establishing change-houses and dust-reduction programs, along with educating personnel "to treat the concentrates with respect, e.g., not to lean or sit on bags of concentrates" apparently improved the situation. So did relying on shorter stints for migrant workers who, in principle, could exceed six-month contracts "only if their radiation exposure were low."[15] High exposures had also been "rectified" at a small monazite mine.[16] The foundational components of nuclearity appeared to be in place: research projects, monitoring instruments, special instructions, job rotation.

The National Physics Laboratory did not remain interested in mine radiation, however, and research on radiation in mines was taken up by S. R. Rabson, a scientist at the Chamber of Mines' Dust and Ventilation Laboratory (DVL). After a tour of US uranium mines, Rabson returned eager to measure radioactivity in South African shafts. Between 1958 and 1961 he surveyed radon in seven gold mines, five of which also produced uranium. He reported peaks that exceeded ICRP and US recommended limits almost seven fold. In some places he found even higher concentrations of radon, but he excluded these peaks from his report, emphasizing instead the *average* concentrations that mostly remained within international limits. Radon levels, Rabson reported, were highest where ventilation patterns took air through old, depleted shafts. Despite taking very few samples and surveying only five of the 27 uranium-producing mines, Rabson concluded that, with a few exceptions, "personnel are seldom exposed to radioactive concentrations in excess of the so-called 'tolerance level.'"

Rabson still wanted to conduct more research, not the least because he knew that American scientists eagerly awaited South African results.[17] It seemed as though he might get his wish in late 1962 when the Chamber established a Radioactivity Panel to supervise such projects. Within a year, however, Rabson's ambitions were frustrated. Medical data proved inadequate to the task of correlating lung cancer with radon exposure. Though "there appeared to be quite a number" of lung cancer cases among workers employed in surface operations, researchers couldn't establish whether this unspecified number exceeded that of the general population. The difficulty of finding enough manpower to conduct in-depth radon surveys also impeded research.[18]

In December 1963, barely a year after its formation, the Radioactivity Panel disbanded.[19] Maintaining nuclearity by continuing research, acquiring more instruments, and training more experts required resources. Rabson's report hadn't accorded much significance to the radon peaks. No one mentioned (or perhaps even noticed) that he hadn't taken enough samples to produce conclusive results. Whatever nuclearity mine shafts might have acquired rapidly dissipated. Rabson shifted the full burden of his attention to uranium treatment plants, which for a time seemed better candidates for nuclearity. In order to understand his work on these plants, we need to backtrack a bit.

TOXICITY TRUMPS NUCLEARITY

The 17 treatment plants in operation by the late 1950s used chemical and metallurgical processes to extract uranium ore from gold tailings. In the course of developing these processes, the director of South Africa's Government Metallurgical Laboratory (GML), Leonard Taverner, had traveled to the US, Canada, and Britain. In 1951, struck by the efforts overseas operators made to protect workers against radiation and toxicological hazards, Taverner proposed a set of health and safety precautions for South African uranium treatment plants.

South African ore was low-grade, so Taverner did not expect gamma radiation to pose a big problem. But uranium in the plants continued to emit radon, which could lead to internal contamination by alpha particles. Uranium also had toxic properties and could cause significant kidney damage if ingested. Other chemicals posed threats too. The ore contained significant amounts of lead, and the sulfuric acid required to leach the uranium out from the gold would burn skin on contact. Taverner felt that a few basic precautions could mitigate the dangers. Radiation levels should be monitored. Workers should wear respirators and protective clothing. Double sets of change-houses should separate clean from dirty clothing to prevent contaminants from going home with workers. Smoking, eating, and drinking in work areas should be prohibited. Pre-employment medical examinations should screen out workers with existing lung and kidney problems. First-aid services should include special equipment and training to deal with uranium poisoning.[20]

Even the simplest precautions, however, struck the Chamber of Mines as "unnecessarily elaborate." Officials worked to banalize uranium, asserting that it was less hazardous than mercury, lead, or cyanide. They claimed that a man could eat up to a pound of uranium oxide "without effect." Chamber officials especially objected to prohibitions on eating and drinking in the workplace, which would force employees to go through the change-rooms to get refreshments. Because the uranium plants operated continuously, this would require hiring "spare men." The industry should avoid "installations and . . . precautions which would seriously affect the economics of the uranium project, and which would, in the long run, be found to be unnecessary."[21]

The industry, already experiencing enough trouble with unionized white labor, saw no need to alarm these workers any further with excessive safety precautions. One official went so far as to warn that "the main danger to the personnel of the plant will be psychological if health hazards are over-emphasised." Expressing faith in the mining groups' awareness that "good industrial hygiene must be practised in any plant and that the uranium plants will be no exception," he concluded that "if there is not an over-emphasis placed on the health hazard involved in uranium production, little trouble can be expected from the personnel."[22]

Taverner defended his recommendations, arguing that they would prove especially important when the inevitable breakdown or spillage occurred. His precautions also "allow[ed] for the fact that natives would be employed in the plants," though he didn't explain how. But Taverner was already losing the battle to none other than Basil Schonland, South Africa's most prominent physicist, who had played an important role in launching South African uranium exports. Schonland had worked closely with atomic scientists in Britain and knew their facilities well.[23] He insisted that "there were no mysterious rays associated with the production of uranium" and that "radiation from the spillage of the liquors handled in the uranium plants would be innocuous."[24] This judgment carried considerable authority, enabling Chamber of Mines officials to dismiss the significance of radon and assert that it was "probably all released during mining operations."

At the very least, precautions could wait while the mining industry sent one of its own to visit overseas uranium plants. Their chosen man was L. S. Williams, chief medical officer for the Goldfields mining company, who seemed impressed by radioactivity's distinctiveness. Williams noted that it could take "twenty years or more to produce any evidence of radiation toxicity." Though he thought that "radiation dangers, *other than those due to Alpha radiation of inhaled dusts* [i.e., radon daughters] are not expected to be of importance," he nevertheless suggested that plants might find a limited gamma monitoring program "instructive." He also recommended ventilation, protective clothing, and, following the example of US plants, monitoring workers' hands and feet for radioactive contamination.[25] But Williams kept the Chamber happy by rejecting American proscriptions on eating, drinking, or smoking. Periodic urine tests would suffice to monitor uranium ingestion without interrupting work or cutting into profits.

Armed with Williams's report, the industry reached an accommodation with the Government Mining Engineer (GME) to postpone formal regulation while a newly constituted Health Committee at the Atomic Energy Board kept an eye on things. After a few years, this committee helped industry develop a code of practice that duly avoided "over-stringent precautionary measures" by prescribing protective clothing and double change-houses.[26] Rather than prohibiting eating, drinking, or smoking, it recommended that "operators be instructed to wash their hands well" before engaging in these activities. It called for "adequate ventilation" but did not specify figures for maximum permissible levels of any contaminants or toxins. Making no mention of radioactivity, it ignored Williams's suggestion to monitor gamma and alpha radiation. Extraction plants could simply tell workers that uranium was "an extremely poisonous material" to be handled "with extreme care."[27]

Not even the Calprods/NUFCOR plant, which contained the highest concentrations of uranium, conducted radiation monitoring in its first three decades of operation. The 1955 code focused instead on the chemical toxicity of uranium. It prescribed monthly urine tests for workers exposed to the highest concentrations of uranium and dust, which included most workers in the uranium oxide plant. Ten percent of the remaining labor force should be randomly sampled every three months for urine testing.[28] Urine-test results for "European" workers should be filed with their medical records but "should be regarded as extremely confidential [and] must not be shown to the employee concerned."[29]

The industry was well aware that unionized white workers could make trouble. In one instance in 1952, two men had attributed "strange symptoms" to "the effects of radiation."[30] The Minister of Mines had publicly dispelled those fears by asserting that tests had shown that "there was no trace of Radon gas" and "no signs of radio-activity of a dangerous level" in the plants. Since then, white labor had apparently remained quiet on the subject.[31] At that time, the industry had more than enough trouble brewing concerning diagnosis, treatment, and compensation for silicosis and tuberculosis.[32] On no account could employers let concerns about radiation provide fodder for concern.

Some plant managers made a good-faith effort to comply with the minimalist code. But many disliked the need for urine tests and double change-houses. Early compliance reports tended to be extremely cursory.[33] Irritated Chamber officials sternly reminded plant managers that if plants didn't persuasively document compliance the GME might institute formal regulations. The rebuke worked to prompt more detailed reporting, though not necessarily better health and safety.[34]

(DIS)COUNTING RACIALIZED CONTAMINATION

To fend off formal regulation by the Government Mining Engineer, the Dust and Ventilation Laboratory decided to conduct its own checks. In 1957, S. R. Rabson began collecting dust samples in plants to check the effectiveness of control measures. At the Calprods plant, he immediately found "exceptionally high counts." Urinalyses confirmed "cases of high contamination" in workers who were somehow ingesting uranium. Wagging the stick of formalized inspections, Rabson persuaded Calprods to improve housekeeping and ventilation at the plant.[35]

Meanwhile, Rabson and his colleagues looked for patterns in the urine data from the 17 extraction plants. They found that "people with beards appeared to have higher counts than those without." But the highest uranium counts of all were recorded in "natives." Rabson attributed this to a lack of "personal cleanliness."[36] The vast racial disparities in working conditions had become so normalized that it apparently didn't occur to him that "personal cleanliness" was difficult to maintain when towels were issued "at 1 towel per 3 Natives," when "all natives of each shift . . . have a common 'clean' and 'dirty' locker for their clothes," or when men had to launder their own work clothes using a courtyard water spigot.[37]

Anyone who fell ill had only himself to blame, as a 1971 NUFCOR (formerly Calprods) manual made clear: "Nobody has ever been sick at NUFCOR from breathing in uranium dust. But, we do know that some Bantu have been careless because we have found uranium in their urine. That is why we check your urine."[38] Black workers certainly felt the surveillance, not just when their urine was tested but at all times. Recalling plant operations in the 1980s, one worker recollected white supervisors looking down at the shop floor from their surveillance post: "You can't believe what a good job it was, just to look on black people. . . . Just watching the black people when they [were] working. If they go to toilet you have to write down, see the time, you write the time. . . . Black people were working very hard, pushing some drums or everything like that. . . . But white people just think up there, you know . . . , looking [at] the black people down there."[39] There was nothing unusual about this division of labor under apartheid, of course. For these workers, urine tests seemed merely another humiliating surveillance tool.

It went without saying that "natives" would never see their test results. Nor would they benefit from "the same stringent medical examinations as applied to European employees." The reasoning behind this had been developed to explain the alleged absence of silicosis in black mineworkers, namely that "natives were essentially migratory workers and, therefore, not subject to the same hazards as permanent European employees."[40] Mine doctors would keep urinalysis results for black workers on site. At the conclusion of a contract, a worker's medical record and fingerprints would be sent to the Witwatersrand Native Labor Association recruiting agency. An addendum to the code of practice explained how uranium health history could be tracked "if such a Native re-engaged for work": "If at the Association's Depot it was found that the fingerprints agreed with those of a Native already in the files, the previous medical record card would be extracted from the file and sent to the mine which had engaged him."[41] The logistical difficulty of quickly matching fingerprints, however, made actual implementation of such a process highly impractical.[42] Nevertheless, in the mid 1960s Rabson expressed continued faith in the system's ability to track the "radiation history" of black workers. For instance, he affirmed that the WLNA would "retain for record purposes only such forms as may indicate occurrence of radiation or uranium toxic effects, *or a radiation history.*"[43] But the code of practice failed to mention radiation, the plants failed to measure radioactivity, and the forms did not have a specific rubric for radiation exposure. Medical interns had 12 minutes to conduct each health examination.[44] Rabson had apparently conflated "radiation history" with the uranium counts measured by urinalysis outside the context of regular health exams. He also appeared to assume that all the data would end up in the same place.

Competing indexing systems made a mockery of meaningful record keeping. Urine records were kept separate from medical records, making correlations impossible.[45] Medical records didn't have a specific rubric for uranium exposure any more than they did for radiation. Nothing indicates that recruitment clerks had training to interpret notations about kidney damage as evidence of uranium exposure or "radiation history." Sometimes, in fact, they couldn't even tell which records corresponded to which workers.[46] "Native employees," one medical advisor complained in 1959, "are designated only by company numbers and if any of these [urine] samples are in excess of the limit allowed, a card is made out for this company number and subsequent records for this worker are entered each month as the samples are received. It now transpires that on the discharge of any of these Natives, his number is transferred to a new entrant— consequently subsequent samples may be shown against the original worker who no longer works in the plant."[47]

The claim that black workers' uranium-related health histories were being reliably recorded represented, at best, a control fantasy. In 2004, an official described the NUFCOR plant's current record-keeping practices in a confidential interview: "We now have a system where . . . we should be able to go back about ten years. Before that, it's very tenuous—there may be some records somewhere, but. . . . Let me put it this way. If some worker, somebody who had worked there in 1970 came and said, 'I've got . . . problems because of an overdose of uranium. . . .' We'd really struggle to find any documentation to refute it."[48]

Once the initial challenge of data collection and interpretation gave way to routine practice, interest waned, especially after Rabson's death in 1969. By 1972 the mining industry's insurance arm had begun to wonder: "What is done rising from all this effort and information? Do we need it?" Surely the Chamber felt "satisfied that the safety precautions in force in the industry are perfectly adequate, and that the time has come to put an end to the medical examinations and urine checks."[49] Before moving to eliminate the health surveillance program, however, the Chamber generated urine count graphs for white and black workers (denoted respectively as "skilled" and "unskilled"). Levels were "satisfactorily low," they proclaimed, and the two curves had roughly the same shape. No one commented that the relative scales of the y axes on the two graphs differed substantially, clearly demonstrating that a larger percentage of "unskilled" workers had high uranium concentrations.[50]

The graphs, furthermore, only documented uranium counts. The separation of urine-test results from medical records made it impossible to determine whether workers had suffered kidney or other damage from their uranium exposure. Nevertheless, the Chamber's medical superintendent displayed no compunction about asserting that any kidney damage he'd detected "appears to have been completely unrelated to the level of uranium excretion in the urine." Undeterred by the absence of statistically meaningful data, he insisted that "clinical and radiological examination of the lungs has not shown any particular pattern which might be attributed to the inhalation of acid mists or of uranium dust. Neither was there any detectable radiological change in the lungs which could be related to ionising radiation."[51] He concluded that "no special trend could be identified as incriminating any of the processes in the Uranium Plants and Refinery as being injurious to health."[52] In the absence of "incriminating" information, the testing program could be reduced or eliminated.

In sum, none of these data shed light on how much uranium-containing dust may have lodged in workers' lungs and released alpha particles through decay into radon and its daughters. No one directly measured radiation at any of the extraction plants. Not even the Calprods/NUFCOR plant

monitored radiation levels. Industry experts convinced themselves that by
treating uranium as a purely toxicological hazard they maintained a sufficiently safe working environment for blacks and whites alike.

MAKING RADON VISIBLE

In 1980, a young British scientist named Shaun Guy replied to an employment ad placed by the South African Atomic Energy Board. The job
involved controlling the dispersal of effluents from the Koeberg nuclear
power plant, then under construction in Cape Town. Guy had written a
master's thesis on the movement of radioactive phosphorus through fresh-
water ecosystems, so he seemed like a perfect fit. But by the time Guy
obtained a visa and arrived in South Africa, nearly two years later, the post
had been filled. He went to work instead in the Atomic Energy Board's
licensing branch, established in 1969 to develop safety standards for South
African nuclear power stations.

The early 1980s were a difficult time to move to South Africa. In 1980,
school boycotts spread from the Western Cape to the rest of South Africa
before being brutally suppressed. The following year, security police assas-
sinated several prominent African National Congress activists. Umkhonto
we Sizwe (known as MK, the military wing of the ANC) responded by
intensifying attacks on military and industrial installations. In 1982, Nelson
Mandela was transferred from Robben Island to the Pollsmoor prison, and
the worldwide campaign to free him began. That same year—not long
before Guy's arrival—MK exploded two bombs in the nearly completed
Koeberg reactor complex, where Guy had expected to work.

The AEB's licensing branch was a small division that included two other
foreign transplants. Phil Metcalf was a British citizen, like Guy; Bert
Winkler was a German émigré from Tanzania who'd acquired South
African citizenship. As reactor construction delays accumulated, Winkler
decided to keep his staff busy by sending them to investigate other components of the nuclear program. The team quickly determined that no
part of the system—not the AEB's headquarters at Pelindaba, not the
uranium enrichment plant, not the Safari research reactor, not the foreign-
owned nuclear-powered ships in South African harbors—was subject to
inspection and licensing.

Taking charge of uranium mining, milling, and conversion, Metcalf realized that South Africa had been producing uranium for over 20 years with no formal regulation of hazards. In the mid 1970s he had traveled overseas to learn about international regulations and monitoring, returning to South Africa full of ideas about ways to monitor underground radon exposure, regulate uranium plants, and manage tailings disposal. The code of practice for the uranium plants, he noted, failed to meet North American or Australian standards in several respects. There were no inspection procedures. The code didn't provide for "competent" radiological protection staff or monitoring. And "the keeping of dose and health registers is not in line with normal practice."[53] Metcalf's early attempts to implement change (which were, by his own admission, somewhat naïve) came to naught.[54]

In 1979, the licensing branch persuaded Ampie Roux, the AEB's powerful president, to write Secretary for Mines W. P. Viljoen about radiation exposure in the mines. Roux, however, didn't seem committed to strong AEB oversight of mining. He maintained that the nuclearity of hazards increased linearly as uranium traveled from mine to mill to conversion to enrichment to reactor. "The intrinsic risk of nuclear damage associated with the early part of the nuclear fuel cycle," he wrote, "increases progressively from the ore mining stage onwards through ore beneficiation to the production of U_3O_8."[55] This sentence could have caused US epidemiologists, French health physicists, and Phil Metcalf himself to gasp in dismay. But it reflected the South African consensus that occluded radon risk in mines. Quintessentially modern, nuclear risk was seen as primarily threatening white technicians and scientists, not mineworkers (whether white or black).

Roux pointed out to Viljoen that "shortcomings" in uranium mining regulation overseas had "led to much criticism in recent times." But, he admitted sympathetically, "the question of possible radiation induced injuries is a very sensitive matter." Any form of regulation or control required "circumspection so as not to cause alarm or provoke other unwarranted adverse reaction within the industry." Sensitive to the problem of labor unrest, the AEB wanted to do only "the minimum necessary consistent with its statutory obligation." Following "proper procedures" would probably lead to the exemption of mines from nuclear licensing.[56]

Roux's sympathetic stance notwithstanding, Viljoen must have found the letter's timing (May 1979) particularly unpropitious. Labor tensions had been building throughout the 1970s. The withdrawal of Malawian workers from the labor pool in 1974 had prompted mining houses to recruit more labor from within South Africa. Faction fights were on the rise, partly as a result of the arrival of new ethnic groups but also in response to changes in job grading that dramatically increased wages for some black miners but not others. To address projected shortages in skilled labor and dissatisfaction among black workers, some mining houses had begun to push for elimination of the color bar and legalization of black unions. The white Mine Workers Union strongly opposed such measures, but in 1977 the government created the Wiehahn commission to review discriminatory labor legislation. The commission issued a series of reports recommending the end of the color bar and the legal recognition of black trade unions. Just weeks before Roux's letter, the MWU launched a strike to protest these recommendations.

No surprise, then, that Viljoen would fear that implementing radon controls might lead the white union to press for "radiation induced disease compensation."[57] Cajoled by Roux's reassurances, he reluctantly agreed to a meeting in October 1979 between the AEB's licensing branch and the Government Mining Engineer to discuss the creation of a committee to control radiological hazards in the mines. But as Shaun Guy noted dryly when he reviewed the record a few years later, "there were no further meetings."[58] The Chamber of Mines argued that "there was no hazard, no indication of hazard," Guy later recalled. "And because the licensing branch didn't have access to [radon] studies at the time, they had no information to go on. They couldn't come at them with a piece of paper and say, hey, this has been measured, this is what we calculate these people get if they work 2,000 hours and something must be done about it."[59] Without data, the power disparity between the Chamber and the licensing branch was too large to overcome.

As apartheid violence intensified and states of emergency multiplied, Guy, Metcalf, and Winkler decided to step up their own battle to regulate the mines using data as weapons. Guy knew that South African data existed; he just had to find them. "I went through the [AEB] library and the archives, contacted people who worked at the AEB who . . . assisted me in getting hold of reports I couldn't ask for myself. So a lot of this

was done underhand. . . . And there were quite serious security implications. . . . You had to sign an official secrets act so some of the stuff
I did was illegal."[60] The Chamber never yielded its secrets, but the Government Mining Engineer's office did. Guy found a hoard of documents,
including Chamber correspondence, that clearly revealed high radon
levels.

Buried among these documents were S. R. Rabson's results from the
early 1960s, complete with raw data showing high peaks in radon levels.
After Rabson's death, the Government Mining Engineer had asked the
Chamber to conduct another radon survey in 1969–70, which had found
"only two" mines with high levels. That confidential report had estimated
that about 180 "skilled" and 1,600 "unskilled" workers had been "exposed
to concentrations above the accepted tolerance level." Because South
Africa had no official standard in 1970, the report referred to the much-
contested US standard.[61] The percentage game rendered this into a perfectly acceptable result, as the number of over-exposed workers represented
just 1 percent of the workforce at the surveyed mines.[62] Comforting, too,
was J. K. Basson's 1971 report, which, as we saw in chapter 6, asserted the
absence of any radon-related lung cancer.

Shaun Guy covered his copy of this documentation—especially Basson's
report—with outraged notations. His interpretation of Basson's impulse to
discontinue research and radon monitoring differed from that of American
epidemiologists, who'd cast the problem in terms of capitalist conflict of
interest. Reading Basson's report as a foreigner in apartheid South Africa,
Guy knew that "a lot of the senior scientists who were involved with the
Chamber and the surveys and writing the epidemiological assessment from
these results were very hostile to the ICRP and their new dose limits. . . . At
that time also there was the whole thing of sanctions and this closing in
and basically there was a lot of hostility to outside organizations which is
a sort of political thing—it's part of the culture."[63]

Guy also noticed problems missed by the Americans. Basson had calculated cumulative exposures "by multiplying the number of shifts worked
underground on the gold mines by the estimated radiation levels for each
mine on which they worked."[64] This statement earned a double question
mark from Guy. First, Chamber officials had measured actual radiation
levels in only about ten percent of the mines. Second, averages were
meaningless. Even within a single mine, radon levels could vary by several

orders of magnitude. Variation had to do with ventilation, and ventilation had to do with race:

If you know anything about working underground at that time . . . even in the '80s . . . most of the work was done by the black guys who were on the face, the stopes. They tended to be in the areas (what they call the return airways) where the air is hotter, right? It's much cooler in the intake airways. So . . . white miners were mostly located for much of the time in the intake airways where their exposure would be less. So if you take the white miners [as] the base line for exposure. . . . That's the wrong benchmark to take, it's a biased mark.[65]

Digging through data from the 1950s and 1960s, Guy saw many instances of substantial radon build-up in working shafts, some reaching 10 times ICRP dose limits.[66] As Guy observed dryly, "dictatorships or autocracies love records 'cause it gives them a feeling of power and control. . . . The Russians and the Communists are just the same and they keep comprehensive records of everything and that's their weakness because once you can get at those records, it's all there, you know."[67]

These data alone might have justified regulatory measures. But the industry had successfully kept formal regulation at bay for over 30 years. It was not about to cave in to a small group of foreign upstarts relying on old data. If anything, argued the Chamber, nuclear regulation of mines seemed less justified than ever. South African uranium production had slumped by the mid 1980s, leading the Chamber to insist that the mines were less nuclear because they produced less uranium. Even gold production had begun to decline.

Guy and Metcalf weren't about to give up. They knew that radon could build up in shafts worked only for gold. And ventilation often circulated air from old working areas where radon could reach very high levels. *Proving* that "hot spots" still existed, though, required new data. Accompanied by two GME inspectors whose help they'd enlisted, Guy and Metcalf met in 1986 with the manager at West Rand, the mine where Rabson had registered the highest radon levels in the 1960s. They slyly proposed to use West Rand as a "model facility with regard to testing survey methods." Initially hesitant, the manager eventually agreed to a short survey provided that it remained "low key [and] confidential." He would have to obtain approval from his board for a longer-term survey, "as it was a 'sensitive' matter given the union 'situation' at present." At that time,

Miners drill into rock at the Sub Nigel East Gold Mine, 1961. (Ron Stone/Fox Photos/Getty Images, used with permission)

neither white nor black workers knew about radon. White workers congregated around intake airways only because they were cooler.[68]

The preliminary West Rand survey showed radon levels 2–5 times the ICRP limits. The licensing branch expressed concern that "workers appear to have been routinely exposed at these and higher levels for the last 30 years."[69] Backed by these data and the GME, Guy and his colleagues embarked on a mission. Over the next two years, they carried out extensive surveys of many Rand mines, an arduous task both technologically and socially. Calibration became a major bone of contention when mine instruments and the licensing branch's radiation detectors offered widely divergent readings. When Guy descended into the shafts without the protection of a GME inspector, he was subject to harassment by managers, foremen, and white workers who worried that regulation would put them out of their jobs.[70] When all was said and done, however, the survey results showed systemically high radon levels.

MAKING MINE SHAFTS NUCLEAR

Obtaining data was only the first step toward regulation. The battles continued as the institutions of apartheid began to crumble in the late 1980s. Laws were rewritten, among them the Nuclear Energy Act. In 1982 the Atomic Energy Board became the Atomic Energy Corporation. In 1988 its licensing branch obtained institutional independence and renamed itself the Council for Nuclear Safety. The CNS was tasked with licensing all nuclear installations, as well as "sites and activities involving radioactive materials." As far as Guy, Metcalf, and others were concerned, they now had purview not just over uranium-producing mine shafts, but also over any mines that had significant radon levels.

The prospect of the Council for Nuclear Safety having such broad regulatory scope alarmed the mining industry, which through the 1990s continued to fight against designating mine shafts as "nuclear" workplaces subject to licensing. Industry officials dusted off and updated familiar arguments: The absence of adequate data, the ambiguities of nuclearity, and the peculiarity of South African conditions all called for delay and more research. For example, since only a few surveys had been conducted, the CNS was "as unaware of the extent of the radiation problems as was the mining industry." Surely more surveys were required to establish feasible licensing procedures.

Furthermore, the Chamber insisted, the wording of existing South African legislation was ambiguous. The legal definition of "nuclear hazard material" included uranium or "any radio-active daughter product thereof," but the legal definition of a "nuclear installation" specifically excluded "any installation, plant or structure which is situated at any mine or works." Surely it would make more sense to designate the Government Mining Engineer rather the Council for Nuclear Safety as radiation regulator for mines, since the GME was well acquainted with the special problems posed by mining. Chamber officials deemed this approach more "in line with international practice."[71]

"International practice" was a double-edge sword, however. Regulatory approaches varied by country. The Chamber of Mines itself rejected certain international standards as unfeasible under "South African conditions," an argument that it had long made with respect to other occupational hazards.[72] Personal radon dosimetry, which had become standard in France by 1990, "could not necessarily be adopted in the South African mining industry because of the large numbers of employees involved."[73] The new radiation monitoring procedures proposed by the CNS, Chamber officials insisted, were "inappropriate for South African mines" because they were "too comprehensive." South Africa, they argued, lacked the equipment and expertise to enact such recommendations uniformly. Of course, experts could be trained and equipment and data obtained. Doing so effectively, though, would require tremendous work, especially given the size of the labor force and the extent of mining operations.

Put differently, producing nuclearity entailed significant time and start-up costs *because* it was distributed among instruments, experts, data, standards, legislation, and so on. To get started on this work, the CNS invited IAEA experts to South Africa to offer training and advice on radiological protection. It established technical specifications for dosimeters and reporting requirements, though conflict continued over instrument calibration and acceptable error margins. It designed protocols for surveying radiation at each mine. All of this served as merely a prelude to the surveys themselves, which could take from three months to two years to conduct.[74] All the while, industry complained. The measures cost too much. The CNS didn't provide enough consultants. Its experts were autocratic, disdainful, and unreasonable. They failed to appreciate the complexities of ventilation. They took the side of labor.[75]

CNS experts saw little reason to trust industry, given its well-established resistance to regulatory authority. No surprise, then, that the CNS sought alliances with labor, particularly the National Union of Mineworkers. Building trust with labor took time, not least because mine managers spread rumors that the CNS wanted to shut down the mines and take away jobs. At one mine with high radon levels, the CNS eventually decided to circumvent management and offer a radiation seminar directly to workers. The National Union of Mineworkers sent representatives from its national health and safety committee, which had just begun paying attention to radiation exposure at the Koeberg nuclear power plant. This broke the ice, and it led to recurring exchanges between NUM representatives and regulatory experts.[76]

LEGITIMACY THROUGH INTERNATIONALISM

The ongoing struggle between the Council for Nuclear Safety and the mining industry over the terms of regulation spilled out into the international arena. In 1990, the International Commission for Radiological Protection announced a 60 percent reduction in the occupational dose limits, from 50 millisieverts to 20 millisieverts per year. The new recommendation included all types of radiation (alpha, beta, and gamma) and applied to all workplaces, including hospitals, reactors, bomb facilities, and mines. It also hewed firmly to the ALARA principle that all exposures should be as low as reasonably achievable given social and economic conditions. As we saw in chapter 6, though, dose limits represented just one step. The implications for different industries and occupations remained to be worked out, as did the modes of implementation.

The new ICRP recommendation caused acute concern among mine operators around the world. What exactly did the new number imply for radon exposure, and for mining in general? The IAEA convened committees to study the question. The CNS and the mining industry both sent representatives, each hoping to influence international standards and gain support for their domestic positions.

In Vienna and Geneva, Chamber experts found to their delight that colleagues elsewhere had already elaborated extensive arguments against enforcing lower radiation limits. Many of these arguments could readily apply to South Africa:

• Nature was unpredictable. Mines differed from other workplaces because they were subject to natural variations in radiation sources. This argument was well suited to South African mines, typically optimized for gold rather than uranium production. Radiation could not "be mastered at the design stage by engineered procedures as . . . in other parts of the nuclear fuel cycle."[77] As we saw in chapter 6, the French claimed to have accomplished just that at their Margnac mine. But by the 1990s that claim was safely stashed in the dustbin of history.

• The ICRP used faulty assumptions. The new limits assumed that people spent 47 years of their lives working. Miners typically worked much shorter periods, because the work was so arduous. Standards should there-

fore be based on a maximum lifetime dose, allowing miners to absorb
higher doses during the periods when they did work.
• The cost of stringent radiation protection was disproportionately high.
French experts argued that radiation protection should be optimized
according to cost-benefit principles.[78] Ventilation to mitigate radon could
cause other problems, such as high air flow that aggravated dust levels.
Spending a lot of money on ventilation, therefore, could be counterpro-
ductive. In a world of limited resources, increased emphasis on radiation
risk would jeopardize the mitigation of conventional hazards that posed a
more immediate threat.
• ICRP recommendations dealt an economic blow to countries that
depended on mining. "Some third world countries," one official statement
held, "are very dependent upon the uranium business as a primary source
of foreign exchange. It is unlikely that some of these mines could rigidly
adhere to the new recommendations and continue to be economically
viable. The economic and social dislocation which would be caused by
the closure of one of these mines would be far worse in its impact than
the radiation risk to the mine workers, even at the higher risk factors
which the ICRP has used in the latest recommendations."[79]

Chamber experts particularly liked this last line of reasoning, which seemed
tailor-made for South Africa. They deployed it repeatedly in their domestic
battles, arguing that for some mines compliance would force closure.

On the international stage, South Africa offered an ideal case for those
arguing that different radiation limits should apply to developing nations.
In 1991, three years before the country's first democratic election ousted
the apartheid regime, South Africa was the sole African country on the
IAEA committee tasked with deriving new "Basic Safety Standards."
Gabon, Niger, and Namibia—all of which were more "dependent
upon the uranium business as a primary source of foreign exchange" than
South Africa—did not participate in drafting the new international
standards.

South African industry experts claimed that the cost of implementing
ICRP revisions would incur serious financial losses (up to 43 percent) and
threaten the jobs of some 90,000 workers, who in turn represented "the
only means of support for nearly one million people throughout the

sub-continent."[80] The mere act of quantification garnered attention. As one participant reported, "South Africa was the only country for which any attempt had been made to quantify the social-economic impact and this was recognized as an important contribution to the argument, *despite the exercise being necessarily a little speculative at this stage*."[81] Such numbers may have been pulled out of thin air, but they supported the argument that the "Basic Safety Standards" should formally incorporate "flexibility" clauses to allow mining operations in third world nations to exceed ICRP limits under certain circumstances.

ICRP and other experts opposed flexibility in the "Basic Safety Standards." So did Phil Metcalf of the CNS, who contributed enthusiastically to the discussions despite the Chamber's attempt to sideline him because of his "extreme views and uncompromising attitudes."[82] Arguments against formalizing flexibility included:

• Maximum annual exposures could not be calculated by simply dividing a maximum lifetime dose by the number of working years. For one thing, the biological effects of radiation exposure varied with age. Workers were more likely to develop cancer later in life if exposed in youth rather than in middle age. For another, it was impractical to expect that radiation exposure records could be kept consistently over the course of a lifetime, especially in "third world" countries. Without accurate records, there was no way to know someone's true lifetime occupational exposure. Though employers should make every effort to keep good records, the best way to control exposure was to keep workplace doses low.
• The mining industry around the world had a history of "whining" about regulation. Its appeals for flexibility were simply more of the same and could not be taken seriously.
• The ICRP's allowance for national social and economic differences was not intended to confer sole power about decisions onto companies, governments, or even regulators. Labor unions should also have a voice, since they represented the people subjected to radiation exposure.

Discussions went round and round. Metcalf and Chamber experts clashed openly, despite injunctions to "avoid conflict within the [South African] team in Vienna."[83] Their differences echoed tensions between uranium mine operators and radiation regulators around the world.

A compromise emerged in 1993. "Special circumstances" could exist, but these had to be clearly limited in duration. National regulators could make exceptions to dose limitations, but they first had to be satisfied that the corresponding practice was "justified." Exemptions should be granted only after regulators had consulted *both* employers and workers.[84]

International consensus thus came down in favor of anointing national regulators as the ultimate arbiters of radiation exposure limits. In South Africa, too, regulators acquired greater formal authority. A 1993 amendment to the Nuclear Energy Act gave the CNS power to license gold mines and other installations where radioactivity levels reached a certain threshold. New licensees included NUFCOR, which now had to comply with CNS requirements, such as keeping clear and consistent radiation exposure records.[85]

NUCLEAR REGULATION

South Africa's transition to democracy provided the occasion for a large-scale inquiry into safety and health in the mining industry. For five weeks in mid 1994, the state-appointed Leon Commission held hearings that covered the full range of health and safety hazards in the mining industry, gathering testimony from unions, the Chamber of Mines, the Government Mining Engineer, the Council for Nuclear Safety, and others. The subsequent report clearly belied the mining industry's claims of adequate ventilation, finding that dust levels had remained high for decades. Although radon and radiation hazards had occupied only a small part of the proceedings, the Leon Commission used radiation in mines as the paradigmatic example of South Africa's failures to conduct solid research and engage in adequate monitoring. Complaining that there was "no Chair of Epidemiology in any South African University," the commission deplored the lack of evidence concerning radiation levels in mines.

The evidence that did exist suggested that in the mid 1990s at least 10,000 mineworkers were exposed to radiation levels above the ICRP's recommended limits. "On average," the report continued, "the 269,000 South African miners employed underground are exposed to ten times as much ionising radiation as medical staff (attributable to their job), three times the average background level to which we are all exposed, and twice the level entailed in working in the nuclear fuel cycle," the latter presumably referring to the post-mine part of the cycle. The matter required "urgent attention" by qualified experts, not by "ad hoc groups or by inexperienced research workers."[86] Rejecting the Chamber's recommendation that radiation regulation should fall to the GME and espousing the NUM's view that the CNS should continue its oversight, the Leon Commission suggested that closer cooperation between the CNS and the GME would benefit workers' overall health.[87]

Clashes over nuclearity per se resumed where the Leon Commission left off. The 1994 elections that brought Nelson Mandela and the ANC to power offered renewed opportunities to dispute the nature of nuclear things. All manner of laws were being rewritten. Once again, nuclear legislation came up for debate.

This time, the CNS spearheaded the initiative and sought even greater autonomy.[88] The mining industry renewed its opposition to CNS authority, suggesting that radiation protection fall under the purview of the Department of Health. Radon, the Chamber contended, was "essentially a health issue and not a nuclear energy issue."[89] Brazenly invoking "South Africa's transition to full democracy," Chamber president A. H. Munro wrote Parliament in 1995 to argue that "the Nuclear Energy Act does not provide for public participation, transparency, or accountability. Instead it puts extensive power and decision-making responsibilities solely in the hands of expert authorities. Furthermore, it also makes no provision for making the essential social judgments in respect of acceptance of certain risks in exchange for benefits to society." In a striking turnaround, the Chamber thus summoned the ICRP as an ally. Munro quoted its 1990 recommendations that "the selection of dose limits necessarily includes social judgements applied to the many attributes of risk. These judgements would not necessarily be the same in all contexts and, in particular, might be different in different societies."[90] Nuclear regulation of mines, Munro insisted, would impede economic and social development in the New South Africa. The Chamber unblushingly accused the CNS of being a "white, male organization" with an inadequate understanding of development challenges.[91] Once again the Chamber's strategies failed—at least on paper. In 1999, the revised Nuclear Energy Act remade the Council for Nuclear Safety into the fully independent National Nuclear Regulator, cementing its authority to regulate radiation in mines.

By this point, regulatory experts had developed a reasonably good working relationship with the black-led National Union of Mineworkers. Nevertheless, as union official Derek Elbrecht told me in a 2004 interview, the NUM still felt that the regulatory agency should have "more teeth— you know, to bite the industry when they go wrong." And industry continued to pit the union against the regulator: "Management actually used the workforce to pressurize the regulator not to close down the shafts" if workers absorbed excess radiation doses, explained Elbrecht, so "the dilemma for us as a union is in how . . . we manage that situation." Some of the union's health and safety officers would have liked to develop independent expertise in radiation exposure—for example, via a task force such as the one they'd set up for silicosis. But, officials explained ruefully, "the union does not have the resources to fund that kind of research." Turnover at the union's national headquarters also posed a challenge, given the amount of time required to get up to speed on matters of radiation exposure. The union and the NNR continued to conduct information sessions on radiation for workers, but these were limited in number because neither organization had sufficient personnel. One union official estimated that even in 2004, fewer than 5 percent of workers exposed to radiation knew what it was.

Measurement, meanwhile, posed a perpetual problem. Down in the shafts, only team leaders carried a dosimeter. But their doses were not representative of those absorbed by other workers: the team leader normally stood "two, three, five metres away," Elbrecht explained, "and he moves around in a bigger circle . . . than the workers at the point of production." Even under the new regulatory regime, therefore, individual workers did not have a reliable record of the doses they'd absorbed. And even though mineworkers could (thanks to Leon Commission recommendations) refuse dangerous duty, Elbrecht noted that workers did not see radiation as dangerous, in part because they couldn't see it at all. Finally, even though the national union leadership tried its best to attend to radiation exposure, local branches didn't see radiation as a "stand-alone issue. It's not seen as the major cause to certain diseases," Elbrecht explained during an interview. It certainly didn't command the attention required by HIV, which by the end of the century had become the most urgent and overwhelming health issue for mineworkers—and many others—in South Africa.[95]

CONCLUSION

For decades, South African radon repeatedly threatened to become visible. On some occasions it peeked through the work of South African scientists. On others it loomed through the queries of international experts. The volatile nature of labor relations under apartheid—including issues such as white miners' anxieties about the color bar, the recognition of black trade unions, and shifting migration patterns—shaped how the mining industry responded to the presence of radon. With occasional help from state scientists, the industry muffled the political menace of the gas by making its physical presence difficult to see. Sometimes this invisibility resulted from deliberate decisions, sometimes from structural suppression, sometimes from both.

Establishing a *credible* dosimetric regime in South African mines—that is, a program that met international scientific standards for sampling and data analysis—required three related perspectival shifts. The first involved seeing mine work in nuclear terms. The second involved seeing existing radon surveys as products of apartheid science. The third involved rejecting assumptions that nuclearity began only with white, mid-level technicians.

As foreign-trained radiation experts with more invested in trusting the ICRP than in upholding the technopolitics of South African mining, Shaun Guy and Phil Metcalf were uniquely well positioned to see the *im*perceptions generated by South African data. Along with other CNS colleagues, they recognized that making radiation visible meant pushing against the powerful alliance between the mining industry and parts of the apartheid state. They remained unimpressed by arguments about "South African conditions" and unfazed by numerical gymnastics that minimized overexposures in miners. They grasped the contradictions that characterized the apartheid's relationship with things "international," whether the powerful lure of international legitimacy, the resentment bred by international condemnation, or the defiance generated by any hint of international interference. And they understood enough about the fractured nature of the state and its workings to realize that establishing a dosimetric regime required producing new data, asserting expertise, and delineating a domain in which that expertise had authority.

Over the course of the 1990s' transition to democracy, the mining industry began to find ICRP philosophy more congenial. As the meaning of "society" changed in South Africa, the industry sought to capitalize on the "social" dimensions of ALARA. On the domestic front, a more inclusive polity gave a new and far more positive valence to arguments that economic development and stringent regulation were incompatible. On the international front, the democratic transition restored the credibility of South African experts, making it possible for them to draw upon—and shape—international debates on radiation mitigation.

Changes have certainly taken place in the South African mining industry's approach to radiation. Working through the Chamber of Mines, the industry has invested several million rand in cleaning up tailings dumps and contaminated scrap. Mines now have radiation protection officers, instruments, and procedures. Instead of denying the danger, AngloGold (the largest remaining mining house) now advertises its radiation monitoring program as exemplary of its commitment to occupational health and safety. AngloGold's 2007 annual report assured readers that in the three mines where radiation exposure "could potentially exceed the NNR [National Nuclear Regulator] annual dose limit,"

extensive radon monitoring systems have been established . . . all workers are issued with continuous personal dosimeters (radon gas monitors), which are analysed on a monthly basis and the reports forwarded to the NNR. . . .

Mitigation measures include efficient ventilation systems and the prevention of leakages from old workings. In addition, extensive administrative limits have been established internally. To prevent over-exposure to radon gas, employees are moved to lower risk underground areas or to surface should they reach the annual dose limit. No over-exposures have been measured in the last three years.[92]

Even for those inclined to take statements in an annual report as transparent truth, however, much remains unspoken and unknown. The National Nuclear Regulator itself, meanwhile, has come under sharp criticism for lack of transparency and for appointing an industry insider to head the agency.[93]

As important as it is to make radiation in mining perceptible and to regulate it, these achievements do not guarantee that exposures are visible to the miners themselves. Retired mineworkers—whether white or black,

South African or foreign—have no way to correlate disease with work history. In addition to accurate retrospective dosimetry, such correlations require medical data far beyond the capacity of South Africa's current health system, overburdened as it is by other occupational diseases and the devastating HIV epidemic. They also require coordination with similarly overburdened medical systems in Lesotho, Mozambique, and other labor-sending areas, where thousands of former mineworkers now rely on their families for care.

In 2004 the first lawsuit by a black mineworker against a South African gold mining company demanded compensation for silicosis. Filed by the British lawyer Richard Meeran, the suit followed his successful litigation against asbestos companies. The case was still in litigation in 2009, when it was estimated that over half a million current and former miners might be eligible for silicosis compensation.[94] It's likely that a significant number of these men were also exposed to high radon levels—probably more than the 10,000 estimated by the Leon Commission. But it will never be possible to reconstruct their full radiation history.

The work of maintaining invisibility, like that of creating nuclearity, has long-lived consequences.

During their 1980s campaign to block the circulation of Namibian yellowcake, Alun Roberts and other members of the Campaign Against Namibian Uranium Contracts sought to enlist the help of the British peace movement. But as we saw in chapter 5, bombs, missiles, and submarines made more compelling targets than uranium mines—especially mines in distant lands. Although the Campaign for Nuclear Disarmament contributed some funds and occasionally mentioned Namibian yellowcake during its rallies, the CANUC movement's support fell short of what Roberts had hoped for. Uranium mining by itself could not generate outrage to match the protests against the deployment of American cruise missiles (as of 1979) or the reactor accidents at Three Mile Island (1979) and Chernobyl (1986). Unlike these events, it would not inspire hundreds of thousands of people to march in London, Rome, New York, or West Berlin.

Australia was the one country where uranium mining did arouse the anti-nuclear movement and become the focus of mass protests. Britain conducted a series of atomic bomb tests there, first in 1952 in the Montebello Islands off the western coast. Many more secret tests followed at Maralinga—home to several groups of Aboriginal peoples—between 1955 and 1963, a time when Aboriginal rights were not yet on the national political agenda. Decades would pass before the full extent of Maralinga's radiological contamination became public.

Some white Australians did march to "ban the bomb" in the 1960s. More got riled up when France began atmospheric testing at Moruroa, a Polynesian atoll about 6,000 kilometers to the east. Environmentalism gathered momentum in 1971 with the establishment of the Australian branch of Friends of the Earth, the group that stole and released the papers documenting uranium cartel operations. But the magic moment for Australian anti-nuclear activists came in 1975 when the government commissioned an inquiry into uranium mining in the Northern Territory.

Australia had provided uranium to the US and Britain from the mid 1950s to 1971, primarily from two mines that had been shut down by the early 1970s and another that had been placed on hold pending a rise in uranium prices. The latter mine reopened in 1974 under the parentage of Rio Tinto Zinc, which also operated the Rössing mine in Namibia. That same year, the Ranger mining syndicate (shortly to become another RTZ uranium mine) sketched out a proposal to extract uranium from the Alligator Rivers region of the Northern Territory, smack in the middle of Aboriginal lands.

Environmental activists launched a campaign against the proposal, recruiting important allies in the trade union movement. To elucidate the social, political, and environmental dilemmas posed by the mine proposal, the governing Australian Labour Party (which had promised to address the burning problem of Aboriginal land rights) initiated the Ranger Uranium Environmental Inquiry, which in two years produced thousands of pages of transcripts. The Ranger Inquiry gave a wide range of constituencies, from Aboriginal leaders to international experts, a chance to express their views on uranium mining, nuclear power, Aboriginal rights, and related topics. The final report issued a qualified recommendation in favor of the mine, albeit with complex caveats leading to the concomitant creation of a national park and of a social and environmental monitoring program, as well as a slower pace of development for mine sites. Although activists did not succeed in stopping the mine, they strongly influenced the conditions of its development. By making uranium the linchpin of anti-nuclear and environmental activism, they also succeeded in placing it at the center of Australian national politics—a position that it has more or less retained.[1]

Elsewhere, mobilization against uranium mining remained outside the anti-nuclear mainstream. Regional opposition to North America mining focused on indigenous rights and compensation claims. In the 1970s, Navajo advocates unsuccessfully lobbied the US Congress to extend black-lung compensation benefits to former uranium miners. Their subsequent lawsuits demanding damages for these miners were also denied. Undeterred, the Navajo allied with other Southwest activists and created a thriving environment for advocacy groups by 1990, the year Congress passed the Radiation Exposure and Compensation Act. Some activists focused on helping former miners and "downwinders" (people who lived near the Nevada Test Site) to file their RECA claims. Others worked to prevent new uranium mines from opening in the Navajo Nation. All, however, operated independently of the Santa Fe anti-nuclear groups that focused on radioactive contamination generated by the Los Alamos facilities where The Bomb had been born.[2]

In the 1980s, Navajo and other Native American advocacy groups began working with First Nation groups who were protesting the effects of uranium mining on or near Native lands in Ontario and Saskatchewan. In 1987 activists submitted a joint statement to the UN Commission on Human Rights arguing that "in none of the uranium-mining projects . . . was there a thorough advance assessment of risks . . . the wishes of the indigenous community were taken into consideration only in those cases where the mine was to be physically located on land recognized as their property. It is difficult to avoid the conclusion that much of the human cost of North American uranium production has been borne, unwittingly and mostly unwillingly, by indigenous peoples."[3] Through the 1980s and 1990s, uranium mining gathered momentum as a global indigenous rights issue. Activists in North America made connections with counterparts in Australia, and with social and environmental justice organizations working on other forms of mining.

Anti-uranium activists may have been outside the anti-nuclear mainstream, but the dangers of radiological contamination did give them a distinctive mission. Some developed expertise in matters radioactive. Others sought ways of translating nuclear exceptionalism into "traditional" idioms. Some Navajo activists, for example, tell a story in which uranium—currently referred to as *leetso* in Diné—was present at the creation. Here's one rendition:

At the time of Creation, the Diné people came to the top, the place of emergence, where they were given instructions and a choice of how they were going to make their living in this world. They were given two yellow substances and asked, "'Which one are you going to take?" . . . This is the corn pollen, and this is the yellow cake, which is uranium—*leetso*, they call it, in our language. Our Diné people chose the corn pollen, the Beauty Way, and that's what they're going to use to sustain their livelihood and use in all their well-being and their prayers and hold it as a significant item for their people. . . .

The yellow substance they did choose, the *leetso*, they were told, would be returned to Mother Earth:

That will be her protector, so leave that with Mother Earth. That's hers. It should never be taken from her. You took the corn pollen way. If for any reason you should go and get that and take that away from her and bring it out, that's when all of the hardships are going to come about. People will begin to suffer.[4]

Indigenous rights activists have pursued similar strategies elsewhere. The Sahtu Dene living near the Port Radium mine in Canada tell of a prophecy that warned them to stay away from the site. Some Australian Aboriginal advocates recount uranium-tinged versions of Rainbow Serpent tales. The cosmic resonance of such stories recalls the salvation/depravity theme of "first world" imagery, rendering nuclear exceptionalism in "fourth world" registers: Nuclearity was never meant to be. It disrupts the cosmological order. It requires reparation and purification.

In 1992 a conference called The World Uranium Hearing attracted activists from six continents to Salzburg, Austria. Indigenous rights advocates, environmentalists, and a handful of medical doctors and scientists who lent the movement their expertise took part in a week of lectures and testimonies on uranium mining. Additionally, talks on Chernobyl, Hiroshima, and Moruroa helped to contextualize and thereby secure uranium's nuclearity. The lexicon of the global indigenous rights movement offered structure and imagery to the event, whose proceedings were published under the title *Poison Fire, Sacred Earth* and dedicated to "all children throughout the world, and to their future free from the physical and psychic contamination which has been caused by disturbing Uranium's timeless sleep in our Earth."

The presence of four African men at the conference signaled the extension of uranium's nuclearity into the southern tip of Africa. One man was a community organizer near RTZ's Palabora mine, which produced uranium as a by-product of copper. Another was a health and safety officer from the National Union of Mineworkers in South Africa. Both averred that before attending the hearings they had never heard about the radiological dangers of uranium. "Really, on our side, there was nothing that we knew before," said the NUM representative, with the concurrence of the community organizer. Both vowed to use their newly acquired knowledge back home (which, as we've just seen, was easier said than done).

The other two African men were Cleophas Mutjavikua, Secretary-General of the Mineworkers' Union of Namibia, and Joe Hangula, chair of Rössing's safety committee. It immediately became clear that Namibia offered a very different set of possibilities than its southern neighbor. Mutjavikua summarized the chain of events that had led an IAEA delegation to spend ten days surveying radiation and occupational safety at Rössing. The group eventually concluded that all was well at the mine, an assessment the MUN rejected:

Our union is questioning whether the IAEA officials really carried out a full independent and objective examination. . . . Accordingly, the MUN is seeking to assemble an independent team to investigate the medical cases, radiation doses, tailings seepage and decommissioning plans of the mine.[5]

The phrasing of this statement makes clear that the MUN had access to a nuclear world that its South African counterpart had barely perceived. To see this nuclear world through Namibian eyes, we must turn to the workers of the Rössing mine.

From the moment it opened, in 1976, Rössing symbolized the capitalist world's complicity in maintaining apartheid in southern Africa. Although illegal by United Nations decree, the mine supplied large quantities of uranium for nuclear weapons and power plants in Europe, Asia, and the US. For the liberation struggle—especially the nationalist South West Africa People's Organization—opposing Rössing's operations offered a means of recruiting allies from the anti-nuclear and anti-apartheid movements outside Namibia. Activists kept the mine in the international spotlight via hearings, publications, and demonstrations, repeatedly invoking apartheid conditions and exposing the transnational web of capital and technology that supported the mine.[6] Mine executives countered these challenges by invoking opposing international authorities. Nevertheless, in the late 1980s activist efforts began jeopardizing Rössing's ability to do business.

Namibian independence came in 1990. Almost overnight, SWAPO reversed its rhetoric and forged strong ties with Rössing management. Accusations of apartheid collusion faded away. So did threats to nationalize the mining industry, as the new government expressed boundless enthusiasm for foreign investment. Rössing executives accompanied newly elected President Sam Nujoma on official visits abroad. Marketing campaigns proclaimed that buying Rössing uranium was akin to aiding a new nation. The postcolonial state fully backed the company's new slogan: "Working for Namibia." But what about those who worked for Rössing?

on facing page: A Rössing worker posing in a bucket loader used to haul ore out of the open pit. (courtesy of Rössing Uranium Limited)

Aerial view of open uranium pit at Rössing, 2004 (Photo: Paul Edwards)

Rössing workers, as we'll see in this chapter, had a very different experience from that of the other uranium miners we've met. From its inception, their mine constituted a highly visible nexus of liberation, labor, anti-nuclear, and transnational politics. They became active participants in these politics by staking their political claims to technological practices. During the 1980s, contesting the racial division of labor led some Rössing workers to refashion surveillance technologies into methods for trade union action. This tactic enabled SWAPO-affiliated union leaders to build support among the workforce even as SWAPO activists abroad called for an embargo on Rössing uranium that could have shut down the mine.

When national independence failed to produce radical changes in working conditions and labor hierarchies, union leaders looked abroad for help. They engaged external expertise to help them understand, measure, and regulate exposure to workplace contaminants, especially radiation and dust. Rössing's first company doctor had staked his career on developing medical nuclear expertise, and in some respects he had turned the mine into a nuclear workplace. The international expertise invoked by the labor

Breaking rocks in the open pit. (courtesy of Rössing Uranium Limited)

union broadened Rössing's nuclearity, making it meaningful to workers and visible on the global stage.

Appealing to external scientific authority carried the political promise of international accountability. It offered workers political channels that bypassed the boundaries of the nation-state. But engaging in science meant accepting its structures. Rössing labor leaders ultimately discovered that technopolitical power staked to nuclearity could be limiting as well as liberating.

THE WEIGHT OF COLONIALISM

From the beginning, Rössing's management had to balance pressure from the South African colonial state to cooperate with repressive policies against pressure from RTZ to spruce up the mine's international image. Starting in the late 1970s, Rössing proclaimed itself a "deracialized" workplace where workers would be evaluated, ranked, and paid based on performance and skill rather than race. Housing nonetheless remained segregated. Black and colored employees lived in Arandis, a company town in the middle of the desert just a few kilometers from the mine. White employees commuted to the mine from their subsidized housing in Swakopmund, a coastal tourist destination some 70 kilometers away. Such contradictions reflected deep tensions between managers who genuinely sought to challenge apartheid and those who fought to maintain their privileges by upholding its practices. These tensions had consequences for every dimension of work and life at Rössing.

Although the southern African system of contract labor was on its way out by the mid 1970s, the Rössing mine relied on migrant workers during its start-up phase.[7] In 1976 and again in 1978, employees went on strike to protest their living and working conditions.[8] These strikes coincided with a wave of technical problems that, as we saw in chapter 5, made it difficult for the mine to meet its early contractual obligations. During both strikes, Rössing readily restored order thanks to the South African riot police, trucked in to augment the mine's own security forces. Still, management could read the writing on the wall. By 1978, high-level political negotiations made the end of South African occupation appear imminent. The company decided it needed a stable, non-migratory labor force to run the mine's complex machinery. It hoped that better workplace relations could help it meet its productivity goals.[9]

In 1977–78, hoping to head off further strikes, improve production, and bolster its image, Rössing's top management began to explore a new labor policy based on the "complete abandonment of racial discrimination."[10] Via an internal newsletter, new structures for evaluating job grade and pay, a commitment to a permanent labor force, and a variety of social programs, management tried to persuade workers that they were now working under capitalism rather than apartheid, and that these were two radically distinct social forms.

Workers, however, had a hard time seeing the difference. Many white foremen actively opposed reform. Management had adopted the Paterson scheme, which coded rank and skill level to determine remuneration. But low turnover hampered promotion options for black and colored employees, who remained on the lower rungs. The location and quality of housing were formally tied to rank instead of race, but since rank remained racially segregated in practice, not much changed. White Namibian engineers, metallurgists, chemists, and medical doctors trained in South Africa, which also provided most of the mine's technological equipment (not the least because this facilitated maintenance and the acquisition of spare parts).[11]

Most significantly, management hung onto the most potent symbol of oppression: the Rössing security force. Headed by Bill Birch, a former member of the South African police in Rhodesia, the force was equipped with submachine guns, automatic rifles, tear gas, grenades, and other weapons. Police dogs remained a common sight, especially on paydays. At one time, guards on horseback had patrolled the site perimeter, but that practice proved unsustainable when the horses collapsed in the intense heat of the Namibian summer. Workers suspected that Rössing security collaborated with the South African police by identifying political agitators, and Bill Birch himself freely admitted to me that he'd held regular meetings with the police to share intelligence.[12]

Despite management's early efforts at reform, workers continued to feel the weight of the colonial state both on and off the mine site. Yet they also felt changes. Some black and colored workers did get promoted. Training programs did develop skills and literacy. The schools, clinic, social programs, and sports venues in Arandis far surpassed the facilities available to non-whites elsewhere in Namibia. And even though significant pay disparities between ranks (and hence between races) persisted, black and colored men still made more at Rössing than they could have made just about anywhere else in Namibia.

LABOR MOBILIZATION AND THE TECHNOPOLITICS OF SURVEILLANCE

"Deracialization" took a new turn in 1979 when Rössing management created the Loss Control division, which was responsible for accident assessment and prevention. Headed by a white manager, the unit was staffed by four black and colored men—the first non-white employees to receive salaries and benefits instead of hourly wages.

As its name made clear, Loss Control's primary goal was to improve productivity by eliminating losses in work time and equipment.[13] Anticipated improvements in occupational safety eventually followed, and in 1987 management sought certification from the South African National Occupational Safety Association and the British Safety Council. Trumpeting these certificates in internal newsletters and external brochures, management hoped that the legitimacy offered by external recognition would help counter international opposition.[14]

Both technologically and socially, Loss Control functioned as a surveillance unit. Safety officers inspected work areas for ventilation, lighting, fire protection, leaks, and other problems. They conducted accident investigations, wrote safety guidelines, verified compliance, and reported violations. Their findings could ultimately lead management to impose sanctions. This gave the Loss Control officers some authority over the foremen, most of whom were white.[15]

Some white superintendents violently resented the new officers. Willem van Rooyen, a colored member of the initial Loss Control team who eventually became a top manager himself, remembers being physically assaulted by one foreman. His assailant was fired despite the objections of some mid-level managers. The general manager's "dictatorial style" in the early 1980s had benefits in such circumstances, van Rooyen wryly observed:

[The general manager] was very focused on setting standards . . . and actual compliance. It was . . . pre-independence and it was . . . harsh in a way that people knew quite extensively, clearly: if you are not obedient and play [by] the rules, you are out of the company. There are no softy issues or like what we may call today fair hearings. . . . You comply or you don't comply. So that regime was . . . actually very conducive to what the safety section and the environmental section were trying to do at the time.[16]

Ironically, the authoritarian habits of occupied Namibia—and the disciplinary workplace measures they legitimated—enabled van Rooyen and his colleagues to impose safety practices on whites as well as non-whites. Their surveillance became a means of slowly and subtly shifting the racial locus of technical authority.[17]

Surveillance also had technopolitical effects that escaped management's intentions. Loss Control officers bore witness to petty injustices and worker grievances. They watched men wrestle with the gigantic machines to drill, blast, and move rocks in the open pit. They heard the deafening, bone-shaking crushers that ground the rocks. They saw corrosion ceaselessly attack the pipes in the solvent extraction plant, smelled the fumes rising from the vats, and breathed the dust that pervaded the site. They learned first hand of the toxicity risks of Final Product Recovery, which roasted and packed uranium ore for shipment. Their reports on all this inscribed the links between hazardous conditions and racial inequalities.[18]

Still, safety surveillance might not have become political action had it not been for Asser Kapere, one of the first four Loss Control officers. Involved in the resistance struggle since childhood, Kapere was a longtime SWAPO member. He itched to organize residents of Arandis and nearby communities, but he decided that "before I become involved . . . in the whole community, I should start to politicize the department [where] I was working." The Loss Control division provided fertile ground:

The only [other] person in that department who came as a SWAPO member was Paul Rooi. . . . Together [we] mobilized the [others] . . . there was a lady from UK who used to be the secretary to the manager of loss control. . . . She also became a SWAPO member. Eventually the [white] manager of Loss Control himself was Alf Butcher, he also became, not a member, but a sympathizer. So . . . if I wanted to go to political meetings or so on, I would tell him . . . "tomorrow I cannot come to work, comrade!"[19]

Kapere, Rooi, and other Loss Control officers used their mobility to radicalize other workers. They transformed their daily tours of the site's technological activities into occasions for trade union mobilization. Support grew steadily, and in early 1986 they notified management of their intention to establish a union.[20]

Beyond the confines of the mine, momentum for independence was resurging. The transitional government had begun to support non-party-affiliated trade unionism in the hope that such organizations would offer a political base independent of SWAPO.[21] Rössing management suspected that supporting a union before independence could pay off afterward. But Kapere's and Rooi's SWAPO affiliations—not to mention the anxiety many white employees expressed at the prospect of independence—made things tricky.

During his first meeting with union leaders, personnel manager Charles Kauraisa insisted that the company supported the union but noted "the nervous attitude of the State" toward SWAPO. The union, he warned, should "be very careful not to prompt an unreasoned response from any part of the State bureaucracy." Union leaders, meanwhile, complained about supervisors "displaying unacceptable fear and interfer[ing] with the rights of their subordinates to join a prospective Trade Union."[22] Kauraisa insisted that management would countermand such interference.

Labor leaders kept noticing undercover security officers at union meetings, however, and suspected that the mine's security service routinely "colluded with the police."[23] In July 1987, Kapere's arrest during a police raid that injured 15 people in Arandis confirmed this suspicion. Acquiescing to worker demands, Rössing publicly condemned the police actions. Dr. Zed Ngavirue—the company's first black chairman[24]—met with the Minister of Justice to demand Kapere's release.[25] The union gained formal recognition later in 1987, merging with the soon-to-be-national Mineworkers Union of Namibia the following year.[26]

The union's first official business at the mine concerned workplace racism. Existing disciplinary mechanisms, the union argued, "embodie[d], if not the provision, then at least the spirit of . . . apartheid and racial discrimination." Most notably:

Where the supervisor is white and the worker black, the merits of the case are largely subservient to skin colour and the overriding consideration remains the view/opinion/version of the white supervisor. Many of these supervisors still consistently display a "baasskap" [sic] mentality. . . .[27]

Furthermore, union leaders remained unimpressed by the company's progress toward the "Namibianisation" of supervisory positions.[28] The technological knowledge of black workers remained undervalued relative to that of whites: "The majority of whites in middle management are promoted regardless of qualifications whereas in stark contrast, black semi-skilled workers (performing the tasks of skilled workers) have to obtain qualifications higher than the legal requirement to make any advancement whatsoever." Workers questioned management's true commitment to racial

Undated photo of acid tank in yellowcake plant. (courtesy of Rössing Uranium Limited)

equality.[29] Meanwhile, management worked to quell rising panic among some whites at the prospect of independence.

The Rössing union's affiliation with the MUN had political and financial benefits, but it also created dilemmas. MUN General Secretary Ben Ulenga, who had recently been released from the apartheid state's notorious Robben Island prison, embraced the liberation struggle's strategy of recruiting international allies. He traveled to Europe frequently to rally trade union support. In these settings Ulenga followed the SWAPO line by issuing repeated demands for an international boycott of Rössing uranium. Back in Namibia, mine management tattled on Ulenga, cautioning workers that a successful boycott would shut down the mine and produce widespread job loss.[30] Staunch SWAPO supporters defended the primacy of the nationalist struggle and declared themselves ready to accept any consequences.[31] Others weren't so sure. Rumblings about forming a breakaway union surfaced.[32]

MUN leaders realized that retaining local support required delivering concrete solutions to specific workplace problems. In 1989, with independence around the corner, union organizers returned to their technopolitical origins by focusing on health and safety issues.[33] They began by demanding increased worker input into health and safety decisions.

Meaningful participation required more knowledge about employees and their workplaces. Loss Control reports testified only to acute accidents. Readings for environmental contaminants such as dust and radiation fell under the direction of the company doctor. Making these data meaningful—for example, linking readings of dust or radiation with workplace exposures and health outcomes—required access to individual medical and personnel files.

During and after Namibia's transition to independence, such knowledge became a subject of conflict between management and workers. In order to understand tensions around the production and use of knowledge about health and exposures at Rössing, however, we must backtrack.

MEDICAL NUCLEARITY AND DISEASE ONTOLOGIES

In 1979 Rössing hired a freshly trained white Namibian doctor named Wotan Swiegers to build a clinic and devise occupational health guidelines. Swiegers began by visiting uranium facilities in Canada, South Africa, and Britain to learn about their practices and obtain an occupational health certificate. Radiation exposure and uranium toxicity particularly interested Swiegers because they offered scientific and technological challenges that would distinguish his work from that of other southern African mine doctors. His expertise, unlike theirs, would be specifically nuclear.[34]

At Rössing, Swiegers ordered equipment for performing a battery of annual tests, including spirometry, x-rays, sputum cytologies, and urine sampling. In cooperation with the South African Bureau of Standards, he set up a system to monitor the monthly radiation exposures of workers in the Final Product Recovery (FPR) area. And—in stark contrast to uranium-producing mines in South Africa—he developed exposure guidelines based on ICRP recommendations. Swiegers thus developed the knowledge, will, and authority to enact nuclearity. He justifiably took pride in his program: "When I visited Canada, for instance, or even France—and South Africa for certain, you know—we were streaks ahead."[35]

When Phil Metcalf of the AEB licensing branch (whom we met in chapter 8) initially toured Rössing, he was especially struck by problems in the solvent extraction section of the plant. Workers were shoveling sludge with no protective clothing, and there were no guidelines setting limits on exposure levels. The second time he visited, he found that Rössing had gone "totally over the top."[36] Whereas yearly lung-function tests and x-rays had been deemed unnecessary elsewhere, Swiegers had plumped for thorough data collection, even when it taxed the patience of management.[37] For Swiegers, this was "pioneer work" that, though ordinary "in a first world country," was "damn difficult to do . . . in a third world country."[38] For Swiegers, Rössing's location in Namibia meant that the simple act of treating the mine as a nuclear workplace made his work exceptional.

Management may have expected the doctor to emulate a long tradition of corporate and colonial medicine centered on maintaining worker productivity, but Swiegers understood that productivity depended on trust.[39] Both at the time and in retrospect, he defended his technological choices as trust-building exercises: "If people see that what you do is . . . high tech and if you're trying to do it properly, . . . it gives them belief in the system." To the annoyance of line managers, he insisted on transferring workers when their exposures exceeded international standards, and on keeping them out of high-radiation zones until their exposure averages dipped back to acceptable levels. He felt that such measures, which followed international "best practice," resolved the tensions experienced by "any occupational health physician, this whole problem of dual loyalty."[40]

Nevertheless, Swiegers didn't always recognize how deeply trust was bound up with larger social, political, and epistemological issues. For example, the annual spirometry that tested lung function required workers to exhale into a machine following prescribed patterns. Failing to reproduce the patterns precisely could lead to conflicts. Controversy was exacerbated when the lung expert whom Rössing retained from Stellenbosch University in South Africa claimed that black people had weaker lungs than white people. One worker remembered the incident well:

That was a very very big bone of contention during those years, you know. . . . Generally in this country you do not get black people smoking. . . . It's only nowadays you are getting young people smoking, but elder people in black communities do not smoke. . . . [Yet] Professor Joubert [stated] that black person's lungs are weaker than white and that's why black people are getting lung problems through detection of lung function machines.[41]

Workers were ordered to wear uncomfortable protective equipment, with little explanation of its purpose. They did not receive copies of their medical tests or monthly exposure results. All of this engendered further mistrust. Swiegers's successor, Jamie Pretorius, later acknowledged that the medical service's explanations of "the concept of radiation," or of "why must you wear a respirator, or . . . a film badge," or of why workers had to supply urine or blood specimens were "paternalistic."[42]

Workers had no choice but to accept such dictates when the alternatives were disciplinary measures or getting fired. Acquiescence did not mean that they trusted experts and their high-tech machines. They saw monitoring practices as inseparable from the discipline inherent in the colonial/corporate system. As SWAPO member and trade unionist Harry Hoabeb later explained, in the 1980s "we saw each other as white and black. Oppressor and the oppressed."[43] This dynamic permeated how workers saw the medical service, especially when the doctors refused to give workers access to their medical records. For Pretorius, the refusal was a matter of "medical ethics . . . the data in there belongs to Rössing [and] therefore this information was not to be divulged to any person."[44] But Hoabeb and his trade union colleagues rejected this rationale:

If you go to your annual medical check-up, that's the most important information that is in your file. . . . But during those times, we don't have a proper definition for "mine-related diseases." . . . What would we call occupational disease? So mostly what was in the files were chronic diseases and so forth . . . , but nothing that would define as occupational disease. So also in that regard really I do not see the point of medical confidentiality.[45]

Trade unionists challenged the notion that disease ontologies existed separately from social context. In turning that challenge into an instrument of political action, they returned to transnational political circuits established during the liberation struggle.

REACHING FOR THE NUCLEAR WORLD

With sanctions seriously affecting business, top management at Rössing eagerly anticipated independence. Soon after the accord was signed in 1988, Rössing arranged a series of meetings with SWAPO. The party leadership reassured the company that they saw it as a key player in the Namibian economy.[46] Rössing proceeded to give "utmost priority" to "establishing good relations with the future leaders of an independent Namibia."[47] Executives attended social functions in Windhoek; hosted a series of lunches for SWAPO leaders, UN officials, senior diplomats, and other business and civic leaders; and invited all these people to tour the mine. In an internal memo from 1989, public affairs manager Clive Algar advised his colleagues on diplomacy:

Sometimes we—and I am as guilty as anybody—tend to show our satisfaction at Rössing's successes in various fields but this may be the wrong psychological approach when dealing with future cabinet ministers whose whole raison d'être is change and improvement of Namibia. Our theme throughout should be not only what we have achieved but what remains to be done, and in speaking about such aims we should make it clear that we are open to suggestion and comment. *This of course hardly applies in the technical area* but is very relevant to the whole human aspect of Rössing.[48]

This tactic worked. Shortly after independence, top company executives accompanied President Sam Nujoma and several of his ministers on a week-long visit to the US, thereby "consolidat[ing] what were already sound and friendly relationships." In June, Dr. Leake Hangala of the Ministry of Mines and Energy joined Rössing's Board of Directors as the new government's nominee.[49] Most reassuring, the new administration formally announced that it would not nationalize the mining industry.[50]

Meanwhile, Rössing invoked international authorities to keep "technical areas" safely outside the political orbit. Anticipating new regulatory legislation, executives preemptively met with experts from the International Atomic Energy Agency and other international nuclear organizations. Armed with their support, Rössing persuaded government officials that special nuclear regulation was not necessary for the new Namibia. Ordinary mining legislation "could quite comfortably accommodate" the regulation of uranium production. Namibia, said the company, would thereby follow the example of Niger and Gabon.[51]

The company took similar steps to shape environmental legislation. Clearly, safety certificates from South African organizations would do little to legitimate Rössing's practices in the eyes of the new government. So the company commissioned a Canadian consulting firm to write a review of Rössing's occupational hygiene and environmental control practices and a proposal for Namibian legislation on such practices. The reports vetted Rössing's existing code of practice as meeting international standards. They recommended that the Namibian state formally adopt this code, along with a licensing and inspection system for enforcement[52]— suggestions that the ministry of mines eventually adopted.[53] International expertise thus helped the company gain firm political footing with the new administration.

This strengthening of the company's relationship with the new state posed challenges for trade unionists. Even before independence, Namibian labor leaders had begun to see tensions between the logic of nationalism and their immediate workplace concerns.[54] Before the transition, such tensions were managed through spatial demarcation. The nationalist demand to boycott Rössing's uranium unfolded at the UN and in other international venues. Contests over the racialization of technical skill and authority occurred in the workplace. Spillover from one domain to the other was minor and manageable. After independence, however, a postcolonial state run by nationalists needed Rössing to help power its economy. The state's embrace of private corporations compromised the democratic promise of the labor movement.[55]

To meet these challenges, Rössing's MUN turned outward to international sources of authority. Building on their pre-independence successes, labor leaders focused on matters of occupational health. Asser Kapere fired a warning shot in an interview with the *Namibian* in which he linked radiation levels at Rössing with "inexplicable ailments" suffered by Arandis residents. The MUN demanded a broad epidemiological investigation of health problems among workers and their families. When union leaders were approached by Greg Dropkin, a British activist who'd joined CANUC in the late 1980s, they seized the occasion.[56] Now working for PARTiZANS, which campaigned against RTZ's global mining operations, Dropkin wanted to investigate dust and radiation exposures at Rössing. The ensuing publicity, he suggested, might ultimately enable workers to claim compensation. Union leaders gladly provided Dropkin with internal company documents and worker testimonies.[57]

The resulting 1992 report, *Past Exposure*, struggled to reconcile PARTiZANS' agenda with the union's. PARTiZANS advocated closing all RTZ operations. The union, already combating a recently announced retrenchment plan that threatened several hundred jobs, didn't want the mine to close. *Past Exposure* managed to find common ground in the retrospective reconstruction of radiation doses. It made the most of the sparse data Dropkin had obtained on dust and radiation levels, extrapolating cumulative exposure figures which it then compared against international standards. The report concluded that dust levels in some areas considerably exceeded Rössing's own standard and that dust levels of airborne uranium exceeded ICRP guidelines. Additionally, workers in Final Product Recovery had experienced particularly high radiation levels before 1982 and still had significant exposures.[58]

Insisting that the allegations were false, company executives suggested that state officials invite independent experts from the IAEA to inspect the mine. The government readily assented.[59] In September 1992, the IAEA and the International Labour Organisation sent a team to take radiation readings at the site. The state initially touted the inspection as an effort to mediate between the company and the union. Accordingly, the MUN submitted a list of concerns for the team to investigate and named three international experts that it wanted on the team. IAEA experts rejected these suggestions, explaining that they were on "an independent technical mission undertaken on behalf of the Namibian Government." They would address union concerns "time permitting," but they first had to fulfill their own mandate.[60]

The five-member team spent two weeks taking readings on dust, radiation, and other contaminants. Their report concluded that the mine's medical surveillance program and facilities were "outstanding," that

Poster for anti-RTZ activism, London, May 1981. (from collection of Laka Foundation)

Rössing's data were "reliable," and that radiation levels were "very low, much lower than current international limits." It acknowledged that "grievances exist about some cases of illnesses, including lung cancer, which are thought to be related to occupational radiation exposure. However," it continued, "such cases can only be addressed in comparison to national vital statistics, which do not seem to exist in Namibia at the present time."[61]

In other words, the lack of control data made any broad epidemiological study impossible. Furthermore, IAEA experts betrayed marked impatience with the union:

. . . the Union were most upset that the mission did not concentrate solely on their perceived problems.

Many health and safety issues raised by the MUN could be resolved through the establishment of a joint Occupational Health Committee. While the Rössing management is quite prepared for it, the MUN does not seem to move into this area. Also, the local branch of MUN seems to lack specific site knowledge, occupational hygiene information, and standards for Occupational Health Committee members.[62]

The team expressed no awareness of the colonial histories and structural inequalities that explained the lack of control data and impeded the union's ability to gain information. Union leaders, for their part, expressed dismay that the team didn't gather its own medical data: "[T]hey did not take . . . examinations, properly draw blood. . . . They just checked with the facilities of the company and they drew their report."[63]

Rössing was thrilled with the IAEA report. It sent copies to the international press, organized a panel discussion on Namibian television, and printed a pamphlet (titled *Past Exposure Exposed*) aimed at reassuring shareholders and customers. All this, however, only delegitimated the report further in the eyes of union leaders, who refused to accept its conclusions.[64]

In response, the IAEA and the ILO invited a group of MUN representatives to visit uranium mines in Canada in 1993. The trip was sobering. Appalled at the sight of open drums of yellowcake leaking onto the floors, Harry Hoabeb acknowledged that Rössing had "far better" standards.[65] Another delegate expressed astonishment that most of the Canadian workers remained un-unionized.[66] And everybody was dumbfounded to see white people mopping floors.[67]

Nevertheless, bad conditions elsewhere did not exonerate Rössing. MUN representatives returned from Canada knowing much more about international practices, but no less determined to find an independent assessment of their own workplace.[68] "The government then challenged us: 'if your boys are rejecting this international body, we are challenging you to come with an alternative to say that this report is false.'"[69] Union comrades in South Africa recommended a medical researcher at the University of Cape Town. His fees proved too high, but he suggested Reinhard Zaire, a Herero-speaking medical student working in Germany. Zaire could study Rössing as part of his research, and the union would reap the benefits. MUN leaders liked this plan; the fact that Zaire was a black Namibian helped them trust him.[70] In late 1992, they contracted Zaire to conduct an "epidemiological evaluation of the Mineworkers at Rössing (with specific references to cancers)."[71]

Management didn't take kindly to this initiative. Jamie Pretorius refused to turn over medical records and denied Zaire access to the site. Undaunted, and more suspicious than ever, the MUN arranged for Zaire to take blood and urine samples in secret to prevent the company from meddling with the results.[72] The company enlisted its government connections to probe Zaire's intentions and slow him down. In April 1993, Dr. Nestor Shivute, the Director of Primary Health, told Clive Algar that the Ministry of Health had indeed given Zaire permission to conduct oncology research. But since "the MUN had thought that this research would provide them with an opportunity to 'nail Rössing,'" Shivute declared himself "ready to withdraw the Ministry's permission for the research."[73] He helped Pretorius obtain a copy of Zaire's research protocol.[74] After the mine doctor identified several procedural problems, the Ministry of Health and Social Services revoked Zaire's research permission.[75] Zaire ignored these strictures. Clearly, though, the company's careful cultivation of allies in the postcolonial state was bearing fruit.

Reinhard Zaire's study soon became entwined with other events. In December 1993, Edward Connelly, a former Rössing worker living in Britain and diagnosed with laryngeal cancer, enlisted Richard Meeran, an environmental lawyer who had prosecuted asbestosis cases against South African mines, to sue Rio Tinto for damages in British courts. Rio first tried to have the case dismissed on the grounds that Namibian courts offered the "natural forum" for the case. Pending a decision on this challenge, Jamie Pretorius prepared a thick report documenting Connelly's work history.[76] In 1997 a British television news program featured interviews with Connelly and two black employees: Petrus Hwaibe (who'd contracted anaplastic anemia) and Petrus Naibab (diagnosed with non-Hodgkins lymphoma). Both men declared themselves ready to follow Connelly by suing Rössing in Britain. British courts dashed their hopes in 1998 by striking down Connelly's suit. But another suit—brought by the widow of Peter Carlson, who had worked at Rössing from 1977 to 1984 and had died of esophageal cancer in 1995[77]—was already in the works. These events clearly signaled the stakes of Zaire's study.

Zaire took his blood and urine samples from Rössing back to Berlin, where he ran tests to determine blood counts, hormone levels, chromosomal aberrations, and more. In 1995, he began presenting his results at hematology conferences in Europe and the US. "Namibia," he declared, "provides a clear test case for the effects of low-dose long-term uranium exposure due to the clean air quality and the lack of other industries with negative health effects." His results showed that miners had "a significant reduction in testosterone levels and neutrophil count" relative to the control group."[78] He asserted that Rössing employees suffered from a higher rate of disease than the general population. But he placed the most emphasis on his results for chromosomal aberrations:

Most remarkably, multi-aberrant cells such as "rogue" cells were observed for the first time in miners which were formerly known only from short time high dosage radiation injury, e.g., from Hiroshima atomic bombing or the Chernobyl accident. We conclude that the miners exposed to uranium are at an increased risk to acquire various degrees of genetic damage, and that the damage be associated with an increased risk for malignant transformation.[79]

Zaire thus made two related claims. First, he asserted that working at Rössing increased the risk of cancer. Second, he claimed to have found the first concrete evidence that exposure to low-level radiation had genetic effects.[80] Zaire saw his intervention in nuclear terms: a career-making scientific breakthrough.

By early 1996, Zaire had posted these texts on the Internet. With help from colleagues at Stellenbosch, Jamie Pretorius dissected the findings and

found many potential problems. The absence of control data made epidemiological comparisons impossible. Zaire lacked accurate radiation exposure profiles for employees, so he couldn't correlate exposures with
chromosome abnormalities. Increased chromosomal aberration didn't necessarily lead to increased malignancy.[81] Rössing's general manager forwarded this analysis to Rio Tinto's headquarters in London.[82] After an
article on the study appeared in the German press, Rössing alerted its
lawyers in Windhoek to stand by.[83]

By July 1996, when Zaire presented his work to the MUN, Rössing
was well prepared. Pretorius attended the meeting and challenged Zaire
at every turn.[84] According to one management observer, Zaire "could not
give credible answers" to the doctor's questions. The observer drew hope
from the "body language" of the union leadership "that some doubt
must have been cast on the validity of Mr. Zaire's findings."[85] Quite on
their own, however, MUN officials were growing disenchanted with
Zaire. They had expected more from his study than an apparent increased
risk of cancer, a less powerful conclusion than increased *incidence* of
diseases with clear workplace origins. They pushed Zaire for stronger
conclusions.

Obtaining better results would require access to medical records and
other company data. Rössing proposed that the union and the company
engage additional independent experts to conduct a verification study of
Zaire's findings. The company would make more data available, and Zaire
could participate. Everyone agreed to this plan, and they began searching
for mutually acceptable experts.[86] To rectify the absence of national epidemiological data, which both Rössing (in 1996) and the IAEA (in 1992)
had invoked as grounds for skepticism, MUN officials wanted the health
study to go national. Rather than isolating health effects of radiation, they
wanted the study to identify *all* the occupational health hazards posed by
work at *all* Namibian mines.[87]

Along the way, Zaire's behavior grew increasingly disturbing. He had
begun holding press conferences without notifying the union. He made
contradictory statements.[88] Union leaders gradually realized that Zaire had
developed an agenda independent of their mandate. In one venue, Zaire
claimed that the study had been commissioned by the German activist
group Akafrik in 1992. In another, he said that he himself had originated
the project.[89] He accepted "financial support" for his research from

anti-nuclear groups. "Once you start getting involved with external people," one worker said bitterly, "what do they want? . . . You get help and whether it means something to you in the end you [are] not sure. . . . You are under obligation for certain things and then you need to be very clever in dealing with these external people because some of them [have] other interest groups that support them or fund their projects but you['re] not clear of what is the project. . . . So that's . . . an international problem, especially if you're from the third world countries."[90]

Zaire began to demand outrageous consulting fees, promising more "comprehensive" health data in return.[91] But MUN leaders had grown disillusioned. Anti-nuclear groups had "come along and hijacked" their study.[92] They suspected that Zaire was trying to blackmail them, notably by telling rank-and-file members that the union's leadership was withholding medical information from them.[93]

Zaire had also tried to blackmail management, demanding a total of 120,000 German marks in return for refraining from lucrative television appearances. He requested another 25,000 British pounds to refrain from appearing as a witness in the Connelly case.[94] By 1998 it was clear that the attempt to find a credible independent expert had taken a disastrous turn. In the words of one union official: "The value of Zaire's study is very great but his personal ambitions, his personal greed . . . overshadowed his scientific work."[95] And then the man himself went missing: "[W]e couldn't trace him later—no emails, no phones. . . . Then government start[ed] looking for him. . . . That was quite a bad angle."[96]

The Zaire affair badly derailed the union's attempt to recruit international expertise. Ironically, it and the Connelly case further strengthened the ties between Rössing and the state. A former industrial relations officer later commented:

I think Zaire helped us to open up. I mean he was a pain in the flesh but he kept everybody awake every day—he was on radio, he was everywhere. . . . The government was saying, "What's going on, Rössing, explain yourselves!" . . . We had to talk to all the ministers that mattered, we had a very strong governmental liaison. . . . One of the issues was low dose radiation long-term exposure, the Connelly case . . . that was very publicized, it was always running headline news in *The Namibian*. Every week there was somebody writing about the Connelly case, Rössing being asked to explain and people saying, "well, you know, can Rössing employees who have suffered cancer also pick up class action against Rio Tinto?"[97]

Rössing hoped that the verification study would put the bad publicity to rest.[98] The union also had high expectations for the verification study, hoping that it would confirm and perhaps strengthen Zaire's findings. But they took nothing for granted and became determined to learn as much as possible on their own:

We had to start reading up on research protocols. [For] any human fluid or tissue, there are certain international criteria. These things we had never known before and then afterwards we start learning about these things. . . .[99]

The company and the union together selected two experts to conduct the study: David Lloyd (a British scientist) and Joe Lucas (the American scientist who had originally designed the chromosome aberration tests used by Zaire). Lucas and Lloyd initially proposed a three-part research plan. First, they would seek to "confirm or refute the Zaire data" by searching for chromosomal aberrations among the ten highest-exposed miners using a control group of equal size. Second, they would evaluate clonal expansion, "an initial step in all cancers and leukemias," in miners. Last, they would "identify and evaluate any cancer patients for exposure at the Rössing plant." The second and third phases would have gone well beyond Zaire's work, though not far enough to address union concerns. Rössing balked at the extra cost, however, claiming that its agreement with the union specified only a verification of Zaire's published results. The study was limited to the first point. Zaire's research legacy thus endured well after he had disappeared.

US research protocols dictated that an Institutional Review Board oversee the project. To forge political consensus, this first-ever Namibian IRB included Rössing managers and doctors, union representatives, politicians (including Asser Kapere, who by then had left Rössing to join the government), and community residents. Lucas and Lloyd collected blood samples, analyzed them independently of each other, and found no chromosomal aberrations. In 2001 the Namibian IRB accepted their conclusions.

Officially, the MUN also accepted the results of the verification study. It dropped efforts to obtain independent assessments of workers' health. Individually, some workers expressed relief at the results and confidence that Rössing ran the safest possible workplace. Others seemed disappointed that the verification study dashed their hopes for compensation. Still others

who noted that chromosomal aberrations were only one possible conse-
quence of exposure continued to fear negative health effects from working
at Rössing. Overall, however, they saw little recourse. "Those people,"
Harry Hoabeb concluded, "were internationally renowned people and
where do you go after those people? Where do you go? You are not going
to have any leg to stand on if you dispute those—those type of bigheads,
you know. Where do you go? The only way that you had to go is probably these anti-uranium people and they will tell you definitely another
story and you are going to become confused. This Greenpeace, *neh*? You
go to Greenpeace, they will tell you another story."[100]

The verification study may have foreclosed the question of chromosomal damage, but it couldn't stop speculation. Many of the workers I
spoke with mentioned someone they knew—or had heard about—who
had cancer, or a brain tumor, or leukemia. A former public relations officer
put it to me this way in 2004: "There is a bit of a concern at the moment,
you know . . . a lot of people . . . are dying at the moment. . . . I'm not
suggesting that they are dying because of anything to do with tumours,
cancers. There's just this very funny feeling . . . every time a cancer case
comes up. . . . You get a bit concerned if you know what I mean."[101]
When she heard these stories, she couldn't help thinking back to the eight
years she'd spent at the mine. Was she next in line? Were the cancers triggered by uranium exposure, or did they merely follow the typical pattern
of an aging population? In the absence of a historical cancer registry, no
one knew.

CONCLUSION

The Rössing mine offered a powerful symbol to the Namibian liberation
struggle. In representing the transnational character of apartheid and corporate power, it provided a concrete expression of the argument that South
African colonialism was not merely a Namibian problem, proving that
boycotts could be effective forms of international action. But the political
significance of Rössing went beyond symbolism. The material specificities
of the uranium mine mattered not just for political economy but also for
labor politics.

The company appealed to international sources of authority—the
IAEA, the ILO, and others—to certify the safety of its workplaces. Both

before and after independence, these bodies offered supra-national legitimation, a way of bypassing the state by appealing to universal standards.
Their invocation of science and technology took matters into domains
beyond the reach of politics.

Workers, however, had learned that neat separations between technology and politics did not exist. Loss Control employees found that technical
inspections they carried out were not simply tools for corporate surveillance and profitability; they could also open up political opportunities.
More broadly, in balancing nationalist strategies (such as boycotting Namibian uranium) and local imperatives (such as maintaining employment and
improving the workplace), Rössing trade unionists maintained their political stature and relevance by advocating access to medical and administrative
knowledge.

Power inhered in technology and knowledge. Workers absorbed this not
as an abstract point, but as lived experience. When independence failed to
bring the sweeping changes they had imagined, workers found creative
ways to apply the lesson. They appealed to authorities outside the new
state and its political party. They cultivated alliances with European activists
and hired an expert of their own. Along the way they participated in the
production of knowledge about their workplaces and their bodies, invoking to their advantage the "universal" scientific and technological standards
of exposure, contamination, and safety. Convinced that the knowledge they
produced would ultimately demonstrate that Rössing didn't measure up
to these standards, they invested their hope in scientific authority.

Rössing workers then discovered the limits and vulnerabilities of their
strategies. The problem was not merely the choice of a disappointing
expert. It also had to do with the financial, social, and epistemological
structures of science itself. The verification study was scientifically impec-
cable because of its narrow frame as *nuclear* research, focused only on the
effect of low-level radiation exposure for chromosomal aberrations. Broadening it to include an investigation of cancers and their etiologies required
resources that the company was unwilling to commit and the labor union
did not possess. In any case, such an expanded study would have kept the
work framed in specifically nuclear terms. Expanding the work to include
all contaminants at all mines, as the union desired, would have required
not only far greater financial resources, but also a national knowledge

infrastructure, beginning with a cancer registry to make the data collected at mines meaningful.[102]

Yet another problem was that external allies had their own agendas. Again, Reinhard Zaire's cupidity was only part of the difficulty. European environmentalists had much to offer when it came to making and interpreting measurements, but their ultimate aim was to shut down RTZ mines, or nuclear facilities, or both. Environmentalists and unionists could make common cause for a while, but in the end they ran up against a fundamental contradiction in their goals. Attempts to resolve the paradox, as Harry Hoabeb said, seemed only to cause confusion. Environmentalists might "tell you another story," but the punch line of their narrative was unacceptable.

This leaves us with a question that, like the concerns of Rössing's workers, transcends nuclearity: How can the fundamental conundrums of technopolitical governance—balancing present employment with future health, regulating industrial activity, building national scientific infrastructures, developing independent expertise—be addressed in postcolonial Africa?

Uranium was not born nuclear. It was not born nuclear in the US or Europe, where ceramic and glass manufacturers first used it as a coloring agent. Nor in Madagascar, where villagers in the Androy used it as weights for fishing lines and ammunition for slingshots. Nor in South Africa, where gold mine operators initially discarded it as waste. Used to color ceramics in the US, fashioned into weights for fishing lines in Madagascar, and discarded from South African gold mines as waste, uranium began its life among humans as a metal of marginal value. Even in the bedrock of Gabon, where highly concentrated uranium sparked a self-sustaining fission reaction 2 billion years ago, it did not achieve *nuclearity*—an expression of technopolitical power, a product of social and cultural contestation.

Nor did uranium, once it became nuclear, remain so once and for all. The apocalyptic fever that followed the explosions over Hiroshima and Nagasaki, the fervor for salvation precipitated by "Atoms for Peace," the clarity of the world order fantasized by the Nuclear Non-Proliferation Treaty, the horror triggered by the prospect of "rogue" nuclear weapons: all this certainly makes nuclearity *seem* self-evident, automatic, indisputable and immutable. There is no doubt—none whatsoever—that we could destroy humanity by detonating a few hundred atomic bombs, or that even one explosion would kill thousands of people and bring buildings and infrastructures crashing down. Nor is there any doubt that nuclear power plants have provided electricity to millions, or that spent nuclear fuel will remain radioactive for millennia, or that medical isotopes can diagnose and treat diseases, or that radon daughters can cause cancer.

on facing page: Uranium worker in Niger, 2005. (Pierre Verdy/AFP/Getty Images, used with permission)

No doubt. Yet as this book has shown, none of this, in fact, makes nuclearity self-evident. Let's review what we've learned, and explore some contemporary implications.

NUCLEARITY REQUIRES WORK

Nuclearity requires instruments and data, technological systems and infrastructures, national agencies and international organizations, experts and conferences, journals and media exposure. When (and where) nuclearity is densely distributed among these elements, it can offer a means of claiming expertise, compensation, or citizenship. It can serve as a framework for making sense of history, experience, and memory. When (and where) network elements are absent, weak, or poorly connected, nuclearity falters, fades, or disappears altogether, failing to provide a resource for people claiming remediation or treatment.

As we saw throughout part II, the nuclearity of uranium mines in Madagascar, Gabon, South Africa, and Namibia had to be uniquely made in each place, overcoming resistance that was sometimes systemic, other times strategic. Nuclearity, if it emerged at all, was often fractured and lumpy. Its temporal and spatial unevenness was wrought by variations in the pace and pattern of decolonizations, the waxing and waning of Cold War logics, and the very nature of radiation, which lingered in soil and water and human bodies long after industrial activities had ceased. Even partial nuclearity depended on regional connections, national politics, and transnational networks. It also, simultaneously, required local work. Once established, it could dissolve over time. Nuclearity requires maintenance.

Achieving and maintaining nuclearity depended, in part, on the willingness and capacity of states to exert sovereignty in nuclear terms. The ability of people to use these terms to make citizenship and health claims also mattered. As we've seen, there were considerable variations on these points among different African states as well as within them. The Malagasy state did not have time to consider nuclearity as a tool of statecraft. Uranium operations there were too minor to garner attention, shutting down within a few years of independence. Niger's President Diori sought to put a price on nuclear exceptionalism in order to increase state revenues, but he was ultimately foiled by political upheaval. Gabon's President Bongo was happy to capitalize on the desire of Iran and other states for nuclearity, but in his dealings with France he found it more useful to treat uranium as a

banal commodity. The Gabonese state, furthermore, displayed little inclination to regulate workplace conditions in nuclear terms. As a result, it took
decades for workers to find ways of using nuclearity to make citizenship
claims, and the success of those claims remains very much in doubt.

South Africa's apartheid leaders embraced nuclearity as a tool of diplomacy, but not as a means of regulating radiation risk. Nuclearity served
both the Namibian liberation struggle and Rössing's labor union in the
wake of independence. But it backfired as a technopolitical tool by nar-
rowing workers' options rather than increasing them. As we saw in chapter
9, this had to do partly with the structures of scientific knowledge production and partly with the unwillingness of the postcolonial state to confront
one of its largest economic engines. Another factor, as we'll see shortly,
was the challenge of building independent regulatory capacity when
the only local source of expertise lay with current or former Rössing
employees.

Establishing exposure standards requires expertise, as does effective
enforcement. Revolving doors between industries and their regulators exist
all over the world. In countries with large scientific, technological, and
legal infrastructures, the problems this poses for effective regulation can be
mitigated to varying extents by independent experts and courts. We saw
an example in South Africa when a small group of determined outsiders
succeeded in building a nuclear regulatory system that, albeit imperfect,
included mines.

Infrastructures in the other African countries we visited, however, did
not match those in South Africa. Heavy reliance on uranium to power
these nations' economies (and often line the pockets of their leaders) lends
these impediments to regulatory capacity even greater weight. Such challenges are by no means confined to uranium. They permeate the regulation
of extractive industries throughout Africa, shaping the practices of governance and the limits of citizenship.[1] Understanding and confronting these
challenges, however, requires attending to the particularities of time and
place. History matters.

THE CHALLENGE OF GOVERNANCE

Uranium communities in Niger have discovered all this in recent years
through Aghir In'Man, a local NGO headed by Almoustapha Alhacen, an
employee of the Somaïr uranium mine. Understanding the importance of

visibility, Alhacen invited French NGOs to Arlit in 2003 to measure radiation levels wherever they could. In market stalls, they found radioactive
scrap metal. At community taps, they found contaminated water. Alhacen
called attention to suspicious invisibilities such as retired workers suffering
from ill-defined diseases, and to the curious absence of *any* cancers from
mine hospital reports.

Areva insisted that "cancers are extremely rare. During 40 years of
mining, not one case has been detected that was thought to have been
caused by exposure to ionizing radiation. Cancer is an illness found mainly
in Western countries with elevated pollution levels and high consumption
of rich food, tobacco and alcohol."[2] Locals, however, did not see cancer
as a disease of the developed world. They feared that "elevated pollution
levels" existed right under their feet. Like workers in Gabon and Namibia,
Nigériens suspected that medical doctors employed by the mine had
neither the inclination nor the expertise to find occupational diseases.
More NGOs were formed. Tchinaghen is currently allied with the ongoing Tuareg rebellion, which demands a share of uranium wealth. ROTAB
focuses on issues of financial transparency and accountability.

In December 2009, a French television documentary on the health
effects of uranium mining prominently featured Niger. Greenpeace International subsequently reported numerous instances of rogue radioactivity
in the desert towns that supported the two gigantic Nigérien mines. Areva
had closed one well and removed contaminated scrap metal from the
markets of Arlit and Akokan after NGOs first raised the alert in 2003–04.
But Greenpeace scientists found high levels of alpha radiation and uranium
in some remaining wells. In Arlit, water used to dampen unpaved roads
for protection against high desert winds left uranium residues. Waste rock
used in road and building construction contributed to elevated radiation
levels. The supply of contaminated scrap seemed endless. The 2010 Greenpeace report quoted one Arlit resident who described the quotidian strategies for coping with poverty by refashioning mine trash into daily
necessities:

People buy scrap metal here to cover the houses and certain materials are used
domestically: they sell carts, ploughs. . . . All this is done with the scrap from the
mine! Axes, knives. . . . A lot of things are made right here! . . . This is the cover
of a soda barrel that comes from the sulphuric acid workshop at Somaïr. Here, in

town, people use it to collect water. They bring this downtown and sell it to women [who] then boil the water for washing.[3]

Greenpeace couldn't resist a conspiratorial explanation for France's search for uranium in Africa. "One of the original incentives," it asserted, "was the growing concern over health in France." The report quoted the conviction of one Arlit resident that "the white men, when they were here, they knew."[4] It did not mention that uranium mining had continued in France and that French employees in both the metropole and Africa had also fallen sick.[5] In reducing Arlit's history to racist neo-colonialism, the report missed the transnational implications of struggles over occupational health compensation in France, where experts, labor, and civil society organizations battle to broaden the scope of *Tableau 6*, the schedule of radiation-induced occupational diseases eligible for compensation.

In Gabon, Areva touts a compensation scheme that serves as both historical accounting and future promise. In July 2010 the company's chief medical officer, Alain Acker, explained to the weekly *Jeune Afrique* that before mounting Mounana's health observatory the company had to catalog the treatments and patient outcomes for all diseases encountered at the COMUF hospital since the 1950s. "We also identified all the former mineworkers who are still alive, over 1000 people, their living conditions, and the exposure that they received," Acker continued, suggesting that Areva *had* kept all such records. So, asked *Jeune Afrique*, how would workers be compensated in case of contamination? "If diseases appear that are linked *in a certain and direct manner* to the exploitation of the mine," Acker replied, "we commit ourselves to bearing the costs following the French model for occupational disease compensation."[6] According to Areva, causality is readily determined, and metropolitan models of occupational disease compensation are generous, fair, and accessible.

However inadvertently, NGOs working in Africa risk nurturing these fantasies with facile invocations of neo-colonialism and simplistic oppositions between black and white health. That *nuclearity* often discriminates along axes of class, race, and geography becomes abundantly clear when we view it as a category through which power relations are constituted and refracted. *Radiation*, however, doesn't discriminate. Just ask the families of white workers who tried to sue Rössing in British courts, or former

French employees of Somaïr, Cominak, or COMUF who are currently claiming compensation.[7]

Greenpeace's 2010 Niger report couldn't resist repeating the punch line from its other campaigns: "Dangerous and dirty nuclear power has no role in our sustainable energy future." We might recall the words of Rössing's Harry Hoabeb from chapter 9: "This Greenpeace, *neh*? You go to Greenpeace, they will tell you another story." And his colleague's caution: "Once you start getting involved with external people—what do they want? . . . You get help and whether it means something to you in the end you [are] not sure." In northern Niger, electricity is a privilege, and the extreme vulnerability of people and ecosystems render tensions between economic development, public health, and environmental degradation particularly acute. The Greenpeace dream of a global non-nuclear future probably has little resonance there. It's very unlikely that Somaïr and Cominak workers would choose to shut the mines down altogether.

Nevertheless, local and international NGOs *are* serving to build the nuclearity of Niger's uranium mines. They supply instruments, experts, and media that can help make nuclearity politically, economically, and medically useful for everyday Nigériens. For instance, Areva responded to the Greenpeace report by initiating a radiation survey of Arlit to identify contaminated dwellings. It also launched a program to control dust by paving the town's streets.[8] Yet nuclearity in Niger remains fragile, its power uncertain. Then there's the problem of accountability, and the related problem of regulation. Niger typically ranks at the bottom of the UN's development index (182nd out of 182 in 2009, right below war-torn Afghanistan). Under these conditions, fashioning a robust regulatory system with the necessary complement of experts, instruments, and laboratories presents colossal challenges.

In Niger, obstacles to building regulatory capacity are compounded by widespread corruption. For decades, most observers agree, uranium revenues flowed primarily into fancy buildings in Niamey and the private bank accounts of government officials. Under these circumstances, can *any* Nigérien state agency maintain the independence required for credible expertise? Consider, furthermore, that Niger's Centre national de la radioprotection (CNRP), a state regulatory body for radiological hazards, was not established until 1998—three decades after the creation of the Somaïr.

Over a decade later, one of its officials apparently told Greenpeace that the CNRP still didn't possess instruments that could measure radon levels. Anticipating that two (two!) such detectors would arrive soon, he expressed faith in the expertise and transparency of radiation protection at the mines.[9]

How much leverage can nuclearity offer in a place regularly ravaged by drought and famine? "Participatory" processes and institutions can make radiation exposure and other health problems visible and actionable. But they also run the risk of foreclosing discussion by generating overly narrow research, creating an illusion of adequate stakeholder representation. Recall from chapter 7 the agreement signed by Areva, Sherpa, and Médecins du Monde to create health observatories overseen by a pluralist group of experts in Gabon and Niger. Neither Sherpa nor MdM has in its ranks people with the nuclear expertise to inform and verify the work performed by the observatories. Nor does the pluralist group include anyone sensitive to local power dynamics.

The observatory arrangements allow state officials rather than labor unions or local civil society to represent workers and residents. Yet Mounana residents still mistrust the state and the radioactivity measurements generated by its radiation protection board.[10] They obtained a modest victory in March 2011 when Areva agreed to demolish 200 radon-infused houses. But the resumption of uranium prospecting in southeastern Gabon leads area residents to suspect renewed collusion between the corporation and the state. Until tensions over knowledge production and public representation are addressed, any nuclearity produced by the observatories is unlikely to have significant political, medical, or economic utility for workers and communities.

Similar dilemmas confront Madagascar, which was not part of the agreement between Areva and French NGOs. Uranium exploration has resumed there too. In 2007–08, a survey reported elevated radiation levels in and around the sites abandoned four decades earlier by the CEA. Unsurprisingly, the Ministry of Energy and Mines had not known of the high levels because the Malagasy state had no involvement in uranium extraction in the 1960s. The old quarries are now filled with water, which local residents drink. The team "strongly recommended" prohibiting human use of the water and instituting a monitoring program.[11] In the wake of intense political upheaval and several deadly tropical storms, it's hard to imagine contaminated water in a still-marginal part of the country receiving much attention.

Namibia now faces new challenges posed by the massive expansion of its uranium industry and by its declared intention to open a nuclear power plant around 2025. In late 2005, Rössing refurbished its operations and opened a new quarry. The following year, the Australian company Paladin opened a mine 50 kilometers south of Rössing in the Namib Naukluft Park. Areva has joined the action with its Trekkopje mine, and others are in line to follow suit. These companies have banded together to form a national Uranium Institute "dedicated to the safe and environmentally friendly exploration and mining of uranium in Namibia." Another tag line declares: "We are protecting the Namibian Uranium Brand."[12] Russia, for one, sees considerable promise in that brand. In May 2010, Russian president Dmitri Medvedev and Namibian president Hifikepunye Pohamba agreed on an investment of up to a billion dollars. Resurrecting Cold War, anti-imperial geographies, both heads of state invoked Soviet support of SWAPO as they signed.[13]

Debates over the form and substance of Namibian nuclearity are in full swing. The prospect of nuclear power prompted the Namibian state to appoint its first Atomic Energy Board in February 2009. Chaired by Wotan Swiegers, Rössing's former medical doctor, the new board intends to discuss the "advanced application of nuclear technology" in Namibia and issue recommendations concerning radiation protection. Following the IAEA's example, however, the corresponding legislation specifically excluded uranium ore from its definition of "nuclear materials." Licenses for uranium mines remain under the purview of the mining ministry, while environmental management falls to the Ministry of Environment and Tourism. Legal experts draw attention to the disjointed character of this regulatory system, which have led to ministerial overlaps in some domains, and absence of oversight in others.[14] Activists for Earthlife Namibia and its South African counterpart insist that "the continued treatment of uranium mining like any other mining activity is problematic" and call for legislation specifically dedicated to regulating uranium.[15]

Both within and outside Namibia, uranium production has caught the attention of civil society. In 2008, the Namibian Labour Resource and Research Institute (LaRRI) published a report on legislative control and occupational health in uranium mining. Apparently unable to interview employees of Paladin's new mine, LaRRI focused on Rössing. The workers cited in the report praised the company's safety record but made clear that they mistrusted its doctors and harbored suspicions about long-term health effects. Unverified stories still circulated about high rates of diseases among retired Rössing workers. One employee suspected that the company offered early retirement to sick workers, who then learned of their illnesses only after independent medical consultations. A union representative echoed laments about the difficulty of linking diseases to uranium exposure: "How can we give proof? What was reported by Dr Zaire is crucial information, which nobody wants to come forward to say if it is true."[16] The LaRRI report described neither Zaire's work nor the subsequent verification study, painting instead a picture of occupational health concerns that had not changed in three decades.

These ahistorical narratives remind us of the complexity and fragility of nuclearity's distribution. The lingering echoes of Zaire's report represent the hope that nuclearity will lead to better conditions, more transparency, and post-employment compensation. Overlooking the conflict and limitations generated by Zaire's work and the genuine improvement of working conditions since the 1970s, however, makes it easy to forget that, without a national medical infrastructure to generate occupational health data, the power of nuclearity to facilitate such outcomes is limited. Nuclearity has to be actively maintained and updated in order to offer any sort of traction. Workers, even unionized ones, possess few resources to do so.

Unlike Greenpeace, LaRRI did not advocate abolishing nuclear power. Workers, it explained, were "not anti development, nor do they want to see themselves and their loved ones languishing in poverty." Rather, Namibia needed to ensure that companies displayed "more honesty and transparency . . . on issues of radiation." One worker suggested legislation that would place a 20-year cap on employment in uranium mines and ensure full medical coverage thereafter. Another proposed routinizing compensation claims. LaRRI itself called for improvements in tailings disposal. Such measures were "especially crucial if Namibia is to become one of the top uranium producers in the world. All eyes will be on Namibia, because an evaluation of a country's performance can be judged on how a country treats the most vulnerable."[17]

Other African countries have only just embarked upon their uranium adventures. Malawi's first uranium mine opened in 2009. Plans are

underway for mines in the Central African Republic, Mali, Tanzania, and
Zambia. The challenges of robust technopolitical governance in these
countries have barely begun.

PRICING THE URANIUM BOOM

The driving force behind the African uranium boom, of course, is demand.
More precisely, expectations of demand. By the turn of the century, the
market devices traced in chapter 2—the OECD Red Books, the Nuexco
spot price, and others—had become well-entrenched forecast engines.
They predicted that existing mines would not meet long-term reactor
requirements once stockpiles of yellowcake were drawn down. But skittish
investors, skeptical about the future of nuclear power, proved reluctant to
finance mine development. When I visited the Rössing mine in early 2004,
the spot price of uranium oxide—by then firmly entrenched as an indicator throughout the industry—hovered around $16 per pound. That was
up from $7 in 2000, but still alarmingly low. There was talk of closing the
mine. Workers worried about losing their jobs.

Then concerns about climate change coalesced. The nuclear industry
embraced global warming as an apocalyptic scenario with enough prob-
ability and immediacy to trump nuclear exceptionalism. The industry
offered visions of a "nuclear renaissance" free of greenhouse gases. Nuclear
power began to seduce African governments: Ghana, Niger, Nigeria,
Senegal, Sudan, and others expressed desire for nuclear reactors to address
energy needs.[18] The industry projected exponential growth worldwide.
French president Nicolas Sarkozy made nuclear power a linchpin of
France's foreign policy, signing nuclear cooperation agreements with China,
Libya, Morocco, Turkey, Saudi Arabia, and others.

Currently engaged in a massive expansion of its nuclear power program,
China needs some 10,000 tons of uranium oxide to attain its 2020 target.[19]
In stepping up uranium exploration to meet this demand, Chinese com-
panies have targeted Niger, Namibia, and Zimbabwe, as well as Mongolia,
Uzbekistan, and Kazakhstan. The broader Chinese challenge to French
economic dominance in Francophone Africa has led some commentators
to pronounce the death of la Françafrique and the birth of la Chinafrique.[20]
But China is not the only country eyeing African resources. India, Japan,

and Canada have also invested heavily in new uranium projects. Many in Africa's political elites relish this competition.

In Niger, Chinese companies landed in the middle of the struggle between Tuareg rebels and the national government over resources and rights in the north. In July 2007, the main Tuareg rebel group, the Mouvement des Nigériens pour la Justice (MNJ), kidnapped a Chinese uranium executive and accused his company of financing the government's weapons purchases. A few weeks later, Nigérien president Mamadou Tandja accused Areva of financing the MNJ and expelled its head of operations. Areva denied supporting the MNJ, which had attacked its outposts and employees.[21] As the situation simmered down, an emboldened Tandja demanded higher yellowcake prices from Areva. His government awarded the company a license to mine an enormous new deposit at Imouraren, much to the chagrin of the company's Chinese competitors. Although he met with greater success manipulating uranium production than Diori had in the 1970s, Tandja himself ultimately fell to a coup d'état in February 2010.

Areva continues to control most of Niger's uranium production, though Chinese companies have made inroads. Somina, a joint venture with majority Chinese capital, has begun to mine the Azelik deposit in Niger's Agadez region. Even before the first ton of uranium emerged from the ground, workers and NGOs protested low pay, bad housing, dangerous working conditions, and poor sanitation. According to one report, Azelik has become known throughout northern Niger as "Guantanamo."[22] As of this writing, no one has mentioned radiation there.

Oceans away, investors in North America, Europe, and Australia enthusiastically embraced uranium's potential for profit. Fantasies about a "nuclear renaissance" began to push up the spot price. From $16 in January 2004, the price of a pound climbed to $26 in 2005, $37 in 2006, then $75 by January 2007. In May 2007, the New York Mercantile Exchange opened the first futures market for uranium. Had uranium finally become completely banal, traded on the same terms and in the same ways as any other commodity? In June 2007, the month after futures trading began, the spot price hit $136 per pound before tapering down to its 2010 low of about $40 per pound.

Wall Street traders salivated at the prospect of uranium futures. At long last, they crowed, there would be "liquidity in the uranium market" and "visibility into pricing," a promise particularly attractive to the non-initiated.[23] Skeptical industry insiders, those who had created the market devices that managed visibility, saw the futures market as a challenge to their power. Futures contracts didn't necessarily reflect the trading of physical uranium. They offered a way to hedge bets, not buy yellowcake directly. The CEO of TradeTech (Nuexco's new incarnation) warned darkly that "the NYMEX futures market being purely a financial instrument, it runs the danger of diverging significantly from the physical market."[24] A world in which uranium was as banal as the housing mortgages then being repackaged into derivatives and collateralized debt obligations seemed far indeed from the intricate politics of the northern Nigérien desert.

The relationship between uranium's "futures market" and its "physical market" remained murky to many observers. When the financial system crashed and it came to light that the bankrupt investment firm Lehman Brothers held 450,000 pounds of yellowcake among its unsold assets, few understood the significance. Nevertheless, the lead sentence in many media reports was guaranteed to attract attention: Lehman was "sitting on enough uranium cake to make a nuclear bomb."[25] Sensationalist, certainly. The yellowcake remained parked in the plants that had produced it. But the tag line signaled the enduring tensions between banality and exceptionalism. With escalating concerns over Iranian enrichment and the looming menace of "nuclear terrorism," how could uranium—whether futures or actual yellowcake—be subject to such casual commerce?

The IAEA had already reconsidered its position on the (non)nuclearity of uranium mines and mills. As we've seen, early IAEA inspections were limited to verifying states' declarations concerning their "nuclear material and activities." But in the early 1990s, revelations that inspections had missed secret Iraqi and North Korean weapons programs prompted the agency to expand its purview. Its 1997 "Additional Protocol" specified a set of technopolitical practices intended to increase inspectors' ability to detect *undeclared* nuclear activities. Though by no means the primary targets of the new measures, uranium mines and yellowcake plants became subject to international oversight for the first time.[26] Adoption of the Additional Protocol remained voluntary, but a state subjecting itself to these more intrusive inspections gained a more definitive proclamation about the (non) existence of weapons programs or intentions. By reformulating the parameters of nuclearity, inspections could better exonerate a nation from evil intentions.[27]

TRIGGER LISTS, TERRORISM, AND TRAFFICKING

Published in 1974, the first "trigger list," a catalog of things nuclear enough to activate safeguards and inspections, was brief. For instance, it mentioned isotope separation plants—a keystone of uranium enrichment—but offered only a general description of their components.[28] In the mid 1970s, enrichment plants were rare, the commercial potential of different designs subject to debate (as we saw in the case of South Africa). There was also the assumption that nuclear expertise belonged to the "first world." Noting that Pakistan's uranium enrichment plant was cleverly assembled from components obtained in many different countries, physicist David Albright writes that "minutely focused regulations were not considered [in the 1970s] because, in part, of the first-world belief that third-world countries like Pakistan wouldn't be able to construct a full-scale nuclear facility."[29]

By the time of the Indian and Pakistani bomb tests in 1998, IAEA trigger lists devoted over 20 pages to itemizing plant components. The catalog included such items as shut-off valves and rotary shaft seals "especially designed" to handle uranium hexafluoride gas. It even specified the tolerances and diameters of such things.[30] This increased specificity, however, didn't resolve much. Unless a nation signs an Additional Protocol requiring it to supply information on the location, status, and production output of uranium mines and yellowcake plants, the IAEA cannot legitimately demand more information and inspections. Niger signed one in 2004, the year after George W. Bush's uranium-from-Africa claim. It came into force in May 2007, just as the MNJ had begun to translate Tuareg demands into uranium politics. Mine *inspections*, however, remain a very low priority for the IAEA, which is perpetually strapped for funds.

The Additional Protocol may have re-nuclearized uranium mines in principle, but in practice yellowcake rarely counts as a "nuclear supply." Most tellingly, the trigger lists used by the Nuclear Suppliers Group continue to exclude yellowcake, saying only that "natural uranium . . . should be protected in accordance with prudent management practice."[31] The NSG clearly states that its primary purpose is to "facilitate the development of trade,"[32] in part by defining the boundaries between licit and

illicit transactions in light of the NPT regime. In the course of this boundary work, the NSG arrogates for itself the right to make exceptions. For
example, in 2008 its members "agreed to exempt India from its requirement that recipient countries have in place comprehensive IAEA safeguards covering all nuclear activities."[33] This renders nuclear sales to India
legitimate despite the country's abstinence from the NPT regime, enabling
reactor component manufacturers in the US and Europe to profit from
the Indian nuclear renaissance. Redrawing the contours of licit trade will
doubtless prove essential for these manufacturers in the wake of the 2011
Fukushima accidents, which will likely dampen nuclear power expansion
in the US and some parts of Europe.

In any case, neither the Additional Protocol nor the NSG stopped
Abdul Qadeer Khan, the metallurgist who ran Pakistan's uranium enrichment project and other systems related to nuclear weapons, from operating
the largest and most successful nuclear smuggling operation in history. In
2004, Khan confessed to selling nuclear weapons designs and materials to
North Korea, Iraq, and Libya. Investigations revealed that his network
included nodes in Switzerland, Malaysia, and South Africa.

In response to growing concerns about nuclear terrorism, political
instruments aimed at curbing proliferation and securing "nuclear materials"
continue to multiply.[34] Notably, UN Security Council Resolution 1540
"imposes binding obligations on all States to establish domestic controls
to prevent the proliferation of nuclear, chemical and biological weapons,
and their means of delivery, including by establishing appropriate controls
over related materials."[35] Adopted in 2004, UNSCR 1540 spawned a new
set of committees and reporting structures. But national reports to the
UN's 1540 committee were anything but standardized at first. Reflecting
the broad spectrum of legislation and capacity of the UN's member states,
early reports ran the gamut from long narratives generated by countries
such as the US and France to the three sentences offered by Namibia,
which boiled down to this: "Namibia does not produce weapons of mass
destruction and therefore cannot provide material support to States or
non-State actors to produce or obtain such weapons."[36]

After the theft of 170 kilograms of uranium oxide from its Final
Product Recovery section in September 2009, Rössing lost any sense of
complacency. The company caught the perpetrators, who turned out to
be FPR employees. This wasn't the first attempted yellowcake theft at

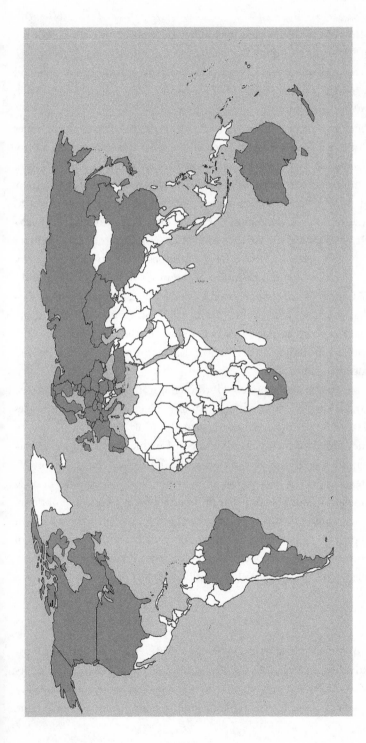

Member states of the Nuclear Suppliers Group. In a classic technopolitical tautology, the NSG includes only (certain) countries in a position to export technologies on its trigger lists. South Africa is the only African member. (source: Wikimedia Commons, en.wikipedia. org/wiki/Nuclear_Suppliers_Group)

Rössing. But this time the new geopolitics of terrorist threat led Rössing to request US help in tightening security measures. Among other things, the company reassured the US government that it had frozen dividend payments to Iran (which still holds 15 percent of the capital), and that it had not, would not send any uranium there.[37]

Nevertheless, concern about the security of African uranium continues to build. In April 2010, an Associated Press story that was picked up by news outlets worldwide reported that most African uranium suppliers were not in compliance with UNSCR 1540.[38] African ore, the story warned, could end up in a "dirty bomb," a device that uses ordinary explosives to spread radiological contamination.[39] Even worse, in the hands of "financially motivated smugglers" African ore could wind up in a "state with an illicit weapons program." (Notice the tacit opposition between the purity of non-proliferation and the pollution of the profit motive.) A few months after the AP story, WikiLeaks released US embassy cables concerning allegations that uranium ore—mined "artisanally" in Congolese sites abandoned a half-century earlier—was being sold on the black market. One embassy official commented that "all of Katanga Province could be said to be somewhat radioactive."[40]

Based on the analysis presented throughout this book, we can see uranium trafficking as the profound systemic effect of historical struggles over nuclearity. From the form those struggles took in African places, we see the unevenness of nuclearity, the range of its meanings and manifestations, the tremendous disparities in priorities, needs, and capacities of different states. Trafficking betrays the risks of not treating uranium-bearing ore as a nuclear thing. The fact that even a handful of Rössing employees believed they could sell stolen yellowcake the way others might sell stolen diamonds or gold is a measure of uranium's banalization. The tiny lots of uranium ore marketed by hopeful Congolese entrepreneurs illustrate how context shapes the meaning of licit or illicit trade.

RADIOACTIVE RESIDUES

In order to control illicit traffic in nuclear things, the UN offers technological and administrative assistance to states that request help implementing UNSCR 1540. As with all aid, this comes with strings attached. For example, assistance contracts might include specific pieces of equipment, such as "radiation monitors, purchasable by recipient State after completion of activity, subject to valid export licence from Government of the provider of assistance."[41] In other words, states can purchase radiation detectors—under controlled conditions—for the purpose of detecting rogue uranium at their borders. In this formulation, nuclear security and radiation exposure remain distinct spheres of operation. The tools of visibility remain tightly regulated, difficult to access, and geared to the needs of international experts. Radiation as evidence of security threat attracts much more attention and funding than radiation as cause of ill health.

Consider the Democratic Republic of Congo, where rumors of uranium trafficking often originate. It was cobalt and copper, not uranium, that pulled some 6,000 "artisanal" diggers to the Shinkolobwe site in the late 1990s.[42] Nevertheless, reports of illicit uranium trafficking and child labor began circulating almost immediately. After a fatal mine collapse in 2004, the DRC government closed the mine, evacuated and burned the miners' village, and requested international assistance in evaluating the site. A UN team spent two weeks there, barely enough time to make a preliminary list of problems. Though the team found "no large scale environmental consequences" and "no acute radiological risk," it did conclude that "chronic exposures . . . are expected to have been in excess of the accepted international safety limits for radiation workers" and that underground miners had likely experienced "significant dust and radon inhalation."[43]

Mining corporations, violence, poverty, and corruption have a strong sway over the DRC—especially in Katanga Province, which holds some of the world's richest mineral reserves. Sociologist Marie Mazalto writes that "the weakness in the capacity of the state to control mining areas makes it almost impossible to implement a coherent mining policy, under the effective control of its own institutions."[44] One study estimates that several thousand children worked as diggers in Katanga's copper-cobalt deposits in 2006. A US embassy official in Kinshasa reported in 2007 that, even though "dangerously high levels of radiation" had been found at one such mine, the operator "was suppressing this fact to continue mining operations."[45]

In 2009, Congolese and Belgian public health researchers published initial findings from their study of human exposure to metals in the Katanga region. The results showed high urinary concentrations of metals, including uranium, and especially in children.[46] This suggests that the

Young diggers resting on sacks of radioactive ore, Democratic Republic of Congo, 2004. (© UNEP/OCHA Environmental Unit)

environmental and health consequences of mining were far more significant than suspected by the UN team, which had measured levels in places rather than in people. But public health researchers must tread carefully. As one of them told me, there are powerful interests arrayed in favor of mining as usual.

The recent experience of Katanga diggers sharply illustrates the temporal revenge of radioactivity. Half a century after Shinkolobwe closed, radiation and uranium still contaminate Katanga. Though other issues of human security in the region are more immediately pressing, it's worth remembering that exposures are significantly more noxious for children than for adults. Many diggers may die of other causes before contracting radiation-induced disease. For those who survive, however, exposures can be a ticking time bomb that may overwhelm the severely strained medical facilities. To this day, the Shinkolobwe site has not been remediated. Yet the bigger danger may lie elsewhere. Uranium in Katanga is now considered a mere "impurity" in cobalt-bearing ore. As thousands of informal diggers transport and store that ore, they unwittingly spread radioactive contamination much farther afield. As in the South African gold mines, the conjunction of uranium with other geologically banal minerals readily translates into invisible exposures.

The situation is not hopeless. New networks of regulation and knowledge production have begun to form. The Forum of Nuclear Regulatory Bodies in Africa was launched under IAEA sponsorship in late 2009 with a primary goal of coordinating information and resource exchange among African regulatory agencies. Its mandate includes both nuclear security and occupational health regulation. Led by the Namibian representative, one of its working groups is tasked specifically with attending to radiation safety in uranium mining and milling.

On the NGO end of the spectrum, several organizations have come together under the umbrella of the African Uranium Alliance. The AUA builds on earlier anti-uranium networks, such as those mentioned in chapter 9, but includes many new African NGOs, among them the Nigérien organization Aghir In'Man. Focusing on the current African uranium boom, the alliance juggles familiar tensions. Although fundamentally anti-nuclear in outlook, it aims to educate those who are desperate for work about the possible hazards of mining. It also tries to assist local NGOs in gathering knowledge about the health and environmental aspects of uranium production. Both the Forum of Nuclear Regulatory Bodies in Africa and the African Uranium Alliance seek to bridge communication gaps between Francophone and Anglophone Africa. At this writing, it remains too early to evaluate their impact.[47]

Struggles over nuclearity permeate our radioactive world. Consider that throughout the US hospital patients have absorbed excessive radiation doses during CT scans and other diagnostic procedures because technicians were unaware that increasing the power of the beam to sharpen an image also intensifies radiation to harmful levels. The enthusiasm of hospital administrators and diagnostic instrument makers for corporeal visibility through radiation thus turned nuclearity into an invisible "side effect."[48]

Or take the Mayapuri district of New Delhi, where scrap metal dealers were radiologically contaminated by laboratory equipment discarded by the University of Delhi. Such equipment could just as easily have come from Russia, Canada, Japan, or myriad other places that export their waste to India. The governance of radioactive waste depends on classifying some material as nuclear enough to require special treatment, while other material remains ordinary enough to be disposed of by "conventional" means. Out there in the messy world, however, one person's trash is another one's livelihood, and seemingly ordinary scrap metal can be more lethal than vitrified plutonium stored in hermetic casks surrounded by concrete.

Out in the messy world, limits on exposures and protocols for circulating nuclear things help to manage danger. They serve as resources for workers seeking better conditions, for regulators or activists hoping to curb corporate excesses. They help Interpol agents prevent the smuggling of lethal materials.

But even the best standards are not, cannot, *should not* be considered sufficient to regulate the movements or hazards of nuclear things. It's not just that implementation varies by place, or that standards are always renegotiated and revised. It's also that the power of nuclear things depends on *both* exceptionalism and banality.

As we have seen in Gabon, Namibia, and South Africa, making radiation exposures visible requires not just special instruments and experts but also ordinary medical infrastructures and statistics. Making those exposures matter politically, socially, and economically requires understanding their place in the ordinary ordeals of existence: the daily tribulation of fetching water, the need to battle multiple diseases simultaneously, the inaccessibility of adequate health care, the fight against corruption, the struggle for political recognition. The power of nuclear things operates within the complex matrix of these travails.

The relative urgency of these needs does not mean that nuclearity can remain a "first world" luxury. The ICRP's ALARA philosophy is often seen as a license to use divergences in living standards to justify differences in valuing human life. But this interpretation only perpetuates social and economic inequalities. In effect, it constitutes a bet that people (especially, but not only, African workers) will die of something else before they develop cancer or other radiation-induced diseases. It ignores the fact that radiation exposure can make the ordinary travails of existence more acute. It leaves no room for the possibility of recruiting nuclearity in service of better governance.

Most African nation-states have more pressing preoccupations than radiation exposures or the unregulated circulation of radioactive materials. Their problems tax meager state resources even without the added burden of corruption. Major exporters of uranium—Namibia, Niger, and now Malawi—count on mining revenues to address these urgent needs. A house built of mine tailings might keep out the torrential downpours of Gabon's rainy season better than one built of wood. But if buildings that continually emitted radon were roundly condemned and demolished in Grand Junction, Colorado, they should not have counted as adequate shelter in Mounana, Haut Ogooué. The delayed nature of radiation-induced cancer should not serve as justification to spend less money on the mechanisms and infrastructures of visibility, prevention, or treatment.

Fuel, we often hear, contributes only a small amount to the cost of nuclear power. As we've seen, however, uranium prices have not included the full costs of health and environmental monitoring. Mining industry leaders complained repeatedly that tighter regulatory standards would put them out of business, fretted endlessly about government regulation impinging on commercial competition. But even the most responsive companies (like Rössing) can bear only part of the cost of regulation. Standards, after all, are effective only when enforced. And enforcement, in turn, requires states to develop significant technopolitical capacities, distributed into agencies and experts, labs and clinics, measuring devices and diagnostic tools, national registries and international data. How much would electric bills rise in the US or Europe if the price of uranium incorporated the full cost of nuclearity in Africa? That calculation remains to be run.

Nuclearity in one register doesn't easily transpose to another. Geopolitical nuclearity doesn't automatically translate into occupational nuclearity. Yet these domains remain connected. African uranium miners depend on the transnational movement of nuclear things, but that movement also depends on African miners. Ultimately, nuclear security must be considered in tandem with other forms of human security—food and health and environmental and political security.

The power of nuclear things has a price.

APPENDIX: PRIMARY SOURCES AND THE (IN)VISIBILITIES OF HISTORY

Several years ago, a graduate student came to me to discuss a research project in contemporary African history. After we'd gone over his archival options, I asked how he planned to conduct oral interviews. He sighed, "Don't you think that oral histories are kind of fetishized in African studies?" Moving with scholarly trends, he chafed against automatically pairing African history with oral sources. "I mean, you wouldn't be asking about oral interviews if I were doing a project on France, would you?" His reflex was that of the discipline-bound historian in training: Why bother with oral interviews if there are documents at hand?[1]

When I first began my dissertation on the history of French nuclear power, over 20 years ago, no official archives were open. That project, and the book that ensued, was built on a foundation of oral interviews. Documents filtered in through strokes of luck, scavenged from the closets of nuclear power plants, the offices of engineers, the file cabinets of labor unions. Officialdom let me peek at its archives eventually, but the most interesting material came from elsewhere.

In researching this book, I adopted a similar approach. I visited mine sites in Gabon, Madagascar, South Africa, and Namibia. I interviewed workers, managers, engineers, doctors, and residents. I went underground into mine shafts and rode in the enormous haul trucks at open pits. I plugged my ears as I witnessed giant ore crushers, and tried to remain unfazed when acid fumes attacked my sinuses in yellowcake plants. I wandered through company towns, talked with people in their homes. I dug through closets and file cabinets and storage rooms in search of documentation.

Interviews do not offer an unmediated window on events. But neither do documents. Both types of sources are products of their times and circumstances. Both make certain events visible while leaving others invisible. They are generated for particular audiences, and the stakes and nuances of their narratives are often difficult to discern. I tried to remain attentive to these issues and to use each type of source to inform my interpretation of the other.

As historian Antoinette Burton has observed, the "backstage" of historical production is rarely visible to readers.[2] The coherence of written history masks the inherent unruliness of research. That is particularly true for this book. Throughout the text, for example, I have cited by name only a few of the people I interviewed, allowing their stories to stand in for those of the multitude. Similarly, I have cited only a small portion of the more than 50,000 pages of archival documentation I consulted. Fitting all that evidence into a single book was impossible. Because of the unorthodox character of some of "my" archives and the wide range of oral interviews I conducted, it seems especially important to let readers glimpse the backstage action. A complete list of 138 interviews is included in the bibliography.

First, the official archives. The "nuclear age" has produced endless piles of publications and reports. Official state archives are notoriously secretive about their nuclear materials and take different approaches to releasing information. Industry leaders and policy makers prefer to keep documentation tightly under wraps. "National security" is the mantra of the nuclear age, secrecy its reflex. Of course, archival silences stem as much from choices about *what* to preserve as from overt classification and secrecy.

Of all the places I worked, France was the most guarded. From the CEA down, none of the French state institutions let me see much. Because of my earlier book, I knew what to expect. First, archivists would ignore the query. Then, if I managed to find someone to exert pressure on my behalf, they would respond with a meager list of documents, claiming that this represented the full extent of what they'd found using the search terms I'd provided. *Ah non madame*, you may not search the inventory yourself. *C'est défendu*—not allowed!

All the documents that interested me required a *dérogation*, a formal request to release the documents for consultation. *Dérogation* queries sit on desks until archivists feel like passing them on to the office that issued the documents. There someone must be found to look at the files and determine whether they can be released. Months or years later,[3] the verdict comes back. Good news! Of the twenty cartons you requested, you can see two! We're sorry, but we can't issue a verdict on ten of the cartons, because they're in storage and we just don't have the personnel to dig them out. *Vous comprenez, bien sûr?* Oh yes, I understood.

Hardened by long experience in France, I discovered in Britain a dizzying contrast. The National Archives hold an enormous collection of uranium-related files deposited by the UK Atomic Energy Authority. The inventory is posted online. Every document older than 30 years is freely available, and obtaining permission to consult more recent documents takes a couple of months. Wow! Yet the documents themselves said little about many of the things I cared about—labor, worker health, African perspectives.

Then there was South Africa. When it dismantled its atomic weapons program, the exiting apartheid government destroyed thousands of documents. (This wasn't exceptional—as many have remarked, the apartheid government made destroying records something of a habit.[4]) What remained proved more accessible than French archives, if less so than British ones. The National Archives of South Africa contain a sizable cache of freely available files from the 1970s and earlier. The Atomic Energy Board's successor institution, Necsa, keeps its presumably larger hoard mostly to itself. The chunks that were trotted out for my inspection covered more sensitive material than similar documents from France. Still, a lot was missing.

Adding all this up, state-sanctioned nuclear archives promise much more than they deliver. One must attend carefully to their limitations. I determined that sticking to publicly available material would have produced a nuclear history with the same silences, the same assumptions, and the same protagonists as all the others. I had to break out beyond the archival walls.

In 1998 I spent time in France tracking down engineers, geologists, and other experts who'd run uranium operations in Gabon and Madagascar. Robert Bodu, the metallurgist who'd inspected the Androy mines, gallantly introduced me to his fellow retirees. Some had kept a few documents. One had hundreds of photos lining his suburban basement studio. Dr. Jean-Claude Andrault, who'd built the hospital at Mounana, maintained a Paris apartment and a villa crammed with African art. He showed me an 8-millimeter film from the early 1960s that melded scenes of him slaughtering an animal he'd hunted with him removing a cyst from an African patient. "I juxtaposed those on purpose!" he exclaimed gleefully. "What do you think?" I didn't know what to say.

My next stop in 1998 was Mounana, just as the mine was getting ready to shut down. From the associate director to the workers contemplating their last paychecks, the mood was sour. What would happen once the last barrel of yellowcake was shipped out? The COMUF sponsored seminars about small business entrepreneurship, but employees who'd participated seemed pessimistic. Someone in the tourism ministry had the idea to turn the area into an ecotourism destination, following the example of a nearby manganese mine. At Mounana, a small hotel had been built on a hill overlooking the old quarry, which had been filled with water, stocked with tilapia, and flanked by an open-air restaurant. Ducks in the middle of this "lake" bobbed about but never budged from position. "They're plastic," one waiter laughed. "In the beginning they had real ducks, but the pythons ate them."

At first, the associate director at Mounana, a Frenchman who'd spent his career in African mines, didn't understand my request to "see historical documents." After he hauled out the company's annual reports, I managed to communicate that I wanted to see correspondence, memos, hiring records, and other such material dating back to the earliest days of operation. He escorted me to a storage room filled with papers stacked every which way, some in boxes or binders, others in piles. Could that be what I meant? Yes! In that case, he said, failing to keep the incredulity out of his voice, we have two more rooms like this. Here are the keys. We'll probably toss all this junk when we shut down next year.[5]

Confronting an archive in deplorable condition produces a mixture of joy and despair. It's exhilarating to see unprocessed documents, a relief not to be at the mercy of a bureaucrat. But the prospect of wading through so much material, and no doubt missing something significant, induced a measure of panic. Where to start? From the oldest stack I could find, I pulled out a file. Long-dead termites fluttered to the floor, followed by a cloud of fine dust as the document disintegrated in my hands.

Fortunately, most of the papers were in better condition, though they'd clearly been dumped in batches by clerks clearing their file cabinets over the years. With my husband's help, I sorted through them as best I could, copying anything that looked interesting. Promising threads would break. For example, I found radiation reports for the late 1960s and early 1970s, and for the early to mid 1980s, but material for other years never surfaced. This was the silence of chaos, born of the expectation that such records had no value, or at least no local value. I still don't know whether Areva headquarters has copies of these reports in its French repository.

Meanwhile, a Gabonese computer technician born and raised in the company town helped me find employees who'd been around since the 1960s. These workers (all men) were suspicious. Who was I? What did I really want? My husband's presence helped to persuade people that I was American, far better than being French in this instance. Some men opened up; others didn't. Some asked me to help them send a family member to America; others just seemed happy to tell their story. Some talked freely as they toured me around their workplaces; others sat stiffly in a room and confined themselves to answering my questions.

Some stories—Marcel Lekonaguia's petition to the state, the horrifying floods of 1970—emerged in both documents and interviews. Sometimes interviews suggested ways of reading documents "against the grain." For example, old memos detailed fines that residents would incur if they didn't follow rules in the company town, while in interviews workers complained that the company expected them to perform maintenance on their dwellings despite the fact that the building would revert to the company if they left the job.

Among the other places I visited, Namibia most closely paralleled Gabon, though the contrasts were noteworthy. I arrived at Rössing in January 2004, just three weeks after management had announced that the mine would shut down in 2007 because of plummeting uranium prices and the imminent depletion of the open pit. After a dramatic increase in uranium prices, the company later made a large capital investment to develop a second deposit and keep the mine open. At the time, though, I conducted interviews with many Rössing workers whose anxieties about losing their jobs inflected their recollections of the struggle to learn about radiation and dust exposures, as well as their memories of the Zaire affair. For example, several men said they'd accepted the results of the 2001 verification study because they saw no other alternative. The prospect of losing their jobs made them think about reopening debates over occupational health.

I talked to truck drivers, plant workers, union officials, managers, safety officers, and the former and current mine doctor—and to some of their wives. Unsurprisingly, the transition to Namibian independence was the most important historical marker in these conversations—not because people felt it was a clean break, but because many expressed dismay at the slow pace of change since then.

Savvy about outside "experts" and their agendas, Rössing workers questioned my motivations. After my collaborator (medical doctor Bruce Struminger) arrived, their questions focused on health issues—either theirs or those of the Navajo whom Struminger treated in the US. Some employees asked us to withhold their names, but most didn't. I was surprised at how openly some of them criticized Rössing, even during interviews conducted on company property. Perhaps the threat of unemployment left them feeling that they had nothing to lose. Perhaps they thought our work would help them in some way. Or perhaps—after decades of liberation struggle, union organizing, and contacts with outside experts—outspokenness had become commonplace at Rössing.

Whatever the case, I was even more surprised by the openness of Rössing's management. In contrast to his COMUF counterpart, the managing director in Namibia immediately understood my request for company archives, leading me to a well-lit room in the Swakopmund headquarters where hundreds of boxes were neatly arrayed on shelves. He apologized for the "disorder" (which compared to Mounana was a model of organization). Board meetings, sales contracts, internal memos, operations logs, reports on labor unrest, the company's Zaire files, and much more were at my fingertips, no insects involved. As in Mounana, I received keys to the Rössing archives and could come and go as I pleased. Even more was available at the mine site itself, carefully preserved in a clean, dry warehouse. These were the records of a company that expected accountability.

I believe that the COMUF director gave me access to files because he couldn't imagine that they contained anything of enduring relevance. I suspect that Rössing's director gave me access because the company wanted to display increased transparency, and a foreign academic offered a relatively unthreatening means of doing so.

In both Gabon and Namibia, my research had a certain temporal and spatial coherence. I visited the sites while they were still operational, and I conducted interviews and consulted archives simultaneously. Researching uranium production in Madagascar and South Africa presented very different challenges.

Because uranium operations in Madagascar had ceased in 1967, the mine archives had long since moved to France. The Cogéma inherited them when it took over from the CEA. Thanks to the good offices of Robert Bodu, I was able to see monthly operations reports, metallurgical and other technical documents written by Bodu and his colleagues, and assorted reports on the "social situation" at Ambatomika. I consulted these in a bare room at the uranium extraction facility in Bessines, in the Limousin region of central France. The archivist made clear that he was providing the documents under protest. Geological maps, which might have suggested the location of unexploited uranium deposits, were off limits. I might be an industrial spy, after all, and the Cogéma wanted to preserve its access to Malagasy ore should market conditions prove propitious in the future.

My best efforts at reading these documents "against the grain" revealed little about Malagasy experiences. I had met two Malagasy technicians who had made their careers with the CEA and now lived in France. Their perspectives were those of a rarefied elite. Clearly the only way to hear workers' voices was to travel to the Androy. This was my biggest gamble. There was no guarantee that I would find anyone 30 years after the fact, or that they'd talk to me if I did.

But I got lucky, thanks in no small part to Georges Heurtebize, a French geologist who'd first traveled to Madagascar in the 1960s. Heurtebize had fallen in love with the Androy, quit his job, and decided to make a life there. He'd become an amateur ethnographer of Tandroy society with a reputation for helping researchers. His generosity was overwhelming. With some advance scouting, he identified several former miners before I arrived. He met my husband and me at the Taolagnaro airport and accompanied us for the first few days of research. He helped me find a good translator, even assisting with some of the interviews and translations himself.

I conducted both formal and informal interviews with former miners and their families in the Androy. The uranium facilities had been dismantled, but one man took us to see an old quarry. Some of the roads we traveled—rocky, unpaved tracks where it was impossible to drive faster than 15 kilometers an hour—had been built by the CEA. Villages in the Androy are small and scattered, the roads connecting them poorly maintained by the postcolonial state. We based ourselves in Tranomaro, a town of about 1,000 located a little over 100 kilometers inland from Taolagnaro. Google Maps claims that the journey takes a little over two hours, but it took us closer to eight hours to make the drive in 1998.

Rumors of my presence traveled quickly. Be sure you stay in Tranomaro on market days, one man told me, because people will come to find you. Sure enough, they did. As I mentioned in chapter 7, many saw me as the latest of the *vazahas* interested in rocks. Some hoped I was an advance scout for a reemerging CEA. When they learned that I was American, they thought perhaps a US mining company had sent me to assess the rich mineral content of southern Madagascar. Although we brought no permanent jobs, we were a modest source of goods. We had several huge bags of rice, which we shared. We typically bought chickens or a goat from our hosts, which they cooked and served to the whole village. I reimbursed the expenses of men who traveled to see me on market days, adding a bit extra for their families.

My translator was Malagasy-born M. Abdoulhamide, whose parents had emigrated to the island from present-day Pakistan. He insisted that we call him "Monsieur Babou" just as everyone else did. An occasional worker in the sapphire mines, he had married a Tandroy woman who ran a small eatery in Tranomaro. Their ten-year-old daughter Simona accompanied us on trips to the villages, which helped to break the ice. I also employed a guide, Joseph Ramiha, a former CEA miner who helped us find the villages of other miners he knew. Although the interviews in Tranomaro were small affairs (usually just me, my husband, Monsieur Babou, and the interviewee), the interviews in outlying villages were a well-attended source of local entertainment. Sometimes we were surrounded by a ring of thirty or forty residents. A young man named Itirik spent over an hour trying to teach me how to light a cigarette with a flint and steel. My dismal failure at this task had his entire village in stitches.

Finally, South Africa. I conducted most of my research on South African uranium production during the year I spent there in 2003–04, though I had assistants working in archives and libraries before, during, and after that time. In addition to holdings at Necsa and the state archives, we explored a variety of mining industry archives. The most valuable were collections held by NUFCOR and the Chamber of Mines.

NUFCOR proved remarkably open. After I was vetted by public relations, the managing director welcomed me and arranged for a tour of the plant. I interviewed workers and technicians who'd been employed there since the early 1980s. A few months after that visit, I conducted another round of interviews, aided by Bruce Struminger and his video camera. Meanwhile the manager gave me access to the company's archives. Minutes from board meetings dating back to the Calprods days were still held in the main offices. Everything else was stored in the attic of the old clubhouse. Again I heard my favorite sentence: "Here's the key." Though not as well organized as the Rössing collections, the NUFCOR holdings nevertheless represented a significant step up from the COMUF. Papers were stored in labeled boxes, binders, and file cabinets, almost free of insects. These included sales contracts, plant operations reports, records of labor disputes, and much more.

The Chamber of Mines, however, was reluctant to give me access to its archives. The routine was familiar. My request was ignored. Records were on microfilm, but the reader was broken. My request had to be examined again. And so on. Yet while the archivist kept finding new ways to put me off, I was talking to scientists, engineers, and managers involved with uranium production. One chemist who worked for the Chamber's environmental division was a proponent of historical studies. He gave me access to his predecessor's files, which dated back to the late 1980s. This collection included reports on mine operations, minutes of meetings between the Chamber and the nuclear regulator, Chamber correspondence with the IAEA, and the like. Eventually one of the officials I'd interviewed put in a good word. After I left South Africa, one of my research assistants obtained access to the microfilms and made hundreds of copies.

I complemented written material by talking to experts who'd worked to make radiation in the mines visible. Struminger and I met at length with Shaun Guy, who not only gave us a riveting interview but also provided copies of documentation he'd collected during his time as a regulator. Phil Metcalf had moved on to a job at the IAEA, but he talked to me on the phone at considerable length. Dr. J. K. Basson declined an interview. Struminger and I spoke with a range of other experts in Johannesburg, Pretoria, and Cape Town, and I talked with several officials from the National Union of Mineworkers who'd begun to work on radiation exposure issues.

The manager of the newly created radiation protection division at AngloGold Ashanti arranged a visit to an underground shaft. Although we didn't climb into the narrow stopes or descend to the deepest (and hottest) possible level, we got a modest sensory feel for the long journey down into the heat, humidity, and noise of South African mines. That, in turn, helped me appreciate the stories that mineworkers told me.

I had planned to spend considerable time interviewing underground workers. But this proved much more difficult—and less fruitful—than I'd anticipated. One challenge lay in finding men who'd worked in the uranium-producing shafts. Uranium and gold came out in the same rocks, but not from all the shafts and not continuously. I did find a dozen men who'd worked in the right shafts at the right time. Unsurprisingly, given the history recounted in chapter 8, they had not heard of uranium. As far as they were concerned, they were just mining gold, but they had many interesting things to say about working underground, stories that deeply informed my understanding. Yet it rapidly became clear that if I were to pursue that route I'd be repeating work done by others who'd explored the recent history of South African gold mining with depth and insight. My material on South African mines therefore doesn't match the material on Gabon, Madagascar, or Namibia. Then again, "South African conditions," as we've seen, are quite distinctive.

Besides, let's face it. Mismatched evidence is the most common of epistemological challenges for historians—as it is for the knowledge producers I discuss in this book.

PUBLICATION HISTORY

Portions of chapters 1, 4, and 5 were previously published in "The Power of Nuclear Things" (*Technology and Culture* 51, January 2010: 1–30).
A portion of chapter 1 was previously published in "Nuclear Ontologies" (*Constellations* 13, 2006, no. 3: 320–331) and in "Negotiating Global Nucle-arities: Apartheid, Decolonization, and the Cold War in the Making of the IAEA," in *Global Power Knowledge: Science, Technology, and International Affairs*, ed. John Krige and Kai-Henrik Barth (*Osiris* 21, July 2006: 25–48).
Portions of chapter 3 were previously published in "On the Fallacies of Cold War Nostalgia: Capitalism, Colonialism, and South African Nuclear Geographies," in *Entangled Geographies: Empire and Technopolitics in the Global Cold War*, ed. Gabrielle Hecht (MIT Press, 2011).
Portions of chapters 7 and 8 were previously published in "Africa and the Nuclear World: Labor, Occupational Health, and the Transnational Produc-tion of Uranium" (*Comparative Studies in Society and History* 51, 2009, no. 4: 896–926).
A portion of chapter 7 was previously published in "Rupture-Talk in the Nuclear Age: Conjugating Colonial Power in Africa" in *Social Studies of Science* (32, 2002, no. 5–6: 691–728).
Much of chapter 9 was previously published as "Hopes for the Radiated Body: Uranium Miners and Transnational Technopolitics in Namibia" (*Journal of African History* 51, 2010, no. 2: 213–234).
The previously published material is reprinted with permission.

NOTES

CHAPTER I

1. David Albright, Iraq's Aluminum Tubes: Separating Fact from Fiction, Institute for Science and International Security, 5 December 2003 (http://isis-online.org/isis-reports/category/iraq/).
2. Joseph Cirincione, *Niger Uranium: Still a False Claim*, Carnegie Endowment Proliferation Brief 7, no. 12, 2004.
3. Itty Abraham, *The Making of the Indian Atomic Bomb: Science, Secrecy and the Postcolonial State* (Zed Books, 1998); "The ambivalence of nuclear histories," in *Global Power Knowledge: Science, Technology, and International Affairs*, ed. J. Krige and K.-H. Barth (*Osiris* 21, July 2006).
4. David Albright, "When could Iran get the Bomb?" *Bulletin of the Atomic Scientists*, July–August 2006: 26–33.
5. Lewis Strauss, speech to National Association of Science Writers, 16 September 1954.
6. On interpretations of Chernobyl in France, see Sezin Topçu, L'agir contestataire à l'épreuve de l'atome: Critique et gouvernement de la critique dans l'histoire de l'énergie nucléaire en France (1968–2008), PhD thesis, École des Hautes Études en Sciences Sociales, 2010.
7. Samuel Huntington, *The Clash of Civilizations and the Remaking of the World Order* (Simon and Schuster, 1996).
8. Régis Debray, *Tous azimuts* (Odile Jacob, 1989).
9. Hugh Gusterson, "Nuclear weapons and the other in the Western imagination," *Cultural Anthropology* 14, no. 1 (1999): 111–143.
10. Office of Technology Assessment, *Nuclear Safeguards and the International Atomic Energy Agency* (OTA-ISS-615), 1995, appendix B.
11. D. A. Holaday, "Some unsolved problems in uranium mining," in *Radiological Health and Safety in Mining and Milling of Nuclear Materials: Proceedings*, volume 1 (IAEA, 1964), 51
12. In some respects, my approach here builds on the work of Ian Hacking and others who have used "historical ontology" and analogous concepts to discuss how

names, classification schemes, and other ways of designating what is/isn't in the world—how things "come into being"—are the products of shifting historical processes. Far from referring to some higher order of classification that corresponds in clear ways to the realities we see and is fixed for all time (as some philosophical traditions would have it), Hacking argues that ontologies result from contestations over power, knowledge, and ethics. (Hacking traces the term "historical ontology" back to its casual use in a single essay of Michel Foucault's, but he and others have taken the notion in several other directions.) Christopher Sellers and other environmental and medical historians have further developed "historical ontology" as an approach to analyzing occupational and environmental hazards, emphasizing how non-expert ways of knowing and seeing have shaped how specific industrial hazards come into being as objects of knowledge, perception, and action. Ian Hacking, *Historical Ontology* (Harvard University Press, 2002); Christopher Sellers, "The artificial nature of fluoridated water: Between nations, knowledge, and material flows," in *Landscapes of Exposure: Knowledge and Illness in Modern Environments*, ed. G. Mitman, M. Murphy, and C. Sellers (*Osiris* 19, 2004: 182–200); Christopher Sellers, *Hazards of the Job: From Industrial Disease to Environmental Health Science* (University of North Carolina Press, 1997).

13. For an introduction to one strand of STS scholarship on the distribution of ontology and agency into things, see Bruno Latour, *Reassembling the Social: An Introduction to Actor-Network-Theory* (Oxford University Press, 2005).

14. John Mueller, *Atomic Obsession: Nuclear Alarmism from Hiroshima to Al-Qaeda* (Oxford University Press, 2009).

15. Itty Abraham argues that nuclear systems are inherently, structurally ambivalent, filled with signs and meanings whose contradictions are fundamentally unresolvable. See "Who's next? Nuclear ambivalence and the contradictions of non-proliferation policy," *Economic and Political Weekly* 45, no. 43 (2010): 48–56.

16. V. Y. Mudimbe, *The Idea of Africa* (Indiana University Press, 1994).

17. Michael Adas, *Machines as the Measure of Men: Science, Technology, and Ideologies of Western Dominance* (Cornell University Press, 1990); Daniel R. Headrick, *The Tools of Empire: Technology and European Imperialism in the Nineteenth Century* (Oxford University Press, 1981).

18. Monica van Beusekom, *Negotiating Development: African Farmers and Colonial Experts at the Office du Niger, 1920–1960* (Heinemann, 2001); John L. Comaroff and Jean Comaroff, *Of Revelation and Revolution*, volume 2: *The Dialectics of Modernity on a South African Frontier* (University Of Chicago Press, 1997); Alice L. Conklin, *A Mission to Civilize: The Republican Idea of Empire in France and West Africa, 1895–1930* (Stanford University Press, 1997); Joseph Morgan Hodge, *Triumph of the Expert: Agrarian Doctrines of Development and the Legacies of British Colonialism* (Ohio University Press, 2007); Timothy Mitchell, *Rule of Experts: Egypt, Techno-Politics, Modernity* (University of California Press, 2002); Helen Tilley, *Africa as a Living Laboratory: Empire, Development, and the Problem of Scientific Knowledge, 1870–1950* (University of Chicago Press, 2011).

19. Michael Adas, *Dominance by Design: Technological Imperatives and America's Civilizing Mission* (Belknap, 2006); Arturo Escobar, *Encountering Development* (Princeton University Press, 1994); James Ferguson, *The Anti-Politics Machine: "Development," Depoliticization and Bureaucratic Power in Lesotho* (Cambridge University Press, 1990); Nils Gilman, *Mandarins of the Future: Modernization Theory in Cold War America* (Johns Hopkins University Press, 2004); Michael Latham, *Modernization as Ideology: American Social Science and "Nation Building" in the Kennedy Era* (University of North Carolina Press, 2000).

20. An analysis of this Lo-Zar story can be found at http://www.comicbox.com. Special thanks to Xavier Fournier for scans of the original Lo-Zar and a wealth of references to African uranium in American pop culture.

21. An early (and perhaps lone) exception to this pattern occurred in the lone issue of *All-Negro Comics*, a 1947 omnibus aimed at an African-American audience that included a story about Lion Man, an African scientist commissioned by the United Nations to protect a uranium deposit in Africa (http://en.wikipedia.org/wiki/African_characters_in_comics).

22. Quotes taken from Jack Kirby, *Black Panther* nos. 5, 7, and 8 (Marvel Comics, 1977–78).

23. These dialectics had a long history, as Saul Dubow has shown in *A Commonwealth of Knowledge: Science, Sensibility, and White South Africa 1820–2000* (Oxford University Press, 2006).

24. A. R. Newby-Fraser. *Chain Reaction: Twenty Years of Nuclear Research and Development in South Africa* (AEB, 1979), 64.

25. Here I cannot even begin to review how Africanists have engaged and challenged notions of "modernity." For a pithy survey of this literature, see Lynn Thomas, "Modernity's failings, political claims, and intermediate concepts," *American Historical Review*, June 2011: 727–740.

26. Recent examples: Kairn Klieman, *The Pygmies Were Our Compass: Bantu and Batwa in the History of West Central Africa, Early Times to c. 1900 C.E.* (Greenwood, 2003); Neil Kodesh, *Beyond the Royal Gaze: Clanship and Public Healing in Buganda* (University of Virginia Press, 2010); David Schoenbrun, *A Green Place, A Good Place: Agrarian Change and Social Identity in the Great Lakes Region to the 15th Century* (Heinemann, 1998).

27. Walter Rodney, *How Europe Underdeveloped Africa* (Howard University Press, 1981).

28. John Thornton, "Precolonial African industry and the Atlantic trade, 1500–1800," *African Economic History* 19 (1990): 1–19.

29. Eugenia W. Herbert, *Red Gold Of Africa: Copper in Precolonial History and Culture* (University of Wisconsin Press, 1984); Colleen E. Kriger, *Pride of Men: Ironworking in 19th Century West Central Africa* (Heinemann, 1999); Peter R. Schmidt, *Iron Technology in East Africa: Symbolism, Science and Archaeology* (James Currey, 1997).

30. Madeleine Akrich, "The de-scription of technical objects," in *Shaping technology/building society*, ed. W. Bijker and J. Law (MIT Press, 1994); Clapperton

Mavhunga, The Mobile Workshop: Mobility, Technology, and Human-Animal Interaction in Gonarezhou (National Park), 1850–present, PhD dissertation, University of Michigan, 2008; William Kelleher Storey, *Guns, Race, and Power in Colonial South Africa* (Cambridge University Press, 2008); Antina von Schnitzler, "Citizenship prepaid: Water, calculability, and techno-politics in South Africa," *Journal of Southern African Studies* 34 (2008): 899–917; Luise White, "'Heading for the gun': Skills and sophistication in an African guerrilla war," *Comparative Studies in Society and History* 51 (2009): 236–259.

31. Herbert, *Red Gold of Africa*; Kriger, *Pride of Men*; Schmidt, *Iron Technology*.

32. Michael Burawoy, *The Color of Class on the Copper Mines, from African Advancement to Zambianization* (Manchester University Press [for] the Institute for African Studies, University of Zambia, 1972); Jonathan Crush, Alan Jeeves, and David Yudelman, *South Africa's Labor Empire: A History of Black Migrancy to the Gold Mines* (Westview, 1991); Tshidiso Maloka Eddy, *Basotho and the Mines: A Social History of Labor Migrancy in Lesotho and South Africa, c.1890–1940* (Codesria, 2004); Charles Van Onselen, *Chibaro: African Mine Labor in Southern Rhodesia, 1900–1933* (Pluto, 1976); Francis Wilson, *Labour in the South African Gold Mines 1911–1969* (Cambridge University Press, 1972).

33. Raymond E. Dumett, *El Dorado in West Africa: Mining Frontier* (Ohio University Press, 1999); James Ferguson, *Expectations of Modernity: Myths and Meanings of Urban Life on the Zambian Copperbelt* (University of California Press, 1999); Patrick Harries, *Work, Culture, and Identity: Migrant Laborers in Mozambique and South Africa, c. 1860–1910* (Heinemann, 1994); T. Dunbar Moodie and Vivienne Ndatshe, *Going for Gold: Men, Mines, and Migration* (University of California Press, 1994); Keith Breckenridge, "Migrancy, crime and faction fighting: The role of the Isitshozi in the development of ethnic organisations in the compounds," *Journal of Southern African Studies* 16 (1990): 55–78; Keith Breckenridge, "'Money with dignity': Migrants, minelords and the cultural politics of the South African gold standard crisis, 1920–33," *Journal of African History* 36 (1995): 271–304; Keith Breckenridge, "The allure of violence: Men, race and masculinity on the South African goldmines, 1900–1950," *Journal of Southern African Studies* 24 (1998): 669–693; Keith Breckenridge, "'We must speak for ourselves': The rise and fall of a public sphere on the South African gold mines, 1920 to 1931," *Comparative Studies in Society and History* 40 (1998): 71–108.

34. James Ferguson, *Global Shadows: Africa in the Neoliberal World Order* (Duke University Press, 2006).

35. Frederick Cooper, "What is the concept of globalization good for? An African historian's perspective," *African Affairs* 100, no. 399 (2001): 189–213; Ferguson, *Global Shadows*.

36. J.-F. Bayart, "Africa in the world: A history of extraversion," *African Affairs* 99, no. 395 (2000): 217–267; Cooper, "Concept of globalization," 190.

37. On "technology," see Nina E. Lerman, "'Preparing for the duties and practical business of life': Technological knowledge and social structure in mid-19th-century

Philadelphia," *Technology and Culture* 38, no. 1 (1997): 31–59; Leo Marx, "Technology: The emergence of a hazardous concept," *Technology and Culture* 51, no. 3 (2010): 561–577; Ruth Oldenziel, *Making Technology Masculine: Men, Women, and Modern Machines in America, 1870–1945* (Amsterdam University Press, 2004); Eric Schatzberg, "Technik comes to America: Changing meanings of technology before 1930," *Technology and Culture* 47, no. 3 (2006): 486–512.

38. Ferguson, *Global Shadows* and *Expectations of Modernity*.

39. L. J. Butler, "The Central African Federation and Britain's post-war nuclear programme: Reconsidering the connections," *Journal of Imperial and Commonwealth History* 36, no. 3 (2008): 509–525; Margaret Gowing, *Independence and Deterrence: Britain and Atomic Energy, 1945–1952* (Macmillan, 1974).

40. Quotes in Alice Cawte, *Atomic Australia, 1944–1990* (UNSW Press, 1992), 41 and in Gabrielle Hecht, *The Radiance of France: Nuclear Power and National Identity after World War II*, new edition (MIT Press, 2009), 62.

41. Sven Lindqvist, *A History of Bombing* (New Press, 2001); Vijay Prashad, *The Darker Nations: A People's History of the Third World* (New Press, 2008).

42. John Dower, *War Without Mercy: Race and Power in the Pacific War* (Pantheon, 1986); Lindqvist, *History of Bombing*.

43. M. Susan Lindee, *Suffering Made Real: American Science and the Survivors at Hiroshima* (University of Chicago Press, 1994).

44. Quoted in Abraham, *Indian Atomic Bomb*, 28.

45. Ibid., 29.

46. For analyses of the speech and the subsequent US campaign, see John Krige, "Atoms for Peace, scientific internationalism, and scientific intelligence," in *Global Power Knowledge*, ed. Krige and Barth; Martin J. Medhurst, "Atoms for Peace and nuclear hegemony: The rhetorical structure of a Cold War campaign," *Armed Forces and Security* 23 (1997): 571–93.

47. This concern resurfaced at the general conference on the statute. According to the leader of the South African delegation, Arab, Asian, and some Latin American delegations raised concerns, parallel to those that they were raising at UN General Assembly meetings, concerning the "'undemocratic' nature of the Statute, and the perpetuation of an elite of 'have' nations which would repeat the inequalities of the first industrial revolution" (Conference on the Statute of the International Atomic Energy Agency, First Progress Report, 20 September to 2 October 1956, 4 October 1956. NASA: BLO 349 ref. PS 17/109/3, volume 2).

48. UN General Assembly Resolution 502 (VI), 11 January 1952.

49. "Final Communiqué of the Asian-African conference of Bandung," 24 April 1955. For more on Bandung, see Christopher J. Lee, ed., *Making a World after Empire: The Bandung Moment and Its Political Afterlives* (Ohio University Press, 2010); Prashad, *Darker Nations*.

50. "Final Communiqué."

51. Bayard Rustin, quoted in Jean Allman, "Nuclear imperialism and the pan-African struggle for peace and freedom: Ghana, 1959–1962," *Souls* 10, no. 2 (2008): 83–102.

52. In 1956, members of the first category were the US, the USSR, the UK, France, and Canada; members of the second were South Africa, Brazil, Japan, India, and Australia. See David Fischer, *History of the International Atomic Energy Agency: The First Forty Years* (IAEA, 1997).

53. South Africa's uranium was located in the same mines that produced its gold. In the decade following World War II, supplying uranium to the US and Britain saved many of these mines from economic collapse and served as conduits for massive foreign investment in the nation's industrial infrastructure. See Thomas Borstelmann, *Apartheid's Reluctant Uncle: The United States and Southern Africa in the Early Cold War* (Oxford University Press, 1993); Jonathan E. Helmreich, *Gathering Rare Ores: The Diplomacy of Uranium Acquisition, 1943–1954* (Princeton University Press, 1986).

54. "International Atomic Energy Agency," Annex to South Africa minute no. 79/2, 28/7/56. pp 10–11. NASA: BLO 349 ref. PS 17/109/3, volume 2. The position of South Africa vis-à-vis the IAEA is thoroughly documented in the BLO 349, BVV84, and BPA 25 series of these archives.

55. I elaborate on this point in "Negotiating global nuclearities: Apartheid, decolonization, and the Cold War in the making of the IAEA," in *Global Power Knowledge: Science, Technology, and International Affairs*, ed. J. Krige and K.-H. Barth (*Osiris* 21, 2006: 25–48). For an even more detailed analysis, see Astrid Forland, Negotiating Supranational Rules: The Genesis of the International Atomic Energy Agency Safeguards System, Dr. Art. dissertation, University of Bergen, 1997. For parallel developments with respect to Euratom, see John Krige, "The peaceful atom as political weapon: Euratom and American foreign policy in the late 1950s." *Historical Studies in the Natural Sciences* 38, no. 1 (2008): 9–48.

56. Donald Sole, "Uranium sales survey: Interim report on Continental Western Europe," 8 June 1959. NASA: HEN 2756 ref. 477/1/17.

57. For example, as reported in AEB Sales Committee, Minutes of the 5th meeting, 24 February 1961. NASA: HEN 2756 ref. 477/1/17 (among many other documents in this series).

58. Krige, "Atoms for Peace."

59. Angela Creager, "Nuclear energy in the service of biomedicine: The US Atomic Energy Commission's radioisotope program, 1946–1950," *Journal of the History of Biology*, 39 (2006): 649–684.

60. John A. Hall, "Atoms for Peace, or war," *Foreign Affairs* 43 (1965), July: 602; Krige, "Atoms for Peace."

61. Sonja Schmid, "Nuclear colonization? Soviet technopolitics in the Second World," in *Entangled Geographies: Empire and Technopolitics in the Global Cold War*, ed. G. Hecht (MIT Press, 2011).

62. Hecht, *Radiance of France.*

63. Avner Cohen, *Israel and the Bomb* (Columbia University Press, 1998); Pierre Péan, *Les deux bombes* (Fayard, 1982).

64. Itty Abraham, "Contra-proliferation: Interpreting the meanings of India's nuclear tests in 1974 and 1998," *Inside Nuclear South Asia* (2009): 106–36; Scott Kirsch, *Proving Grounds: Project Plowshare and the Unrealized Dream of Nuclear Earthmoving* (Rutgers University Press, 2005).

65. "International Atomic Energy Agency," Annex to South Africa minute no. 79/2, 28/7/56. pp. 5–6. NASA: BLO 349 ref. PS 17/109/3, volume 2.

66. Ibid. Plutonium fell into the category of "special fissionable materials." As a highly radioactive, extremely explosive, human-made material, it represented (and continues to represent) the pinnacle of nuclearity, the most exceptional of all things nuclear.

67. A. J. Brink to H. R. P. A. Kotzenberg, "Sale of uranium by France," 14 March 1962. NASA, HEN 2756 ref. 477/1/17.

68. Abraham, *Indian Atomic Bomb*; Forland, Negotiating Supranational Rules; George Perkovich, *India's Nuclear Bomb: The Impact on Global Proliferation* (University of California Press, 1999); Lawrence Scheinman, *The International Atomic Energy Agency and World Nuclear Order* (Resources for the Future, 1987).

69. For an interesting new take on the NPT, see Grégoire Mallard, *Crafting the Nuclear World Order (1950–1975): The Dynamics of Legal Change in the Field of Nuclear Nonproliferation*, Working Paper 10-005, Roberta Buffet Center for International and Comparative Studies, 2010 (available at http://papers.ssrn.com).

70. Article IV of Treaty on the Non-Proliferation of Nuclear Weapons, signed 1 July 1968 in Washington, London, and Moscow (emphasis added).

71. Between 1958 and 1993, the IAEA gave out $617.5 million in "technical assistance." The top ten recipients were Egypt, Brazil, Thailand, Indonesia, Peru, Pakistan, Philippines, Bangladesh, South Korea, and Yugoslavia (OTA, *Nuclear Safeguards*).

72. IAEA, INFCIRC/66/Rev. 2, 16 September 1968.

73. IAEA, INFCIRC/153 (Corrected), June 1972. According to article 112, "*nuclear material* means any source or any special fissionable material as defined in Article XX of the Statute" and "the term source material shall not be interpreted as applying to ore or ore residue."

74. This was the so-called Zangger committee, chaired by Claude Zangger and initially composed of 15 states that were "suppliers or potential suppliers of nuclear material and equipment" (IAEA, INFCIRC/209/Rev. 1, Annex).

75. Notably, two trigger lists developed in parallel, one under the rubric of INFCIRC/209 and one under the rubric of INFCIRC/254. Different nations adhered to different lists. The two lists were "brought into synch" in 1977, but they continued to develop separately.

76. E.g., IAEA, INFCIRC/209/Rev. 1/Mod. 4, 26 April 1999.

77. Gowing, *Independence and Deterrence*; Helmreich, *Gathering Rare Ores*.

78. Early (failed) proposals to use the IAEA an international fuel bank would have placed uranium beyond "the market" in a slightly different way, by fully centralizing its distribution.

79. Michel Callon, *Laws of the Markets* (Wiley-Blackwell, 1998); Michel Callon, Yuval Millo, and Fabian Muniesa, eds., *Market Devices* (Wiley-Blackwell, 2007); Donald MacKenzie, Fabian Muniesa, and Lucia Siu, eds., *Do Economists Make Markets?* (Princeton University Press, 2007); Mitchell, *Rule of Experts*.

80. Mounana interviews, 1998; Antoine Paucard, *La mine et les mineurs de l'uranium français. III: Le Temps des Grandes Aventures (1959–1973)* (Editions Thierry Parquet, 1996), 228.

81. Interview with Jeremy Fano, 18 August 1998 (translator: M. Abdoulhamide).

82. W. C. Hueper, *Occupational Tumors and Allied Diseases* (Thomas, 1942).

83. Holaday, "Some unsolved problems."

84. Robert Bodu, "Compte-rendu de mission à Madagascar," pp. III-6, Direction des Recherches et Exploitations Minières, Mars 1960. Cogéma papers, Razès.

85. IAEA/ILO manual (panel from 1973, manual from 1976).

86. "Draft: South African Energy Poicy: Discussion Document: Comment," 2 October 1995, 10–11, CoM-ES.

87. Interview with Shaun Guy, 12 July 2004.

88. Interview with Dominique Oyingha, 17 July 1998.

89. Cited in Robert Proctor and Londa Schiebinger, *Agnotology: The Making and Unmaking of Ignorance* (Stanford University Press, 2008), 1. See also Allan M. Brandt, *The Cigarette Century: The Rise, Fall, and Deadly Persistence of the Product That Defined America* (Basic Books, 2007); Gerald Markowitz and David Rosner, *Deceit and Denial: The Deadly Politics of Industrial Pollution* (University of California Press, 2002); Robert N. Proctor, *Cancer Wars: How Politics Shapes What We Know and Don't Know About Cancer* (Basic Books, 1995); David Rosner and Gerald Markowitz, *Deadly Dust: Silicosis and the Politics of Occupational Disease in Twentieth-Century America* (Princeton University Press, 1991).

90. Sheila Jasanoff, "Bhopal's trials of knowledge and ignorance," *Isis* 98, no. 2 (2007): 344–350; Linda Nash, *Inescapable Ecologies: A History of Environment, Disease, and Knowledge* (University of California Press, 2007); Rosner and Markowitz, *Deadly Dust*; Christopher Sellers and Joseph Melling, eds., *Dangerous Trade: Histories of Industrial Hazard Across a Globalizing World* (Temple University Press, 2011).

91. Julie Livingston, *Improvising Medicine in an African Oncology Ward* (Duke University Press, in press).

92. For one example among many, see Jock McCulloch, *Asbestos Blues: Labour, Capital, Physicians and the State in South Africa* (Indiana University Press, 2002).

93. Michelle Murphy, *Sick Building Syndrome and the Problem of Uncertainty: Environmental Politics, Technoscience, and Women Workers* (Duke University Press, 2006). On how such issues relate to radiation exposure, see Adriana Petryna, *Life Exposed: Biological Citizens after Chernobyl* (Princeton University Press, 2002); on how they relate to nuclear issues more broadly, see Joseph Masco, *The Nuclear Borderlands:*

The Manhattan Project in Post-Cold War New Mexico (Princeton University Press, 2006).
94. Petryna, *Life Exposed*.
95. Anna Lowenhaupt Tsing, *Friction: An Ethnography of Global Connection* (Princeton University Press, 2005).

CHAPTER 2 AND PROLOGUE

1. http://www.orau.org/ptp/collection/consumer%20products/consumer.htm.
2. Itty Abraham, "Rare earths: The Cold War in the annals of Travancore," in *Entangled Geographies*, ed. Hecht.
3. Gowing, *Independence and Deterrence*, 369.
4. This is well-trodden historical terrain. See Robert Bothwell, *Eldorado: Canada's National Uranium Company* (University of Toronto Press, 1984); René Brion and Jean-Louis Moreau, *De la mine à Mars: la genèse d'Umicore* (Lanoo, 2006); Gowing, *Independence and Deterrence*; Helmreich, *Gathering Rare Ores*; Jacques Vanderlinden, *A propos de l'uranium congolais* (Académie royale des sciences d'outre-mer, 1991).
5. Cited in Gowing, *Independence and Deterrence*, 372.
6. These events can be traced via British documents held in TNA: AB16/3131; AB16/3292; AB16/2514; AB16/2516, and other boxes.
7. Paucard, *La mine et les mineurs de l'uranium français*, volumes 1–3 (Éditions Thierry Parquet, 1992), Goldschmidt, *Les Rivalités Atomiques 1939–1966* (Fayard, 1967).
8. Paucard, *La mine et les mineurs*, volume 3.
9. Ibid.
10. Callon et al., *Market Devices*; Donald MacKenzie, *Material Markets: How Economic Agents Are Constructed* (Oxford University Press, 2009), 2.
11. On the "Third World" project, see Prashad, *Darker Nations*. For discussions of US technopolitical imperialism during the Cold War, see Adas, *Dominance by Design*; John Krige, *American Hegemony and the Postwar Reconstruction of Science in Europe* (MIT Press, 2006); Ruth Oldenziel, "Islands: The United States as a networked empire," in *Entangled Geographies*, ed. Hecht. On specifically European fears of American imperialism, see Victoria de Grazia, *Irresistible Empire: America's Advance through Twentieth-Century Europe* (Belknap, 2005); Richard F. Kuisel, *Seducing the French: The Dilemma of Americanization* (University of California Press, 1993).
12. Krige, "The peaceful atom," 9–48.
13. The act allowed private ownership of "nuclear materials," thereby enabling American utilities to hold direct title to their reactor fuel.
14. Thomas L. Neff, *The International Uranium Market* (Ballinger, 1984), 43–49; June H. Taylor and Michael D. Yokell, *Yellowcake: The International Uranium Cartel* (Pergamon, 1979), 33.
15. H. McL. Husted, Uranium Sales, Report no. 11, Johannesburg, 13 October 1965, 17. NASA: MMY65, M3/7 volume 2 (emphasis added)

16. Ibid.

17. Hecht, *Radiance*.

18. OECD Nuclear Energy Agency and International Atomic Energy Agency, *Forty Years of Uranium Resources, Production and Demand in Perspective: The Red Book Retrospective* (2006).

19. For a brief time, the Red Books also accounted for thorium deposits.

20. These various categories changed over time, but those details need not concern us here.

21. OECD and IAEA, *Uranium Resources: Revised Estimates* (1967), 5 (emphasis added).

22. OECD and IAEA, *Uranium Production and Short Term Demand* (1969), 4.

23. OECD and IAEA, *Uranium: Resources, Production, and Demand* (1970), 10.

24. OECD and European Nuclear Energy Agency, *World Uranium and Thorium Resources* (1965), 7; *Uranium Resources*, 1967, 8.

25. OECD and IAEA, *Uranium Production*, 1969, 3.

26. Thanks to Dan Hirschman for this observation.

27. OECD and IAEA, *Uranium: Resources, Production, and Demand* (1975), 11.

28. One of them, George White, also lay claim to international recognition as "one of the principal architects of nuclear power" by virtue of having initiated and managed General Electric's commercial nuclear power business. Nuexco initially aimed to facilitate the loan of reactor fuel as an alternative to fuel leasing arrangements offered by the US AEC.

29. Nuexco Monthly Report, 11-18-68. Its monthly newsletter anonymized and reported "market information" in the form of "bid and ask quotations"—that is, the prices at which lots of uranium were being offered for sale or loan, and the bids that potential buyers were making.

30. There was "no evidence," it said in January 1969, "that a market level has been established at which business can be done" (Nuexco Monthly Report, 1-22-69).

31. Nuexco Monthly Report, 2-18-69, 1.

32. Ibid.

33. Nuexco Monthly Report, 9-29-69, 3.

34. Nuexco Monthly Report, 6-19-70.

35. Nuexco Monthly Report, 12-19-70, 1b.

36. Nuexco Monthly Report, 1-21-71, 3.

37. Nuexco Monthly Report, 12-21-73.

38. Nuexco Monthly Report, 11-13-72.

39. Nuexco Monthly Report, 10-21-70, 1a.

40. "NUEXCO: Nuclear Exchange Corporation," August 1976, reproduced in *Allegations that uranium prices and markets have been influenced by a foreign producers' cartel and other factors which may or may not have been responsible for the sevenfold increase in the price of uranium: Hearings before the Subcommittee on Oversight and Investigations, of the House Committee on Interstate and Foreign Commerce*, 94th Congress (4 November 1976), 103–116.

41. André Giraud, quoted in Paucard, *La mine et les mineurs,* volume 3, 23.
42. Hecht, *Radiance.*
43. Antoine Paucard et al., *La mine et les mineurs de l'uranium français, Tome IV/vol* I (Areva, 2007), 173. Uranex records are not available.
44. NUFCOR: Minutes of Meeting of the Executive Committee of the Board of Directors, 27 November 1968; Minutes of Meeting of the Executive Committee of the Board of Directors, 28 October 1970.
45. This was known as the "split tails" policy.
46. COMUF: COMUF, LC/JM, 12.XII.1973, "Développement du marché de l'uranium," 4.
47. L. C. Mazel to H. F. Melouney, 2 May 1972, reproduced in *Allegations that uranium prices and markets have been influenced* (1976), 174–6.
48. NUFCOR: Minutes of Meeting of the Executive Committee of the Board of Directors, 17 October 1973; Taylor and Yokell, *Yellowcake.*
49. E.g., Nuexco Monthly Reports 10-20-72 and 10-22-73. At one stage, apparently, the EEC even contacted governments of Canada, Australia, France and South Africa asking about price fixing in uranium contracts and noting that any arrangement had to be cleared with the EEC! See NUFCOR: Minutes of Meeting of the Executive Committee of the Board of Directors, 19 July 1972.
50. "Report of the Discussions in Paris on the Uranium Industry, 1–4 February 1972," in *Allegations that uranium prices and markets have been influenced* (1976), 344–362.
51. This at least was how Nufcor preferred to proceed, beginning in late 1972 (for deliveries as of 1978). NUFCOR: Minutes of Meeting of the Executive Committee of the Board of Directors, 29 November 1972.
52. Taylor and Yokell, *Yellowcake,* 90–91.
53. NUFCOR: Minutes of Meeting of the Executive Committee of the Board of Directors, 25 October 1972 and 29 November 1972.
54. Anthony David Owen, *The Economics of Uranium* (Praeger, 1985), 46–47.
55. Shigeru Nakayama and Hitoshi Yoshioka, eds., *A Social History of Science and Technology in Contemporary Japan,* volume 4: *Transformation Period, 1970–1979* (Marston, 2006); Japan Atomic Energy Commission, *White Paper on Nuclear Energy* (in Japanese) (JAEC, 1993). Special thanks to Morris Low for these references, and for his translation of the White Paper figures.
56. NUFCOR: Minutes of Meeting of the Executive Committee of the Board of Directors, 17 October 1973.
57. Paucard, *La mine et les mineurs,* volume 3, 72.
58. And by then, Canadian producers reported that they were having difficulty obtaining government approval to continue participating "in a cartel-like organization" (NUFCOR: Minutes of Meeting of the Executive Committee of the Board of Directors, 1 May 1974; Neff, *International Uranium Market,* 50, 83).
59. Neff, *International Uranium Market,* 50.
60. Taylor and Yokell, *Yellowcake,* 118–124.

61. NUFCOR: Minutes of Meeting of the Executive Committee of the Board of Directors, 1 May 1975.

62. Taylor and Yokell, *Yellowcake*, 124–5.

63. *Allegations that uranium prices and markets have been influenced* (1976).

64. Nuexco Monthly Report, 3-18-75, 1.2.

65. *Allegations that uranium prices and markets have been influenced* (1976), 43.

66. Neff, *International Uranium Market*; Owen, *Economics of Uranium*; Taylor and Yokell, *Yellowcake*.

67. *Allegations that uranium prices and markets have been influenced* (1976), 96.

68. COMUF: n.a., "Principales Informations Nucléaires," Octobre-Novembre 1975, 6 (although unattributed, this document is clearly part of a series of documents produced by Cogéma for its subsidiaries). COMUF archives.

69. *Proceedings of the International Conference on Nuclear Energy Commodities*, London, 25–28 October 1976, 138.

70. Ibid., 139.

71. Ibid.

72. Ibid., 79.

73. Ibid., 83.

74. Ibid., 121–22.

CHAPTER 3 AND PROLOGUE

1. Final Communiqué of the Asian-African Conference, Bandung, Indonesia, 24 April 1955.

2. Lee, *Making a World after Empire*; Prashad, *The Darker Nations*.

3. Nico Schrijver, *Sovereignty over Natural Resources: Balancing Rights and Duties* (Cambridge University Press, 1997), 18.

4. Ibid., 83.

5. Quoted in Schrijver, *Sovereignty*, 98–9.

6. Quotes in Schrijver, *Sovereignty*, 100.

7. Dubow, *Commonwealth of Knowledge*. See also Nancy L. Clark, *Manufacturing Apartheid: State Corporations in South Africa* (Yale University Press, 1994); Renfrew Christie, *Electricity, Industry and Class in South Africa* (SUNY Press, 1984); Deborah Posel, "A mania for measurement: Statistics and statecraft in the transition to apartheid," in *Science and Society in Southern Africa*, ed. S. Dubow (Manchester University Press, 2000). Stephen Sparks takes this theme head on in Apartheid Modern: SASOL and the Making of a South African Company Town, 1950–2009 (PhD dissertation, University of Michigan, 2011).

8. Rössing was expected to contribute up to 10% of Namibia's GDP, so its role in the economy of a future independent nation was considered pivotal. The mine's illegality depended, of course, on interpreting the UNCN's decrees as legally binding—which the US and several other countries did not. Conflicts over the binding nature of such decrees are discussed in Schrijver, *Sovereignty*. The UNCN

held hearings on the export of Namibian contracts in 1980, the proceedings of which are publicly available in United Nations General Assembly, "Report of the Panel for Hearings on Namibian Uranium. Part Two: Verbatim Transcripts of the Panel Held at Headquarters from 7 to 11 July 1980," A/AC.131/L.163 (United Nations, 30 September 1980). Publications in which conflicts over the Namibian contracts played out include Alun Roberts, *The Rössing File: The Inside Story of Britain's Secret Contract for Namibian Uranium* (CANUC, 1980) and Allan D. Cooper, ed., *Allies in Apartheid: Western Capitalism in Occupied Namibia* (St. Martin's Press, 1988). For an excellent overview, see Gretchen Bauer, *Labor and Democracy in Namibia, 1971–1996* (Ohio University Press, 1998).

9. Consul (Commercial) New York to Secretary for Commerce and Industry in Pretoria, 30 November 1962, included in AEB Marketing Advisory Committee, Supplementary Agenda for the 1st meeting for 2 February 1962. NASA: HEN 2756 ref. 477/1/17 (emphasis added).

10. See Gabrielle Hecht, "On the fallacies of Cold War nostalgia: Capitalism, colonialism, and South African nuclear geographies," in *Entangled Geographies*, ed. Hecht.

11. Minutes of Meeting of the Executive Committee of the Board of Directors, 28 November 1967. NUFCOR archives.

12. "A Report on the Exploitation and Marketing of the Uranium Resources of South Africa and South West Africa," Secret, AEB1/75 (Amended), p. 15. NECSA archives.

13. I. F. A. de Villiers and E. W. Hunt to A. J. A. Roux, 22 September 1969. NUFCOR archives.

14. Paul N. Edwards and Gabrielle Hecht, "History and the technopolitics of identity: The case of apartheid South Africa," *Journal of Southern African Studies* 36, no. 3 (2010): 619–639.

15. US Central Intelligence Agency, Directorate of Science and Technology, *Weekly Surveyor*, 4 May 1970, 1. DNSA: DocID 651169.

16. The material on these debates is too voluminous to cite individually. Records of meetings, economic and technical feasibility studies, and other correspondence relating to a potential UF_6 plant for NUFCOR can be found in the UTAC papers from 1968 to 1973, in the Goldfields archives.

17. Raad op Atoomkrag, Uitvoerende Komitee (Produksie), "Notule van die 1ste Vergadering van die Uitvoerende Komitee (Produksie)," 20 Maart 1969, Pelindaba. NASA: EAE 143 ref. EA 2/2/13 vol. 1.

18. T. E. W. Schumann, AEB, "South Africa's policy in respect of safeguards in connection with uranium sales," 11/2/0(S), 29/3(S), not dated. NECSA archives.

19. Some archival documentation concerning the enrichment project survives in the National Archives of South Africa. See, for example, papers in EAE 143 ref EA 2/2/13, vol. 1 and MEM 1/590, ref. 121/2.

20. A. R. Newby-Fraser, *Chain Reaction: Twenty Years of Nuclear Research and Development in South Africa* (Atomic Energy Board, 1979), 91. This was the official

history of the AEB, written in English by its public relations director; it is chock full of nationalist assertions of South African technological prowess.

21. Quoted in ibid., 92.

22. Ibid., 92–94 (emphasis added). Also see "Verklang deur Sy Edele die Eerste Minister," 20 Julie 1970. NASA: MEM 1/590, ref. 121/2.

23. "Background notes on some aspects of Atomic Energy in South Africa," not dated (1970), CW/96/014. TNA: EG7/117.

24. See, for example, M. H. Morgan (British Embassy, Cape Town) to Arculus (Science and Technology Department, Foreign and Commonwealth Office), 1 August 1970. TNA: EG 8/198. Numerous, complex efforts to figure out South African technology are documented in these files.

25. D. E. Lyscom (Industry, Science and Energy Department) to P. M. Laver (British Embassy, Cape Town), 2 April 1973. TNA: EG 8/198.

26. H. S. Weeks (Ministry of Defense) to J. Thompson (Department of Trade and Industry), 20 June 1972, "South Africa: Potential Nuclear Capability," Ref: 438/70. TNA: EG8/198.

27. Ibid.

28. A. J. A. Roux, "Uraanverrykingskorporasie van Suid-Afrika, Beperk. Die Suid-Afrikaanse Verrykingsproses en Buitelandse Reaksie en Samewerking tot en met die Europese Kern-Konferensie, Parys, tussen 21 en 25 April 1975"; Steag, "Draft: Uranium Enrichment Corporation of South Africa," Essen, April 21, 1975; André Giraud to A. J. A. Roux, not dated (1975); "Draft Protocol of agreement between AEB/UCOR and CEA," 2.4.1975. NASA: MEM 1/590 ref 121/2. Also documented in NASA: EAE 143 ref EA 2/2/13 vol. 1 and TNA: EG 8/198 and EG7/138.

29. J. C. Paynter and H. E. James, *Feasibility Study of the Commercial Production of Uranium Hexafluoride in South Africa*, report 924, National Institute for Metallurgy Research, 16 March 1970; NUFCOR, report of 18th meeting of UTAC, 14 May 1970, 4. Goldfields archives, UTAC papers.

30. A. J. A. Roux, "Uranium enrichment," paper delivered at a meeting of the South African Academic for Science and Art, 19 March 1971. TNA: EG8/198. This talk was published in the July 1971 issue of AEB's in-house magazine, *Nuclear Active*, and was subsequently excerpted in other publications.

31. A. J. A. Roux to A. W. S. Schumann, Liaison between AEB and NUFCOR, 3.6.74, part of NUFCOR/UTAC Circular no. 26/74, 5 July 1974. Goldfields archives, UTAC papers.

32. R. E. Worroll, Report on Interview with Dr. A. J. A. Roux in Pelindaba, 17 June 1974, part of NUFCOR/UTAC Circular no. 26/74, 5 July 1974. Goldfields archives, UTAC papers.

33. Hecht, "Negotiating global nuclearities."

34. Minutes of Special Meeting of the Executive Committee of the Board of Directors, 4 August 1971. NUFCOR archives.

35. R. E. Worroll, Report on Interview with Dr. A. J. A. Roux in Pelindaba, 17 June 1974, part of NUFCOR/UTAC Circular no. 26/74, 5 July 1974. Goldfields archives, UTAC papers.

36. Minutes of Meeting of the Executive Committee of the Board of Directors, 19 June 1974. NUFCOR archives.

37. R. E. Worroll, Report on Interview with Dr. A. J. A. Roux in Pelindaba, 17 June 1974, part of NUFCOR/UTAC Circular no. 26/74, 5 July 1974. Goldfields archives, UTAC papers.

38. R. E. Worroll, Report on Interview with Dr. A. J. A. Roux in Pelindaba, 17 June 1974, part of NUFCOR/UTAC Circular no. 26/74, 5 July 1974. Goldfields archives, UTAC papers.

39. These estimates figured in AEB 1/75, "A Report on the Exploitation and Marketing of the Uranium Resources of South Africa and South West Africa," Pelindaba, 4 June 1975, JWS/JH/MK. Necsa archives.

40. Ibid.

41. AEB, "Guidelines for a Uranium Marketing Policy for the mid-1980s," Ex.Co. 3/75, Necsa archives.

42. A. W. S. Schumann to A. J. A. Roux, 16 March 1976. Necsa archives.

43. Anthony Jackson, *Qalindaba* (unpublished manuscript, courtesy of the author); interview with Anthony Jackson, 6 April 2004; A. Roux to B. G. Fourie, "Samesprekings met die VSA insake waarborge," 1981–09–16; "Besprekingspunte vir die VSA-afvaardiging se besoek, tentatief 21–26 Oktober 1981," 1981-09-16. NECSA archives.

44. The South African mining giant Anglo American had gone so far as to acquire the mining rights before deciding that it didn't have the technical expertise or up-front capital to set up the mine successfully. RTZ, however, had extensive experience with open-cast mining elsewhere.

45. Technically, the contract was with Riofinex, RTZ's South African affiliate at the time. "Linked to the main contract there was a provision that should the new mine, then in an early stage of development by RTZ, fail to be 'commercially viable' (in the opinion both of RTZ and of the Authority) by December 30th 1968, then a fallback contract with Rio Algom in Canada of 4,000 tons over the same period and on the same principal terms should come into effect." F. Chadwick (UKAEA) to M. I. Michaels (MinTech), 8 December 1969. TNA: EG7/29.

46. F. Chadwick (UKAEA) to M. I. Michaels (MinTech), 8 December 1969. TNA: EG7/29.

47. "Uranium supplies for the Nuclear Power Programme," Memorandum by the Minister of Technology PN68(1), 22 February 1968; F. Chadwick (UKAEA) to M. I. Michaels (MinTech), 8 December 1969. TNA: CAB134/3121, EG7/29.

48. These internal debates are documented in UKAEA archives held at The National Archives of the United Kingdom (TNA): FCO 45/749, FCO 45/750, FCO 45/751, EG7/117, and CAB 148/111.

49. M. I. Michaels (MinTech) to C. Cunningham, 4 March 1970. TNA: EG7/29.

50. M. I. Michaels (MinTech), "Uranium Ore from South West Africa," 26 February 1970. TNA: EG7/29.

51. Cabinet Committee on Overseas Policy, Draft: "Uranium supplies for the nuclear power programme," Note by the Minister of Technology, March 1970. TNA: EG7/29.

52. Up to £5 million. M. I. Michaels (MinTech) "Uranium Ore from South West Africa," 26 February 1970. TNA: EG7/29.

53. M. I. Michaels (MinTech), "Uranium Supplies from South West Africa," 2 March 1970. TNA: EG7/29.

54. Ibid.

55. OPD (70) 7TH Meeting held 6/5/70, "Uranium supplies for the nuclear power programme." TNA: EG7/29. ; Michael Stewart (PM/70/60), "Uranium in South West Africa," not dated.

56. Denis Greenhill to Mr. Wilson, 8 June 1970. TNA: FCO45/75, 82826.

57. Sir Val Duncan to Denis Greenhill, 8 June 1970. TNA: FCO45/75, 82826.

58. Barbara Rogers, *Namibia's Uranium: Implications for the South African Occupation Regime* (UN Council for Namibia, 1975), 15.

59. J. B. Johnston, "RTZ and South West Africa," June 1970. TNA: FCO45/75, 82826.

60. Ibid.

61. Denis Greenhill to Mr. Fingland, 24 September 1970. TNA: FCO45/75, 82826.

62. UNCN, "Report of the Panel for Hearings on Namibian Uranium. Part Two: Verbatim Transcripts of the Public Meetings of the Panel Held at Headquarters from 7 to 11 July 1980."

63. Thynne to Eadie, 4 April 1974. TNA: EG 7/139.

64. P. M. Foster, Central and Southern African Dept (FCO), to Mr Aspin, 29 April 1974. Draft answer to Mr John Biffen's question about Rossing Uranium contract. TNA: EG 7/139.

65. Ibid.

66. Secret memorandum, "Namibia: Uranium Supplies," revised October 1974. TNA: FCO: 96/414.

67. Confidential memorandum, "Uranium from Namibia," not dated but circa mid 1975. TNA: FCO: 96/414.

68. Martin Reith, Central and Southern African Dept, FCO to W. E. Fitzsimmons, Dept of Energy, 10 October 1974. TNA: EG 7/139.

69. Martin Reith, Central and Southern African Dept, FCO to Department of Industry, 10 October 1974. TNA: EG 7/139.

70. Quoted in Alastair Macfarlane, Labour Control: Managerial Strategies in the Namibian Mining Sector, PhD thesis, Oxford Polytechnic, 1990, 420.

71. Telex from Crosland to UKMIS New York, 10 May 1976 (TNA: FCO 45/1940); Minutes of Sales Meetings: 17 February 1981, 13 April 1981, 14 July

1981 (RUL archives); Jun Morikawa, *Japan and Africa: Big Business and Diplomacy* (Witwatersrand University Press, 1997), 80–81 and 154–156.

72. Martin Reith, "Note for the Record: Namibia Uranium," 23 February 1976. TNA: FCO 45/1940.

73. "Record of a meeting between the Secretary of State for Foreign and Commonwealth Affairs and Mr. Sam Nujoma, President of SWAPO, at the Foreign and Commonwealth Office on 11 June at 4.15 pm," 1975. TNA: FCO 96/414.

74. Martin Reith to Private Secretary, "Namibia—Uranium," 23 February 1976. TNA: FCO 45/1940.

75. James Callaghan to Prime Minister, PM/76/21, 16 March 1976. TNA: FCO 45/1940.

76. And indeed his *History of Resistance in Namibia* (published in London, in Addis Ababa, and in Paris in 1988, two years before formal independence) made no mention of uranium or Rössing.

77. James Callaghan, "Namibia Uranium," 12 March 1976. TNA: FCO 45/1940.

78. W. J. Vose (British Embassy Pretoria) to J. M. O. Snodgrass, Esq. (British Embassy Cape Town), 8 March 1976. TNA: FCO 45/1940.

79. Martin Reith, Confidential Note for Record, 17 May 1976. TNA: FCO 45/1940.

80. Ibid.

81. Martin Reith to A. J. Coles Esq (Head of Chancery, Cairo), 21 May 1976. TNA: FCO 45/1940.

82. Martin Reith to R. W. Whitney, 11 June 1976. TNA: FCO 45/1940.

CHAPTER 4 AND PROLOGUE

1. Frederick Cooper, *Africa since 1940: The Past of the Present* (Cambridge University Press, 2002), 77.

2. Ibid.; Frederick Cooper, *Decolonization and African Society: The Labor Question in French and British Africa* (Cambridge University Press, 1996).

3. François-Xavier Verschave, *La Françafrique: Le plus long scandale de la République* (Stock, 1999); Pierre-Noël Giraud, *Géopolitique des Ressources Minières* (Economica, 1983).

4. Pierre Péan, *L'Homme de l'ombre: éléments d'enquête autour de Jacques Foccart, l'homme le plus mystérieux et le plus puissant de la Ve République* (Fayard, 1990); Verschave, *La Françafrique*; Monsieur X. and Patrick Pesnot, *Les dessous de la Françafrique* (Nouveau monde, 2008); Jacques Foccart, *Journal de l'Élysée* (Fayard and Jeune Afrique, 1997–2001).

5. Péan, *L'Homme de l'ombre*; Florence Bernault, *Démocraties ambiguës en Afrique centrale: Congo-Brazzaville, Gabon, 1940–1965* (Karthala, 1996); Jean-François Obiang, *France-Gabon: pratiques clientélaires et logiques d'état dans les relations franco-africaines* (Karthala, 2007).

6. Schrijver, *Sovereignty*.

7. Hecht, *Radiance*.

8. And indeed the temptation was too strong to resist: the CEA went ahead and bought some South African ore. In these early nuclear days, too much uranium seemed like a good problem for France to have. But CEA negotiators insisted on secrecy: France shouldn't be seen (by the postcolonies) consorting with the apartheid state on uranium matters. AEB Marketing Advisory Committee, Minutes of the 6th meeting, 23 March 1964; H. McL. Husted, "Uranium Sales," Report no. 7, 8 November 1963, p. 10; NASA: HEN 2757 ref. 477/1/17/2.

9. Robert Bodu, *Les secrets des cuves d'attaque: 40 ans de traitement des minerais d'uranium* (Cogéma, 1994), 64–67.

10. CEA/DP/DREM, Groupement Afrique-Madagascar, "Notice d'Information destinée aux Européens susceptibles de partir pour l'Afrique ou Madagascar" (1.2.63). Cogéma archives.

11. Cooper, *Decolonization and African Society*; Frederick Cooper and Randall M. Packard, eds., *International Development and the Social Sciences: Essays on the History and Politics of Knowledge* (University of California Press, 1997).

12. COMUF, "Plan décennal," 21 juin 1971, JC/JL.W/DT, "libérera" quote on p. 7. COMUF archives.

13. JC/JF, "Conséquences financières d'une réduction d'activité de Mounana," 11 juin 1970; COMUF, LC/JM, "Développement du marché de l'uranium," 12.XII.1973. COMUF archives.

14. COMUF, "Plan décennal," 21 juin 1971, JC/JL.W/DT. COMUF archives.

15. JC/JF, "Conséquences financières d'une réduction d'activité de Mounana," 11 juin 1970; COMUF, "Plan décennal," 21 juin 1971, JC/JL.W/DT; J. Peccia-Galleto to C. Guizol, 6 juillet 1971. COMUF archives.

16. UF/JC/JF, "Dispositions prises au passage de M. Guizol les 29 et 30 juillet 1969," 1er août 1969. See also CG/mc,0577/70, "Compte-rendu fait par M. de Courlon de nos entretiens du 28 juillet 1970—Programme quadriennal 1970–1974 du 21.7.70," 20 août 1970. COMUF archives.

17. Bodu, *Les secrets des cuves*, 78.

18. Verschave, *La Françafrique*; Foccart, *Journal de l'Élysée*; Obiang, *France-Gabon*; Douglas Yates, *The Rentier State in Africa: Oil-Rent Dependency and Neo-Colonialism in the Republic of Gabon* (Africa World Press, 1996); François Ngolet, "Ideological manipulations and political longevity: The power of Omar Bongo in Gabon since 1967," *African Studies Review* 43, no. 2 (2000): 55–71.

19. Delphine Mauger, "Building the *Gabon Nouveau*: Technology and the construction of the Bongo-state, 1973–1986," seminar paper, University of Michigan, 2003. On postcolonial technological nationalism, see Gyan Prakash, *After Colonialism: Imperial Histories and Postcolonial Displacements* (Princeton University Press, 1995); Brian Larkin, *Signal and Noise: Media, Infrastructure, and Urban Culture in Nigeria* (Duke University Press, 2008). On technological nationalism, see Hecht, *Radiance*; Edwards and Hecht, "History and the technopolitics of identity."

20. The French ambassador made a note of these moves in Ambassade de France au Gabon, Jean Ribo à Michel Jobert, 18 mars 1974, 31/DAM. ANF: AG 5 (FPU) SD 78.

21. Quoted in Ambassade de France au Gabon, Jean Ribo à Michel Jobert, 30 novembre 1973, Synthèse 18–73, période du 1 au 30 novembre 1973. ANF: AG 5 (FPU) SD 78.

22. Robert Edgar Ndong, Les multinationales extractives au Gabon: le cas de la compagnie des mines d'uranium de Franceville (COMUF), 1961–2003, PhD thesis, École Doctorale Sciences Sociales, Université Lumière—Lyon II, 2009, dir. Serge Chassagne, 160.

23. COMUF, "Plan Quinquennal 1975–1979," 31.05.74. COMUF archives.

24. Ndong, "Les multinationales extractives," 77–81.

25. COMUF, "Plan Quinquennal 1975–1979," 31.05.74; "Plan Quinquennal 1976–1980," 9.4.1975; "Plan Quinquennal 1977–1981," 14.6.76; "Plan Quinquennal 1978–1982," 4.07.77. COMUF archives.

26. Ndong, "Les multinationales extractives," 159–164.

27. André Salifou, Le Niger (L'Harmattan, 2002); C. Raynaut, "Trente ans d'indépendance, repères et tendances," Politique Africaine, 38 (1990); Cooper, Africa since 1940; Jean-Paul Azam et al., Conflict and Growth in Africa, volume 1: The Sahel (OECD, 1999).

28. Foccart, Journal de l'Elysée, tome I, 743–744.

29. Foccart, Journal de l'Elysée, tome II, 343–344.

30. The history in this and the next few paragraphs is pieced together from documents contained in the following: ADF: DAM 2816, Commission franco-nigérienne de coopération (juillet 1967–décembre 1969); Jacques Baulin, Conseiller du Président Diori (Editions Eurafor-Press, 1986); Foccart Journal de l'Elysée; Paucard, Les mines et les mineurs, volume 3.

31. Pompidou to Diori, 21 novembre 1969, ADF: DAM 2816.

32. Accessible documents in the ADF reference this first payment, but they only go to late 1969. Other sources suggest that subsequent payments may not have been made, but the details are too murky to form a coherent explanation.

33. Born Jacques Batmanian, Baulin sought French nationality during his time advising Félix Houphoët-Boigny, which Foccart claims he personally denied (Journal de l'Elysée, tome III, 123–4).

34. Baulin, Conseiller, 111–119. As Areva's official history makes clear, the CEA found this interprétation particularly outrageous.

35. Pierre Taranger at the meeting of the Comité de l'Énergie Atomique on 7 February 1974, quoted in Paucard et al., La mine et les mineurs, volume 1, 156.

36. See several documents in ANF: AG 5 (FPU) SD 147.

37. Ambassade de France au Gabon, Jean Ribo à Michel Jobert, 1 mars 1974. Synthèse 2–74, période du 1 au 28 février 1974. ANF: AG 5 (FPU) SD 78.

38. Baulin, Conseiller, 101–119.

39. Salifou, Le Niger, 202.

40. Baulin, *Conseiller*, 116.

41. Cited in Baulin, *Conseiller*, 117.

42. Account based on Baulin, 118.

43. Foccart, *Journal de l'Élysée*, tome V, 599.

44. "Niger: le prix de l'uranium," *Politique-Hebdo* (4 avril 1974), quoted in François Martin, *Le Niger du Président Diori: 1960–1974* (L'Harmattan, 1991), 373.

45. Raynaut, "Trente ans d'indépendance," 13.

46. Richard Higgott and Finn Fuglestad, "The 1974 coup d'état in Niger: Towards an explanation." *Journal of Modern African Studies* 13, no. 3 (1975): 396; Raynaut, "Trente ans d'indépendance"; Salifou, *Le Niger*.

47. Qaddafi had refused the central premise of nuclear exceptionalism expressed in the NPT, declaring in 1974 that "the future will be for the atom. . . . Atomic weapons will be like traditional ones. . . . And we in Libya will have our share of this new weapon." Quoted in John Yemma, "Will 'new boys' joining nuclear club be responsible?" *Christian Science Monitor*, 4 May 1981.

48. Peter Blackburn, "Niger's Koutoubi says he's 'prudently optimistic' about future of world uranium market," *Nuclear Fuel* 11, no. 14 (1986): 7.

49. This is *not* the transaction to which the Bush administration was referring in 2002–03.

50. Figures culled from information compiled by the Nuclear Threat Initiative (NTI) and posted in its country profiles at http://www.nti.org/. See also International Institute for Strategic Studies, *Nuclear Black Markets: Pakistan, A. Q. Khan and the Rise of Proliferation Networks* (IISS, 2007).

51. Thomas Gilroy, "Niger cuts off Libyan uranium supply as fear of internal disruption spreads," *Nuclear Fuel* 6, no. 3 (1981): 4.

52. Quoted in Yemma, "Will 'new boys'," 8.

53. "France denies uranium sales," Associated Press, 3 January 1980.

54. In addition to the obvious reasons, the state had taken out huge loans for infrastructural development, using projected sales as a guarantee. Salifou, *Le Niger*; Emmanuel Grégoire, *Touaregs du Niger: Le destin d'un mythe* (Karthala, 1999).

55. Michel Pecqueur, quoted in Douglas Glucroft, "Cogéma registered its first loss last year but says it prepared well for a dry spell," *Nuclear Fuel* 7, no. 15 (1982): 7.

56. Described and quoted in Ian Skeet, *OPEC: Twenty-Five Years of Prices and Policies* (Cambridge University Press, 1988), 131.

57. P. Ampamba-Gouerangue to Ministre d'Etat chargé des Mines, de l'Energie et des Ressources Hydrauliques, Objet: Coopération avec l'Iran, not dated but clearly reporting on a meeting that took place on or just before 10 February 1976. COMUF archives.

58. E. A. M'Bouy-Boutzit (Ministère des Mines, de l'Énergie et des Ressources Hydrauliques) to Directeur Général de la COMUF, 29 October 1975, Objet: Livraison d'uranium à l'Iran. COMUF archives.

59. Edouard Alexis M'Bouy-Boutzit, "Confidentiel-secret note à la haute attention de Monsieur le Président de la République," 6.1.76, p. 2. COMUF archives.
60. Memo from P. Ampamba-Gouerangue to Ministre d'Etat chargé des Mines, de l'Energie et des Ressources Hydrauliques, Objet: Coopération avec l'Iran, not dated but clearly reporting on a meeting that took place on or just before 10 February 1976. COMUF archives.
61. "Compte rendu des négotiations entre le Gouvernement Gabonais—COMUF—et l'Organisation Iranienne de l'Energie Atomique (O.I.E.A.)," 20/02/76. COMUF archives.
62. Ibid.
63. Edouard-Alexis M'Bouy-Boutzit to Monsieur le Président de la République, draft, not dated but clearly written between February and June of 1977. Also "Aide-mémoire sur la négociation COMUF/OIEA," no author, not dated but covers main negotiating points from October 1975 to 31 May 1977; P. Kayser to M. Peccia-Galletto, note 14/DC—PK/jl, 14 février 1977. COMUF archives.
64. H. E. Dr. Akbar Etemad, "Banquet Address," in *Proceedings of the International Symposium on Uranium Supply and Demand, London, June 1977* (Mining Journal Books, 1977), 252–253.
65. "Compte-rendu d'entretien avec M. M'Bouy-Boutzit," 19 juillet 1977; "Entretien du 5 août 1977 avec M. Ampamba," 8.8.1977, Ph.K/JCo. COMUF archives.
66. "Agreement between Compagnie des Mines d'Uranium de Franceville "COMUF" (Gabon) and Atomic Energy Organization of Iran," unsigned, undated contract draft. COMUF archives.
67. Telex, Kayser to Basset, 8 Septembre 1978, UF/N. 446/DC—PK/JL. COMUF archives.
68. Memo, Kayser to Basset, 29 Septembre 1978, Note 114/DC—PK/jl. COMUF archives.
69. "Entretien avec Monsieur M'Bouy-Boutzit," 12 et 19 Octobre 1978. COMUF archives.
70. "Entretien du 5 août 1977 avec M. Ampamba," 8.8.1977, Ph.K/JCo. COMUF archives.
71. Ibid.
72. On Italy, see "Compte-rendu d'entretien avec M. M'Bouy-Boutzit," 19 juillet 1977; on South Korea, see "Entretien avec Monsieur M'Bouy-Boutzit," 12 et 19 Octobre 1978; COMUF, Conseil d'Administration, Procès-verbal (minutes of meetings), 1982–1990, COMUF archives. Also Paucard et al., *Les mines et les mineurs*, volume 2, 220 passim.
73. Foccart, *Journal*, tome III, 196.
74. Ibid., 204–5, 207, 217.
75. Maurice Delauney, *Kala-Kala: De la grande à la petite histoire, un ambassadeur raconte* (Robert Laffont, 1986), 313.
76. Paucard et al., *Les mines et les mineurs*, volume 2, 222.

77. This was Vincent Berger. The quotation is from an obituary: F. Paolini, "Vincent Berger," *Mines: Revue des Ingénieurs*, May–June 2007 (http://www.annales.org/archives/x/berg.html).

78. Bodu, *Les secrets des cuves*.

79. Description and quotes taken from *COMUF Panorama* (April–June 1982).

80. Ndong, "Les multinationales extractives," 84.

81. Paulin Ampamba-Gouerangué, "Uranium in Gabon," in *Uranium and Nuclear Energy: 1982. Proceedings of the Seventh International Symposium held by the Uranium Institute, London, September 1982* (Butterworths, 1983).

82. Ampamba-Gouerangué, "Uranium in Gabon," 125.

83. For example, in 1984 France bought 615 tons at 575 FF/kg, Belgium bought 127 tons at 598.6 FF/kg, and Japan bought 158 tons at 687 FF/kg. COMUF, Conseil d'Administration, Procès-Verbal: séance du 7 décembre 1984. COMUF archives.

84. COMUF, Conseil d'Administration, Procès-Verbal: séances du 2 décembre 1983, 6 décembre 1985. COMUF archives.

85. Verschave, *La Françafrique*.

86. COMUF, Conseil d'Administration, Procès-Verbal: 30 novembre 1989. COMUF archives.

CHAPTER 5 AND PROLOGUE

1. The "founding patrons" included Julius Nyerere, Seretse Khama, Agonsstinho Neto, Kenneth Kaunda, and Olusegun Obasanjo. The "sponsors" included Olof Palme and Coretta Scott King.

2. *Nuclear Collaboration with South Africa: Report of United Nations Seminar*, London, 24–25 February 1979.

3. Edwards and Hecht, "History and the technopolitics of identity."

4. African National Congress, *The Nuclear Conspiracy: FRG Collaborates to Strengthen Apartheid* (PDW-Verlag, 1975).

5. Ibid., 11–14.

6. Ibid., 17.

7. Quoted in Zdenek Červenka and Barbara Rogers, *The Nuclear Axis: Secret Collaboration between West Germany and South Africa* (Julian Freedmann Books, 1978), 89, 76.

8. For a summary of the controversy and some of the evidence, see Jeffrey T. Richelson, *The Vela Incident: Nuclear Test or Meteoroid?* (National Security Archive Electronic Briefing Book 190) (available at http://www.gwu.edu).

9. Some of the most prominent examples: Dan Smith, *South Africa's Nuclear Capability* (World Campaign Against Military and Nuclear Collaboration with South Africa, 1980); Peter Davis, director, "South Africa: the Nuclear File" (film, 1980); J. D. L. Moore, *South Africa and Nuclear Proliferation* (St. Martin's Press, 1987);

Ronald W. Walters, *South Africa and the Bomb: Responsibility and Deterrence* (Lexington Books, 1987).

10. Cohen, *Israel and the Bomb*; Sasha Polakow-Suransky, *The Unspoken Alliance: Israel's Special Relationship with Apartheid South Africa* (Pantheon, 2010).

11. Ferguson, *Global Shadows*; Mitchell, *Rule of Experts*; Carolyn Nordstrom, *Global Outlaws: Crime, Money, and Power in the Contemporary World* (University of California Press, 2007); Janet L. Roitman, *Fiscal Disobedience: An Anthropology of Economic Regulation in Central Africa* (Princeton University Press, 2005); Willem van Schendel and Itty Abraham, *Illicit Flows and Criminal Things: States, Borders, and the Other Side of Globalization* (Indiana University Press, 2005).

12. Telephone interview with Barbara Rogers, 20 January 2010.

13. Barbara Rogers, "Foreign Investment in Namibia," United Nations Council for Namibia, September 1974, 28.

14. Barbara Rogers, "Namibia's Uranium: Implications for the South African Occupation Regime," June 1975.

15. Ibid., 17, 18.

16. Ibid., 19.

17. On the history of the NSC and its relationship with the anti-apartheid movement, see Chris Saunders, "Namibian solidarity: British support for Namibian independence," *Journal of Southern African Studies* 35, no. 2 (2009): 437–454.

18. "What is CANUC? A Brief Overview," not dated; "Principle [*sic*] Areas of Work and Proposals for CANUC in 1981," not dated, BCLAS: MSS AAM 1131.

19. Alun Roberts, *The Rössing File: The Inside Story of Britain's Secret Contract for Namibian Uranium* (Campaign Against the Namibian Uranium Contracts, 1980).

20. Anti-apartheid activists in the West saw this recognition as a victory over the Pretoria regime, failing to see that within the liberation struggle it empowered SWAPO over other organizations. Bauer, *Labor and Democracy in Namibia*; Colin Leys and John S. Saul, *Namibia's Liberation Struggle: The Two-Edged Sword* (Ohio University Press, 1995).

21. "SWAPO statement by Comrade Theo-Ben Gurirab, at Uranium Hearings, July 7, 1980." BCLAS: MSS AAM 1169.

22. "Security Scheme—All copies of scheme distributed on 29 Nov. 1978 to be destroyed by shredding" (no author or company letterhead, but references to individual employees and divisions make clear that the document was authentic), BCLAS: MSS AAM 1169.

23. "SWAPO statement by Comrade Theo-Ben Gurirab."

24. Martin Reith, "Note for the Record: Namibia Uranium," 23 February 1976. TNA: FCO 45/1940.

25. United Nations General Assembly, "Report of the Panel for Hearings on Namibian Uranium. Part Two: Verbatim Transcripts of the Panel Held at Headquarters from 7 to 11 July 1980," A/AC.131/L.163 (United Nations, 30 September 1980), 289.

26. Ibid., 293–4 and 313–29.

27. Ibid., 177–182.

28. Ibid., 168–9.

29. Alastair Macfarlane, Labour Control: Managerial Strategies in the Namibian Mining Sector, PhD thesis, Oxford Polytechnic, 1990, 157–62. Also RUL: Board of Directors 41st meeting, 30 September 1976.

30. Managing Director, Report to Directors, December 1979. RAS.

31. Board of Directors 44th meeting, 28 April 1977. RAS.

32. Georg Kreis, *Switzerland and South Africa, 1948–1994: Final Report of the NFP 42+ Commissioned by the Swiss Federal Council* (Peter Lang, 2007), 441–444.

33. Board of Directors 43rd meeting, 17 February 1977. RAS.

34. Managing Director, Report to Directors, May 1980. RAS.

35. These clauses read as follows: "The parties hereto acknowledge and accept that all or any of the concentrates sold here-under shall be used solely for civil purposes and shall be subject to the safeguards requirements of the International Atomic Energy Agency." Again, though, the IAEA didn't enforce such safeguards requirements, so the clause was little more than window dressing. Example taken from RUL: Agreement between Tokyo Electric Power Co., Inc., and RTZ Mineral Services Limited, 30 April 1974 (amended 3 February 1986). RAS.

36. RUL: Board of Directors 80th meeting, 7 December 1983.

37. Minserve negotiated the "world market price" for all Japanese customers in a single block, but the others were negotiated separately. RAS: Minutes of Sales Meeting, 13 April 1981.

38. Board of Directors 62nd meeting, 5 June 1980. RAS.

39. Board of Directors 75th meeting, 16 March 1983. RAS.

40. Board of Directors 74th meeting, 27 October 1982 and 75th meeting, 8 December 1982. RAS.

41. Board of Directors 70th meeting, 2 December 1981. RAS.

42. Board of Directors 75th meeting, 8 December 1982. RAS.

43. Board of Directors 88th meeting, 23 August 1985. RAS.

44. SWAPO, *Trade Union Action on Namibian Uranium: Report of a Seminar for West European Trade Unions Organized by SWAPO of Namibia in Co-Operation with the Namibia Support Committee*, 1982.

45. CANUC, *Namibia: A Contract to Kill. The Story of Stolen Uranium and the British Nuclear Programme* (Action on Namibia Publications, Namibia Support Committee, 1986).

46. Michael Bess, *Realism, Utopia, and the Mushroom Cloud: Four Activist Intellectuals and Their Strategies for Peace, 1945–1989* (University of Chicago Press, 1993); Lawrence S. Wittner, *Toward Nuclear Abolition: A History of the World Nuclear Disarmament Movement, 1971 to the Present* (Stanford University Press, 2003).

47. CANUC, *A Contract to Kill*, 67.

48. Ibid., 75.

49. Ibid., 75.

50. Minutes of Sales Meetings, 16 September 1983. RAS.

51. Minserve, London Office Memorandum, J. H. G. Senior to P. Daniel, 21 February 1986; Minserve, London Office Memorandum, G. R. Elliott to Sir Alistair Frame, 9 March 1988. RAS.

52. Minutes of Sales Meetings, 22 November 1984. Also see Kreis, *Switzerland and South Africa*, 442. RAS.

53. Minutes of Sales Meeting, 16 September 1983. RAS.

54. Board of Directors 90th meeting, 5 March 1986. RAS.

55. RTZ 1984 Annual Report, quoted in Macfarlane, *Labour Control*, 271.

56. Kreis, *Switzerland and South Africa*, 383–85.

57. C. A. Algar, "Sanctions Against South Africa which might affect imports by Rössing Uranium Limited," report, 22 January 1986. RAS.

58. GLS/mje, "Mandatory UN Sanctions," 11 June 1986. RAS.

59. R. B. Carlisle to G. L. Stobart, 31 October 1986. RAS.

60. S.P.C. Stewart to P. Daniel, 20 February 1986. RAS.

61. Ibid.

62. Minserve, London Office Memorandum, J. H. G. Senior to P. Daniel, 21 February 1986. RAS.

63. Minserve, London Office Memorandum, J. H. G. Senior to P. Daniel, 21 February 1986 (emphasis added). RAS.

64. Minserve, London Office Memorandum, J. H. G. Senior to P. Daniel, 21 February 1986. RAS.

65. Board of Directors 93rd meeting, 21 November 1986; Minserve, London Office Memorandum, C. A. Macaulay to C. A. Gibson, 4 September 1986. RAS.

66. Board of Directors 93rd meeting, 21 November 1986. RAS.

67. Minserve, London Office Memorandum, G. R. Elliott to Alistair Frame, 9 March 1988. RAS.

68. Minserve, London Office Memorandum, G. R. Elliott to Alistair Frame, 9 March 1988; Board of Directors 93rd meeting, 21 November 1986. RAS.

69. Diane Harmon, "Report for the September 1, 1987 meeting with Minserve Administration A.G.," 31 August 1987; Diane Harmon to John Senior, 26 September 1987; RUL, Board of Directors 95th meeting, 5 June 1987. RAS.

70. Diane Harmon to John Senior, 26 September 1987. RAS.

71. Board of Directors 95th meeting, 5 June 1987. RAS.

72. NRC, SECY-87-223, "Imports of South African origin uranium," 17 September 1987 (emphasis added). The NRC also agreed that it wouldn't consider Rössing to be a parastatal company. RUL: Diane Harmon to John Senior, 17 September 1987.

73. Minserve, London Office Memorandum, G. R. Elliott to Sir Alistair Frame, 9 March 1988. RAS.

74. Dutch activists, including one parliamentarian, first articulated this argument at the 1980 UNCN hearings (UNGA, "Hearings on Namibian Uranium, 123,

passim.) For more on the Urenco lawsuit from an activist perspective, see David de Beer, "The Netherlands and Namibia: The political campaign to end Dutch involvement in the Namibian uranium trade," in *Allies in Apartheid: Western Capitalism in Occupied Namibia*, ed. A. Cooper (St. Martin's Press, 1988). An overview of the suit can be found in Schrijver, *Sovereignty over Natural Resources.*

75. de Beer, "The Netherlands and Namibia," 127.

76. Minserve, Minutes of a Sales Meeting, 28 February 1986. RAS.

77. Minserve, Minutes of a Sales Meeting, 15 February 1988. RAS.

78. Euratom's reluctance was apparently precipitated by the European Parliament's investigation into the Transnuklear/Mol affair, which evidently brought to light unsavory details about the depth of Europe's dependence on southern African uranium. Euratom officials told Minserve that they were "required to ensure security of supply to the Community and in their view the predominance of Southern African material might jeopardise this." RUL: Minserve, Minutes of Sales Meeting, 13 December 1988.

79. Minserve, Minutes of a Sales Meeting, 29 October 1986. RAS.

80. F. McGoldrick, "Flag swaps," in *Uranium and Nuclear Energy: 1988. Proceedings of the Thirteenth International Symposium held by the Uranium Institute* (Uranium Institute, 1989), 48-9.

81. Minutes of Sales Meetings, 18 December 1985. RAS.

82. Minserve, London Office Memorandum, J. H. G. Senior to Z. Ngavriue and M. Bates, 9 May 1988; RTZ, Internal Memorandum, J. G. Hughes to A. E. Buxton, 4 July 1988. In the end, the presentation in Washington didn't take place, because the timing conflicted with the US presidential election. RAS.

BORDERLANDS

1. Ferguson, *Global Shadows*; Nordstrom, *Global Outlaws*; Roitman, *Fiscal Disobedience*; van Schendel and Abraham, *Illicit Flows.*

2. Murphy, *Sick Building Syndrome*. For how such issues relate to radiation exposure, see Adriana Petryna, *Life Exposed: Biological Citizens after Chernobyl* (Princeton University Press, 2002).

CHAPTER 6 AND PROLOGUE

1. For examples, see Michael A. Amundson, *Yellowcake Towns: Uranium Mining Communities in the American West* (University Press of Colorado, 2002); Howard Ball, *Cancer Factories: America's Tragic Quest for Uranium Self-Sufficiency* (Greenwood, 1993); Timothy Benally, Phil Harrison, Chenoa Bah Stilwell, and Navajo Uranium Miner Oral History and Photography Project, *Memories Come to Us in the Rain and the Wind: Oral Histories and Photographs of Navajo Uranium Miners and Their Families* (Navajo Uranium Miner Oral History and Photography Project, 1997); Peter H. Eichstaedt, *If You Poison Us: Uranium and Native Americans*

(Red Crane Books, 1994); Valerie Kuletz, *The Tainted Desert: Environmental Ruin in the American West* (Routledge, 1998); George T. Mazuzan and J. Samuel Walker, *Controlling the Atom: The Beginnings of Nuclear Regulation, 1946–1962* (University of California Press, 1984); Eric W. Mogren, *Warm Sands: Uranium Mill Tailings Policy in the Atomic West* (University of New Mexico Press, 2002); Norman Moss, *The Politics of Uranium* (Universe Books, 1982); Simon J. Ortiz and David King Dunaway, "An interview with Simon Ortiz: July 14, 1988," *Studies in American Indian Literatures* 16 (2005): 12–19; Judy Pasternak, *Yellow Dirt: An American Story of a Poisoned Land and a People Betrayed* (Free Press, 2010); Robert Proctor, *Cancer Wars: How Politics Shapes What We Know and Don't Know About Cancer* (Basic Books, 1995); Raye C. Ringholz, *Uranium Frenzy: Boom and Bust on the Colorado Plateau* (Norton, 1989); Raymond W. Taylor and Samuel Woolley Taylor, *Uranium Fever; or, No Talk under $1 million* (Macmillan, 1970); Stewart L. Udall, *The Myths of August: A Personal Exploration of Our Tragic Cold War Affair with the Atom* (Pantheon Books, 1994); J. Samuel Walker, *Containing the Atom: Nuclear Regulation in a Changing Environment, 1963–1971* (University of California Press, 1992).

2. W. C. Hueper, *Occupational Tumors and Related Diseases* (Thomas, 1942).

3. L. Teleky, "Occupational cancer of the lung," *Journal of Industrial Hygiene and Toxicology* 19 (1937): 73.

4. Zbynek Zeman and Rainer Karlsch, *Uranium Matters: Central European Uranium in International Relations, 1900–1960* (Central European University Press, 2008).

5. Proctor, *Cancer Wars*, 190.

6. Hon. W. Wirtz, Statement before the US Congress. Subcommittee on Research, Development, and Radiation. Joint Committee on Atomic Energy. *Radiation Exposure of Uranium Miners*. 90th Congress, 1st Session. 1967.

7. Hueper, *Occupational Tumors*.

8. M. Susan Lindee, *Suffering Made Real: American Science and the Survivors at Hiroshima* (University of Chicago Press, 1994); Catherine Caufield, *Multiple Exposures: Chronicles of the Radiation Age* (University of Chicago Press, 1990); Barton Hacker, *The Dragon's Tail: Radiation Safety in the Manhattan Project, 1942–1946* (University of California Press, 1987); Barton Hacker, *Elements of Controversy: The Atomic Energy Commission and Radiation Safety in Nuclear Weapons Testing, 1947–1974* (University of California Press, 1994).

9. For an insider history, see Roger Clarke and Jack Valentin, "A history of the International Commission on Radiological Protection," *Health Physics* 88, no. 4 (2005). For an insider history of radiological standards in the US, see J. Samuel Walker, *Permissible Dose: A History of Radiation Protection in the Twentieth Century* (University of California Press, 2000).

10. Soraya Boudia, "Global regulation: Controlling and accepting radioactivity risks," *History and Technology* 23, no. 4 (2007): 389–406; Soraya Boudia, "Sur les dynamiques de constitution des systèmes d'expertise scientifique. La naissance du

système d'évaluation et de régulation des risques des rayonnements ionisants," *Genèses* 70 (2008): 26–44.

11. International Atomic Energy Agency, International Labour Organisation, and World Health Organization, *Radiological Health and Safety in Mining and Milling of Nuclear Materials: Proceedings*, volumes 1 and 2 (International Atomic Energy Agency, 1964).

12. D. A. Holaday and H. N. Doyle, "Environmental studies in the uranium mines," in *Radiological Health and Safety in Mining and Milling of Nuclear Materials: Proceedings*, volume 1 (IAEA, 1964), 15.

13. *Radiological Health and Safety*, 357.

14. Duncan A Holaday et al., *Control of Radon and Daughters in Uranium Mines and Calculations on Biologic Effects* (US Public Health Service, 1957).

15. V. E. Archer et al., "Epidemiological studies of some non-fatal effects of uranium mining," in *Radiological Health and Safety*, 23.

16. Holaday and Doyle, "Environmental studies in the uranium mines," 13.

17. Archer et al., "Epidemiological studies," 21.

18. D. A. Holaday, "Some unsolved problems in uranium mining," in *Radiological Health and Safety*, 51.

19. Ibid., 54.

20. Robert Avril, Charles Berger, Francis Duhamel, and Jacques Pradel, "Measures adopted in French uranium mines to ensure protection of personnel against the hazards of radioactivity," in *Proceedings of the Second United Nations International Conference on the Peaceful Uses of Atomic Energy, Held in Geneva, 1–13 September 1958*, volume 21: *Health and Safety: Dosimetry and Standards* (United Nations, 1985), 63.

21. Ibid., 64.

22. F. Duhamel, M. Beulaygue, and J. Pradel, "Organisation du contrôle radiologique dans les mines d'uranium françaises," in *Radiological Health and Safety*, 63; Holaday and Doyle, "Environmental studies in the uranium mines,"19.

23. F. Billard, J. Miribel, G. Madeleine, and J. Pradel, "Méthodes de mesure du radon et de dosage dans les mines d'uranium," in *Radiological Health and Safety*, 415.

24. Ibid., 423.

25. Paucard, *La mine et les mineur*, 50.

26. D. Mechali and J. Pradel, "Evaluation de l'irradiation externe et de la contamination interne des travailleurs dans les mines d'uranium françaises," in *Radiological Health and Safety*, 373.

27. R. E. Albert (AEC Division of Biology and Medicine), Memo to files, subject: Medical Services in the South African Gold Fields and the Shinkolobwe Uranium Mine." NV0727618. DOE Nevada Test Site electronic archives.

28. Ibid.

29. These were listed as "factors which influenced cases sent to autopsy by the medical attendants (personal interests and bias etc.), religious grounds for relatives refusing autopsy, type of cases treated in the hospital (e.g., special clinics), etc."

S. F. Oosthuizen et al., "Experience in radiological protection in South Africa," in *Second United Nations International Conference on the Peaceful Uses of Atomic Energy*, 25–31.

30. For an analysis of how population categories have only recently changed in American medical research, see Steven Epstein, *Inclusion: The Politics of Difference in Medical Research* (University of Chicago Press, 2007). On apartheid science, see Dubow, *A Commonwealth of Knowledge*.

31. R. E. Albert (AEC Division of Biology and Medicine), Memo to files, subject: Medical Services in the South African Gold Fields and the Shinkolobwe Uranium Mine." NV0727618. DOE Nevada Test Site electronic archives.

32. C. G. Stewart, and S. D. Simpson, "The hazards of inhaling radon-222 and its short-lived daughters: consideration of proposed maximum permissible concentrations in air," in *Radiological Health and Safety*, 333–357. Stewart headed the ICRP from 1969 to 1977.

33. *Radiological Health and Safety*, 347.

34. This study was conducted in collaboration with the National Research Institute for Occupational Research and a group of researchers at the Chamber of Mines.

35. J. K. Basson et al., Lung Cancer and Exposure to Radon Daughters in South African Gold/Uranium Mines, Atomic Energy Board: PEL 209, Pelindaba, March 1971.

36. R. G. Beverly to J. T. Sherman, subject: Report entitled "Lung Cancer and Exposure to Radon Daughters in South African Gold/Uranium Mines," NV0061126; R. D. Evans to C. R. Richmond, subject: Report on Lung Cancer and Exposure to Radon Daughters in South African Gold/Uranium Mines, NV0061125. DOE Nevada Test Site electronic archives.

37. A. H. Wolff to I. Mitchell, subject: Lung Cancer and Exposure to Radon Daughters in South African Gold/Uranium Mines (no enclosures), NV0061124; M. A. Schneiderman (National Cancer Institute) to Deputy Assistant Administrator for R&D, EPA, subject: Report concerning White South African Gold Miners and Bronchiogenic Cancer, NV0061122; V. E. Archer to A. Wolff, subject: Preliminary Report re: Lung Cancer and Exposure to Radon Daughters in South African Gold/Uranium Mines (Criticisms of Report), NV0061123. DOE Nevada Test Site electronic archives.

38. J. K. Basson et al., "A biostatistical investigation of lung cancer incidence in South African gold/uranium miners," in *Proceedings of the Fourth International Conference on the Peaceful Uses of Atomic Energy*, Geneva, 6–16 September 1971, volume 11 (IAEA, 1972), 13–30.

39. IAEA, *Radon in Uranium Mining: Proceedings of a Panel Held in Washington, DC 4–7 September 1973* (1975), 7, 48.

40. Ibid., 28.

41. Ibid., 54.

42. Ibid., 57.

43. Ibid., 34.

44. Ibid., 12.

45. Example: Thomas O'Toole, "Miner cancer deaths tied to smoking," *Washington Post*, 19 October 1971. This article was reprinted in numerous local papers, and the finding itself was cited frequently—including by McGinley during the 1973 panel on radon in Washington (ibid., 9).

46. Proctor, *Cancer Wars*; Rosner and Markowitz, *Deadly Dust*; Markowitz and Rosner, *Deceit and Denial*.

47. An early publication of this research was Jonathan M. Samet et al., "Uranium mining and lung cancer in Navajo men," *New England Journal of Medicine* 310, no 23: 1481–1484.

48. Two publications from this study had appeared: J. Chameaud, R. Perraud, and J. Lafuma, "Cancers du poumon expérimentaux provoqués chez le rat par des inhalations de radon," *Compte rendu de l'Académie des Sciences* 273 (1971): 2388–2389; R. Perraud, J. Chameaud, J. Lafuma, R. Masse, and J. Chrétien, "Cancer broncho-pulmonaire expérimental du rat par inhalation de radon. Comparaison avec les aspects histologiques des cancers humains," *Journal français de médecine et de chirurgie thoraciques* 26, no. 172 (1972): 25–41.

49. An early English language publication of this research was M. Tirmarche et al., "Mortality of a cohort of French uranium miners exposed to relatively low radon concentrations," *British Journal of Cancer* 67: 1090–97.

50. Proctor traces out these debates, for chemical exposures in particular, in chapter 7 of *Cancer Wars*.

51. Henri Jammet, "Les problèmes de protection posés dans l'extraction et le traitement de l'uranium et du thorium," in *Radiation Protection in Mining and Milling of Uranium and Thorium* (International Labour Office, 1976), 3–10.

52. Massan Quadjovie, "Mesures techniques et administratives de radioprotection dans les exploitations d'uranium de Mounana," in *Radiation Protection in Mining*, 141.

53. Interview with Dominique Oyingha, Mounana, Gabon, 17 July 1998.

54. "Radiation protection in uranium and other mines," *Annals of the ICRP* 1, no. 1 (1977).

55. Boudia, "Global Regulation" and "Sur les dynamiques de constitution des systèmes d'expertise scientifique."

56. "Radiation protection in uranium and other mines," 4.

57. Clarke and Valentin, "A history of the International Commission on Radiological Protection," 7.

58. This is consistent with Ulrich Beck's observation that "determinations of risk are the form in which ethics, and with it also philosophy, culture and politics, is resurrected inside the centers of modernization." Beck, *Risk Society: Toward a New Modernity* (Sage, 1992), 28.

59. Caufield, *Multiple Exposures*, 179–183. Proctor traces ALARA back to the NCRP's rejection of the threshold theory in 1948 (*Cancer Wars*, 158–9).

60. Implications of Commission Recommendations that Doses Be Kept as Low as Readily Achievable. A report of ICRP Committee 4 (ICRP Publication 22, 1973), 7.

61. Ibid., 3.

62. Ibid., 4.

63. Ibid.,16.

64. B. Lindell, "Basic concepts and assumptions behind the new ICRP recommendations," in *Application of the Dose Limitation System for Radiation Protection: Practical Implications* (IAEA, 1977); Revised IAEA Basic Standards for Radiation Protection, Vienna, 5–9 March 1979, 3 (emphasis added).

65. As explained by David Sowby of the ICRP. See Caufield, *Multiple Exposures*, 183.

66. Lindell, "Basic concepts and assumptions behind the new ICRP recommendations," 8.

67. Michael Thorne, quoted in Caufield, *Multiple Exposures*, 183.

68. IAEA, *Application of the Dose Limitation System*, 304–315.

69. J. C. Zerbib, "Les recommandations de la CIPR et les travailleurs," in IAEA, *Application of the Dose Limitation System*, 387–413.

70. Zerbib, "Les recommandations de la CIPR, 404–5.

71. Manuel Gomez, ed., *Radiation Hazards in Mining: Control, Measurement, and Medical Aspects* (American Institute of Mining, Metallurgical, and Petroleum Engineers, 1981).

72. J. Kruger, P. J. Kruger, and A. H. Leuschner, "A comparison of the methodologies of intake measurement and bioassay for assessing exposure to personnel in uranium milling operations," in Gomez, *Radiation Hazards in Mining*.

73. ICRP Publication 32 (1981).

74. These arguments were on display throughout the Technical Committee Meeting on Impact of the New ICRP Occupational Dose Limits on the Operation of Underground Mines, held at IAEA headquarters 3–7 June 1991; they are too numerous to cite individually. CoM-ES archives.

75. S. E. Frost, "Impact of the New ICRP Occupational Dose Limits on the Operation of Underground Mines," April 1991, p. 22, produced for the Technical Committee Meeting on Impact of the New ICRP Occupational Dose Limits on the Operation of Underground Mines, 3–7 June 1991, IAEA. CoM-ES archives.

76. Naomi Oreskes and Erik M. Conway, *Merchants of Doubt: How a Handful of Scientists Obscured the Truth on Issues from Tobacco Smoke to Global Warming* (Bloomsbury, 2010).

77. Frost, "Impact of the New ICRP Occupational Dose Limits," 33.

78. Jay H. Lubin et al., *Radon and Lung Cancer Risk: A Joint Analysis of 11 Underground Miners Studies*, NIH report 94-3644 (National Institutes of Health, 1994).

79. Epstein, *Inclusion*, especially 152 passim. Epstein doesn't define this as a specifically transnational process.

80. Paul N. Edwards, *A Vast Machine: Computer Models, Climate Data, and the Politics of Global Warming* (MIT Press, 2010).

81. Robert Proctor, "Agnotology: A missing term to describe the cultural production of ignorance (and its study)," in *Agnotology: The Making and Unmaking of Ignorance*, ed. R. Proctor and L. Schiebinger (Stanford University Press, 2008), 3.

CHAPTER 7 AND PROLOGUE

1. Hecht, *Radiance*.

2. CEA/DP/DREM, Groupement Afrique-Madagascar, "Notice d'Information destinée aux Européens susceptibles de partir pour l'Afrique ou Madagascar" (1.2.63), 2, 4, Cogéma archives. Henceforth cited as "Notice."

3. Ibid., 22.

4. Ibid., 18–19.

5. Ibid., 14.

6. Ibid., 12, 11, 9.

7. On middles and intermediaries, see Nancy Rose Hunt, *A Colonial Lexicon of Birth Ritual, Medicalization, and Mobility in the Congo* (Duke University Press, 1999); Benjamin N. Lawrance, Emily Lynn Osborn, and Richard L. Roberts, eds., *Intermediaries, Interpreters, and Clerks: African Employees in the Making of Colonial Africa* (University of Wisconsin Press, 2006).

8. Interviews: Lucien and Lucienne Gabillat, 29 January 2001; Geneviève and Jacques Michel, 26 January 2001. R. Bodu, "Compte-rendu de mission mars 1960," DREM, 12/7/60. Cogéma archives.

9. Interview with Paul Rasolonjatovo, 25 and 26 January 2001, France.

10. Interview with Yves Legagneux, 24 February 1998, France.

11. Interviews with Fanahia, 13 and 14 August 1998.

12. Indeed, the Tandroy way with zebus continues to cause consternation among other Malagasies today, as was plenty evident from remarks made to me during my visit to Madagascar.

13. Xavier des Ligneris, sm n. 27, 9/1/61, Directeur des Exploitations à M. le Directeur Général. "Effectifs—Logements—Budget 1961." COMUF archives.

14. Ibid.

15. Examples in "Note de Service à MM. les Responsables du marché," 8.4.63; "Note d'Information à tous les Agents," 19.2.64; "Note de Service n. 49 à MM. les adhérents et invités du Cercle des Employés," 27.5.64; "Rapport d'Activités, Service Social," 1967. COMUF archives.

16. Interview with Juste Mambangui, 13 July 1998.

17. On the Ghanaian response, see Jean Allman, "Nuclear imperialism and the pan-African struggle for peace and freedom: Ghana, 1959–1962," *Souls* 10 (2008): 83–102.

18. Robert Bodu, "Compte-rendu de mission à Madagascar," Direction des Recherches et Exploitations Minières, Mars 1960. Cogéma archives.

19. Hecht, *Radiance*.

20. Fanahia interviews.

21. Interview with Mahata, 16 August 1998 (translators: M. Abdoulhamide and Georges Heurtebize). Although I didn't know it at the time, such questions had their obverse in northern Madagascar, where miners speculated that sapphires were used in bombs. See Andrew Walsh, "In the wake of things: Speculating in and about sapphires in Northern Madagascar," *American Anthropologist* 106, no. 2 (2004): 225–37.

22. Fanahia interviews. Such investments strategies contrast with the "daring consumption" that Andrew Walsh describes for some young men working in the 1990s in the sapphire-mining town of Ambondromifehy in "Hot money" and daring consumption in a northern Malagasy sapphire-mining Town," *American Ethnologist* 30, no. 2 (2003): 290–305. The people I interviewed were, necessarily, long-term inhabitants of the region with deep social networks that bolstered and justified such investments; I do not know how migrant workers spent their wages.

23. Mahata interview.

24. Fanahia interview.

25. Interview with Jeremy Fano, 18 August 1998 (translator: M. Abdoulhamide).

26. Paucard, *La mine et les mineurs*, volume 2, 323.

27. Ibid.

28. Interview with group of women, Madagascar 1998, anonymity requested.

29. Interview with Joseph Ramiha, 12 August 1998 (translators: M. Abdoulhamide and Georges Heurtebize).

30. Ramiha interview.

31. M. Abdoulhamide, during the Ramiha interview.

32. Ibid.

33. Marc Edmond Morgaut, "Mission à Madagascar pour le Commissariat à l'Energie Atomique du 11 au 21 novembre 1958." Cogéma archives.

34. These were typically companies that also extracted other resources, especially mica and sisal. They included the de Heaulme brothers and Lanoue.

35. Fanahia interviews.

36. M. Harel, "Rapport Mensuel 'Exploitation,'" Mois de Décembre 1964," CEA-DP, GAM—Division Sud. Cogéma archives.

37. Robert Bodu, "Compte-rendu de mission à Madagascar," IX-4, Direction des Recherches et Exploitations Minières, Mars 1960. Cogéma archives.

38. Ibid., IX-5.

39. Ibid.

40. Y. Legagneux, "Rapport d'Activité du Service 'Expoitation,'" mai 1955, 21. CEA- DREM, Mission de Madagascar, Division du Sud. Cogéma archives.

41. Paul Texier, "Rapport d'Activité du Service 'Expoitation,'" Juillet 1959, no page number. CEA- DREM, Mission de Madagascar, Division du Sud. Cogéma archives.

42. Email communication from Sandrine Clerc to Gabrielle Hecht, 16 May 2007.

43. The grade of Malagasy uranothorianite could reach up to 20% uranium and 50% thorium. Robert Bodu, "Compte-rendu de mission à Madagascar," Direction des Recherches et Exploitations Minières, Mars 1960; Robert Bodu, "Le Traitement des Minerais Radioactifs à Madagascar de 1946 à 1968," avril 1986, 32. Cogéma archives.

44. IAEA, *Manual on Radiological Safety in Uranium and Thorium Mines and Mills*, 9–10.

45. Xavier des Ligneris to Secrétaire Général, le 8 juillet 1961. COMUF archives.

46. Xavier des Ligneris to Secrétaire Général, le 8 juillet 1961; X. des Ligneris, "Consignes Relatives à la Protection Contre les Dangers dus à la Radioactivité," Mounana le 5 mai 1961, approuvé par le Directeur des Mines du Gabon, Libreville, le 1 juin 1961. COMUF archives.

47. Pierre le Fur, Note de Service 072bis, 3/9/64. COMUF archives.

48. X. des Ligneris, "Rapport—Contrôle des radiations," HR/AP 2076, 5 janvier 1968. COMUF archives.

49. Henri Pello, Service Exploitation, Note d'organization, "Stockage et distribution des film detecteurs de radioactivité," 26/9/66. COMUF archives.

50. X. des Ligneris, "Rapport—Contrôle des radiations," HR/AP 2076, 5 janvier 1968. COMUF archives.

51. X. des Ligneris, "Rapport—Contrôle des radiations," HR/AP 2076, 5 janvier 1968; X. des Ligneris, "Rapport—sur le contrôle des risques radioactifs. Février 1968," YT/AP 2169, 21 mars 1968. COMUF archives.

52. X. des Ligneris, "Rapport—Contrôle des radiations," HR/AP 2076, 5 janvier 1968; X. des Ligneris, "Rapport—sur le contrôle des risques radioactifs. Février 1968," YT/AP 2169, 21 mars 1968; X. des Ligneris, "Reference: Votre UF/JL/JF29/68," HP/MB 210/69, 27 janvier 1968; X. des Ligneris, "Rapport sur le contrôle des risques radioactifs. Mois de mai 1968," YT/LR 2275, 20 juin 1968. COMUF archives.

53. J. de Courlon to X. des Ligneris, 10/3/67. COMUF archives.

54. X. des Ligneris, "Reference: Votre UF/JL/JF29/68," HP/MB 210/69, 27 janvier 1968. COMUF archives.

55. Ibid.

56. Interview with Christian Guizol, 26 February 1998.

57. Numerous interviews and conversations in Mounana.

58. C. Guizol, "Rapport sur le contrôle des risques radioactifs. Mois de décembre 1969," YT/sc 0118/70, 9 février 1970. COMUF archives.

59. In part, the difference had to do with assumptions about how long workers would spend in radioactive environments (8 hours a day for 365 days, or 8 hours a day for 264 days—i.e., discounting weekends and holidays). The discrepancy also involved assumptions about the state of equilibrium between radon and its

noxious daughters. I found no evidence that the COMUF actually measured this equilibrium. Rather, Guizol apparently decided—citing the engineering literature—that ventilation at Mounana was "normal" and that radon remained in its shafts "for less than half an hour after it gets released." C. Guizol, "Rapport sur le contrôle des risques radioactifs. Mois de Décembre 1969," YT/sc 0118/70, 9 février 1970. COMUF archives.

60. C. Guizol, "Rapport sur le contrôle des risques radioactifs. Mois de Mars 1970," YT/sc 0184/70, 27 avril 1970. COMUF archives.

61. Since the archives weren't at all organized (see appendix), it was impossible to find complete records on any single topic. My not finding records of state inspections is therefore inconclusive.

62. "Consigne pour la distribution et l'emploi des explosifs," COMUF, Exploitation de Mounana (not dated, ca. 1959). COMUF archives.

63. Convention passée entre la Compagnie des Mines d'uranium de Franceville et le Représentants des Travailleurs (Délégués du Syndicat UTMG, Section de Mounana), 1 avril 1965; C. L. Durand to Directeur Général à Neuilly, 28 Décembre 1965. COMUF archives.

64. Interview with Marcel Lekonaguia, 21 July 1998.

65. Interview with Dominique Oyingha, 17 July 1998.

66. Ibid.

67. Ibid.

68. Ibid.

69. Conversation with author, 13 July 1998.

70. Ibid.

71. Ibid.

72. Christian Guizol to Directeur Général de la Caisse Gabonaise de la Prévoyance Sociale, 19/10/70, Objet: Allocations familiales de M. Lekonaguia Marcel; Christian Guizol to Directeur Général de la Caisse Gabonaise de la Prévoyance Sociale, 26/10/70, Objet: Monsieur Lekonaguia; J. C. Andrault to Docteur C. Gantin, 27/10/70. COMUF archives.

73. Oyingha interview.

74. Christian Guizol to Directeur Général de la Caisse Gabonaise de la Prévoyance Sociale, 19/10/70, Objet: Allocations familiales de M. Lekonaguia Marcel; Christian Guizol to Directeur Général de la Caisse Gabonaise de la Prévoyance Sociale, 26/10/70, Objet: Monsieur Lekonaguia; J. C. Andrault to Docteur C. Gantin, 27/10/70. COMUF archives.

75. "Rapport Trimestrielle [sic] Radioprotection," Octobre-Novembre-Décembre 1984 (COMUF, Mounana); "Rapport Trimestriel Radioprotection," avril-mai-juin 1986 (COMUF, Mounana), COMUF archives. Apparently workers didn't get sanctioned for keeping the films. Was this thanks to another change in management? Guizol had been replaced by then, and the new director, Henri Basset, was less interested displaying discipline and more focused on steering Gabonese

employees toward management-level positions. Many workers I spoke with agreed with the assessment of one former employee: "That one, he was the good director." Interview with François Mambangui, 31 July 1998.

76. J. P. Laurent, "Compte rendu des visites effectuées par J. P. Laurent du 20 au 28 avril," 25 août 1976. COMUF archives.

77. P. de Bonald, "Rapport de Mission Sécurité à Mounana (COMUF), janvier 1977," 3 février 1977. COMUF archives.

78. Ibid.

79. J. Haas to Directeur COMUF, Objet: Assistance en Radioprotection, Prestation CRPM, 30 mai 1989. COMUF archives.

80. Rapport mensuel Radioprotection, avril 1992. COMUF archives.

81. Interview with Lambert Ondounda, 16 July 1998.

82. Lekonaguia interview.

83. Oyingha interview.

84. Ferguson, *Expectations of Modernity*.

85. Martin Magnana (Gouverneur de la Province du Haut-Ogooué) to Directeur Général de la COMUF, 11 avril 1983, n. 231 PHO/CAB and Note de renseignements n. 15 du 7/04/1983. COMUF archives.

86. COMUF, Direction Générale Adjointe Technique, PT/MJM, 29 avril 1983, "Compte-rendu de la reunion de travail du 25.04.83." COMUF archives.

87. The other countries in attendance were Egypt, Morocco, Niger, Nigeria, Sudan, Zaire, and Ivory Coast. The Gabonese delegation consisted of 22 people representing the ministries of energy, mines, education and research, and public health.

88. J. Moine to P. Zettwoog, 15 septembre 1983, re: Séminaire AIEA sur la radioprotection, Libreville 14–25 novembre 1983. COMUF archives.

89. "Recommandations envoyées à Lynn Marple qui doit présenter au Séminaire Gabon, des conférences sur les résidus miniers" (not dated; filed in the COMUF archives with other correspondence on the 1983 Libreville seminar; emphasis added).

90. Ibid.

91. VJ/JM, "Mission de Monsieur Jug à Mounana du 5 au 13 novembre 1984. Compte rendu concernant la gestion des rejets," 23.11.84; VJ/JM, "Mission de Monsieur Jug à Mounana du 5 au 13 novembre 1984. Compte rendu concernant la partie mines," 30.11.84. COMUF archives.

92. V. Jug to J. Moine, 29.5.86. COMUF archives.

93. Ibid.

94. Ibid.

95. UF/DT-n. 29-VJ/JM, V. Jug to H. Basset, 9 mars 1987, Objet: COMUF/État radiologique site de Mounana. COMUF archives.

96. N. Fourcade to V. Jug, 20.2.87, Objet: État radiologique site de COMUF/Mounana, Envoi de rapport, UF/DT-n. 29-VJ/JM, V. Jug to H. Basset, 9 mars 1987, Objet: COMUF/État radiologique site de Mounana. COMUF archives.

97. N. Fourcade and M. C. Robe, "Synthèse des résultats de mesures de concentration en radon 222 sur le site de Mounana, de 1984 à 1985. Comparaison du site de Mounana avec des site français," Rapport n. 4, COM/002 (2), 23 février 1987 (CEA, IPSN, Laboratoire de Recherche sur la Protection des Mines). COMUF archives.

98. In Gabon, they ranged from 2.5 picocuries in very wet January to 5.8 picocuries in dry July; the average was 3.5 picocuries over the course of the whole year. In France, the report cited readings ranging from 4.7 to 5.9 picocuries at one site, and from 2.0 to 8.0 picocuries at another.

99. Fourcade and Robe, "Synthèse des résultats," 16.

100. Loi no. 11/2001 du 12 décembre 2001 fixant les orientations de la politique de prévention et de protection contre les rayonnements ionisants, Hebdo informations, Journal hebdomadaire d'informations et d'annonces légales, 2002-02-23, no. 451, 22–23 (Gabon).

101. Jules Mbombe Samaki, "Memorandum sur la nécessité de la prise en compte de la veille sanitaire et du dédommagement des anciens travailleurs miniers," Libreville, 25 avril 2005 (private communication). See also reports in the Gabonese press: "Le Collectif des anciens travailleurs miniers interpelle la Comuf," L'union, 3 February 2006; "Les anciens travailleurs miniers de la Comuf réunis en collectif," L'union, 17 February 2006.

102. Jules Mbombe Samaki, "Memorandum sur la nécéssité de la prise en compte de la Veille sanitaire et du dédommagement des anciens travailleurs miniers," private, anonymous communication, Libreville, 25 April 2005.

103. Samira Daoud and Jean-Pierre Getti, "Areva au Gabon: Rapport d'enquête sur la situation des travailleurs de la COMUF, filiale gabonaise du groupe Areva-Cogéma," 4 April 2007 (http://www.asso-sherpa.org/).

104. Ibid., 7.

105. Ibid., 24.

106. Ibid., 11–12.

107. Ibid., 14.

108. "L'observatoire de Mounana," L'union, 1 juin 2007.

109. "Sites miniers: AREVA et SHERPA créent un dispositif d'observation de la santé," press release, Paris, 19 June 2009.

CHAPTER 8 AND PROLOGUE

1. Mining has been a pivotal theme in the history of southern Africa; there is a vast literature and considerable historiographical debate on the subject. The material in this prologue draws on this scholarship, including the following works: Moodie and Ndatshe, Going for Gold; Crush et al., South Africa's Labor Empire; Harries, Work, Culture, and Identity; Van Onselen, Chibaro; Charles Van Onselen, Studies in the Social and Economic History of the Witwatersrand, 1886–1914 (Ravan, 1982); V. L. Allen, The History of Black Mineworkers in South Africa, volumes I–III

(Moor, 2005); Randall M. Packard, *White Plague, Black Labor: Tuberculosis and the Political Economy of Health and Disease in South Africa* (University of California Press, 1989); McCulloch, *Asbestos Blues*; Elaine N. Katz, *The White Death*; Breckenridge, "'Money with dignity'"; Breckenridge, "The allure of violence."

2. Both before and after the institutionalization of apartheid, African men migrated to the mines from near and far. Some traveled relatively short distances, from Pondoland (in what is now the Eastern Cape) or Lesotho. Others journeyed from (present-day) Mozambique, Malawi, Botswana, Swaziland, Zambia, and even Tanzania. Political as well as economic shifts shaped migration patterns, which changed over time. After World War II, recruitment networks spread into tropical Africa. In the 1960s, Zambia and Tanzania recalled their workers in protest against the apartheid regime. In the 1970s, a range of forces—a rise in the price of gold, the intensifying proletarianization of the manufacturing workforce, the withdrawal of more than 100,000 Malawian miners—led to a relatively rapid increase in mine wages and changes in the makeup of the labor force. By the late 1970s, many mineworkers had begun to see themselves as proletarian wage earners rather than as farmers trying to maintain their rural homesteads. See Moodie, *Going for Gold*.

3. South African historians have long debated the origins and necessary conditions of formal apartheid. Building on this scholarship, my working premise is that colonialism, capitalism, and aspirations to "modernity" all shaped (and were implicated in) formal apartheid; and that "industry" and "the state" were not a unitary, monolithic entities but rather complex institutions with numerous links between them and rivalries and contradictions interior to each. For more on the creation and maintenance of apartheid, see P. L. Bonner, Peter Delius, and Deborah Posel, *Apartheid's Genesis, 1935–1962* (Ravan Press and Witwatersrand University Press, 1993); Deborah Posel, *The Making of Apartheid, 1948–1961: Conflict and Compromise* (Oxford University Press, 1991); Dan O'Meara, *Forty Lost Years: The Apartheid State and the Politics of the National Party, 1948–1994* (Ravan Press and Ohio University Press, 1996); Hermann Giliomee, *The Afrikaners: Biography of a People* (University of Virginia Press; Tafelberg Publishers, 2003); Merle Lipton, *Capitalism and Apartheid: South Africa, 1910–1986* (Wildwood House, 1986); William Beinart and Saul Dubow, *Segregation and Apartheid in Twentieth-Century South Africa* (Routledge, 1995); Dubow, *A Commonwealth of Knowledge*; Saul Dubow, *Racial Segregation and the Origins of Apartheid in South Africa, 1919–36* (St. Martin's Press, 1989); Bill Freund, *The African City: A History* (Cambridge University Press, 2007). For a useful introduction to the historiographic debates, see Daryl Glaser, *Politics and Society in South Africa* (Sage, 2001).

4. K. Breckenridge, "Verwoerd's bureau of proof: Total information in the making of apartheid," *History Workshop Journal* 29, no. 1 (2005): 83–108.

5. Interview with Edwin Pheeha and Michael Moabeng, 14 July 2004.

6. Ibid.

7. See Moodie, *Going for Gold* and other works cited in note 1 above.

8. *Report of the Commission of Inquiry into Safety and Health in the Mining Industry*, volume 1 (Department of Minerals and Energy Affairs, Pretoria 1995), chapter 4.

9. Packard, *White Plague, Black Labor*.

10. Livingston, *Debility*.

11. Jock McCulloch, "Counting the cost: Gold mining and occupational disease in contemporary South Africa," *African Affairs* 108 (2009): 561–602; Allen, *History of Black Mineworkers*, volume 2, 171.

12. Francis Wilson, *Labour in the South African gold mines 1911–1969* (Cambridge University Press, 1972); Chamber of Mines annual reports (especially 1966); Newby-Fraser, *Chain Reaction*; David Fig, *Uranium Road: Questioning South Africa's Nuclear Direction* (Heinrich Böll Stiftung, 2004).

13. The other themes that emerged in these interviews—workplace dangers, compound living, migration experience—have been deeply explored by other scholars, notably T. Dunbar Moodie and Vivienne Ndatshe (*Going for Gold*) and Jean Léger ("Talking Rocks": An Investigation of the Pit Sense of Rockfall Accidents Amongst Underground Gold Miners, PhD dissertation, University of the Witwatersrand, 1992).

14. Elana Janson, The Development of the Uranium and Nuclear Industry in South Africa, 1945–1970: A Historical Study, PhD dissertation, University of Stellenbosch, 1995, 84–5.

15. RSJ du Toit, "Experience in the control of radiation at a small thorium mine," in *Radiological health and safety*, 205 *passim*.

16. GME U9/11/2, "Minutes of the Fourth Meeting of the Health Committee of the Atomic Energy Board," 24 August 1955. NASA: GES 2175 ref 179/133.

17. S. R. Rabson, "Radon and Radioactivity in Mines: Summary," 30 November 1962; S. R. Rabson, "Survey of Radon and Radon Daughters: Vogelstruisbult G.M. Co. Ltd.," 30 September 1962; P. Buckley-Jones, "Survey of Radon and Radon Daughters: Luipaardsvlei Gold Mining Co. Ltd.," 30 November 1962, SG papers.

18. W. S. Rapson, Research Advisor to Mr. Gibbs, Government Mining Engineer, 8 October 1964, SG papers.

19. Rabson's project shifted to the jurisdiction of the Medical Panel at that point. Circular No. 19/63, "Record of Meeting of the Radioactivity Panel held on Wednesday, 19 December 1963 in the Chamber of Mines Building," 30 December 1963, SG papers.

20. "A Precis of 'Memorandum Regarding Health Hazards Relating to the Operation of Uranium Leach Plants to be Established on Certain of the Witwatersrand Gold Mines' (by Professor L. Taverner)," 11 September 1951. CoM archives.

21. E. T. Pinkney, "Some Comments on Professor Taverner's Memorandum Relating to Health Hazards," 11 September 1951; Transvaal Chamber of Mines, "Report of Meeting held in the Office of the Government Mining Engineer on Wednesday, 3 October, 1951, at 2:30 pm," 8 October 1951; R. R. Porter, "Review of the Toxicology of Uranium as Applied to Rand Ores," 3 October 1951. CoM archives.

22. E. T. Pinkney, "Some Comments on Professor Taverner's Memorandum Relating to Health Hazards," 11 September 1951; Transvaal Chamber of Mines, "Report of Meeting held in the Office of the Government Mining Engineer on Wednesday, 3 October, 1951, at 2:30 pm," 8 October 1951; R. R. Porter, "Review of the Toxicology of Uranium as Applied to Rand Ores," 3 October 1951. CoM archives.

23. Brian Austin, *Schonland: Scientist and Soldier* (Witwatersrand University Press, 2001); Dubow, *Commonwealth of Knowledge*.

24. Transvaal Chamber of Mines, "Report of Meeting held in the Office of the Government Mining Engineer on Wednesday, 3 October, 1951, at 2:30 pm," 8 October 1951. CoM archives.

25. L. S. Williams, "Report on a Study of Health Hazards in the Production and Handling of Uranium Compounds in England and America," 30 April 1952. CoM archives.

26. GMEU 9/11/1, "Minutes of the First Meeting of the Health Committee of the Atomic Energy Board, held in the Office of the Government Mining Engineer on Tuesday, the 6thJuly, 1954, at 10 a.m." NASA: GES 2175 ref 179/133.

27. Hartebeestfontein Gold Mine, "Safety in the Uranium Plant," n.d. but ca. 1956–57. CoM archives.

28. "Safety Practices in Uranium Plants: Code of Practice," 20 May 1955. CoM archives.

29. Ibid.

30. This example is cited in a recap in The Rand Mutual Assurance Company, Memorandum to the General Manager, Chamber of Mines of South Africa, "Medical examination of Employees in the Uranium Industry," 10 August 1972. CoM archives.

31. "No Radio-Activity Danger in Uranium Mines—Viljoen," extract from *Rand Daily Mail*, 9 October 1952. CoM archives.

32. McCulloch, "Counting the cost."

33. Manager, Luipaards Vlei Estate and Gold Mining Company to General Manager, Transvaal and OFS Chamber of Mines, 14 January 1957. CoM archives.

34. Reports are far too numerous to cite individually—collected throughout uranium plant records for 1956 and 1957 in CoM archives.

35. Director, DVL to Manager, Calcined Products, 14 March 1957. CoM archives.

36. Circular no. 19/63, "Record of Meeting of the Radioactivity Panel held on Wednesday, 19 December 1963, in the Chamber of Mines Building," 30 December 1963, SG papers. The timing of data collection affected the count. Urine collected at the beginning of a work week reflected long-term uranium exposure, "it being assumed that over the week-end the uranium from any short-term exposure will have largely disappeared." Sampling at the end of a week—more "practical" from a work-flow perspective—enabled managers to identify sources of short-term exposure, which in turn could reveal a "harmful operation" and "prevent . . . re-occurrence." Workers exhibiting high counts were re-sampled up to two times. Only if the third sample manifested high uranium levels did the reading constitute

a "high count" for work-flow and record-keeping purposes. So only the third reading would (in principle) prompt action to remove the worker from his task and check his work area. More discussion of urine counts and record keeping in P. H. Kitto, Director, Physical Sciences Laboratory, "Projects Nos. PU/908/67 and PU/912/67," 24 November 1972; A. T. Milne, General Manager, Transvaal and Orange Free State Chamber of Mines, to the Government Mining Engineer, 8 December 1964; J. D. Grieg, "Projects Nos. PU/908/67 and PU/912/67: Analysis of Urine from Employees on Uranium Plants and NUFCOR works, Review of Results 1957–1971," 24 November 1972. CoM archives.

37. S. S. Nosworthy to H. Clark, Daggafontein Mines, 16 January 1957. CoM archives.
38. NUFCOR, "Health and Safety: Respirators or Dust Masks," WLL/MK, 4 March 1971. NUFCOR archives.
39. Pheeha and Moabeng interview.
40. GMEU 9/11/1, "Minutes of the First Meeting of the Health Committee of the Atomic Energy Board, held in the Office of the Government Mining Engineer on Tuesday, the 6th July, 1954, at 10 a.m." NASA: GES 2175 ref 179/133.
41. Transvaal and Orange Free State Chamber of Mines, "Explanatory Memorandum: Safety Practices in Uranium Plants: Code of Practice," 6 June 1955.
42. Breckenridge, "Verwoerd's bureau." This argument is extended in Edwards and Hecht, "History and the Technopolitics of Identity."
43. S. R. Rabson, "Safety Precautions in Uranium Plants: Code of Practice," 28 February 1964, CoM archives (emphasis added).
44. McCulloch, "Counting the cost."
45. The Rand Mutual Assurance Company, Memorandum to the General Manager, Chamber of Mines of South Africa, "Medical examination of Employees in the Uranium Industry," 10 August 1972; Statistician, Memorandum to G. H. Grange Esq., Technical Adviser, 20 Sept 1972. CoM archives.
46. Transvaal and Orange Free State Chamber of Mines, Employees in Uranium Plants, Medical Record (blank form, no date). CoM archives.
47. R. A. Mathews, Medical Advisor, Transvaal and Orange Free State Chamber of Mines, Social Services Department, A/18/4, "Memorandum for the General Manager," 6 August 1959. CoM archives.
48. Confidential interview, April 2004, Johannesburg metropolitan area.
49. The Rand Mutual Assurance Company, Memorandum to the General Manager, Chamber of Mines of South Africa, "Medical examination of Employees in the Uranium Industry," 10 August 1972. CoM archives.
50. Statistician, Memorandum to G. H. Grange Esq., Technical Adviser, 20 September 1972. CoM archives.
51. C. C. Freed, Medical Superintendent, Memorandum to Manager, The Rand Mutual Assurance Company Ltd., "Uranium Examinations," 6 November 1972. CoM archives.
52. Ibid.

53. LB/35/6/10, J. O. Tattersall to J. W. L. de Villiers, 18 July 1980, re: Chamber of Mines Code of Practice on Safety. SG papers.

54. Interview with Phil Metcalf, 17 June 2004.

55. A. J. A. Roux to W. P. Viljoen, 16 May 1979, internal ref. LB/35/6/10, SG papers.

56. Ibid p. 2.

57. S. Guy, "A review of files at the Government Mining Engineer concerning radiation in mines and works," 26 August 1986, SG papers.

58. Ibid, 7.

59. Interview with Shaun Guy, 12 July 2004.

60. Ibid.

61. Walker, *Containing the Atom*.

62. J. D. Grieg, R. Rolle, V. B. Dods, B. G. Hampson, and J. P. du Plessis, "Report on a survey of Radon and Radon-daughter concentrations in Twenty-Five Gold Mines," Research Report No. 36/70, October 1970. Goldfields archives.

63. Guy interview.

64. Basson et al., Lung Cancer, 12.

65. Guy interview.

66. "Results of Radon Daughter Sampling in Bird Reef," West Rand Consolidated Mines, Ltd., Mine Office, West Rand, 13 December 1973, SG papers.

67. Guy interview.

68. S. Guy, "Memorandum: Meeting at West Rand Consolidated with the Mine Manager, 24 February 1986." LB/35/6/10/8. SG papers.

69. AEC Licensing Branch, "Report of the Underground Survey for Radon Daughters at West Rand Consolidated Mine, 5 March 1986," 23 May 1986, LB/35/1/13; LB/35/6/10/8, SG papers. (By this point, the AEB had changed its name to Atomic Energy Corporation of South Africa.)

70. Guy interview.

71. JW/MT/TH1, "Special Meeting of the Technical Advisory Committee on Nuclear Licenses," 16 March 1990; "Meeting with Representatives of the Government Mining Engineer and the Council for Nuclear Safety," 20 December 1990; SG/ak, "Minutes of the Meeting held on 30 August 1990 at the Council for Nuclear Safety's Offices in Verwoerdburgstad"; T. R. N. Main to P. J. Hugo, 29 May 1991, re: Nuclear Energy Act, 1982: Application to Mines. CoM-ES archives.

72. McCulloch, "Counting the cost."

73. JW/MT/TH1, "Special Meeting of the Technical Advisory Committee on Nuclear Licenses," 16 March 1990; "Meeting with Representatives of the Government Mining Engineer and the Council for Nuclear Safety," 20 December 1990. CoM-ES archives.

74. J. M. Stewart to J. C. Greef, "Low Level Radiation Exposure in Gold Mines: Information for the Medical Bureau for Occupational Diseases," 27 February 1991; J. O. Tatersall to A. van der Linde, "International Atomic Energy Agency Training Course on Radiological Protection in the Mining and Milling of Radioactive

Ores," 24 April 1991; B. C. Winkler to J. Burrows, "Training Manual for the Chamber of Mines Certificate in Radiological Protection—Screening," 13 December 1991; D. G. Wymer to R. Meade, "Radiation Control in South African Gold Mines—Investigation into Waterborne Radiation," 3 October 1991; COMRO, "Record of the Meeting with AEC on the Contract Research Programme held on 15 July 1992 at COMRO," File 231; P. E. Metcalf to D. Wymer, "Visit of Mr J Uranium Ahmed—Head Radiation Safety Section, International Atomic Energy Agency," 25 August 1992. CoM-ES archives.

75. A few examples among many include: John Edwards to Peter Hunter, "Areas of Concern with respect to the CNS," 8 November 1993, Randfontein Estates Gold Mining Company; ZB Swanepoel to Penuell Maduna, "Matters Concerning the CNS," 4 August 1998. CoM-ES archives.

76. Metcalf interview; interview with Derek Elbrecht, 15 May 2004.

77. This argument was made by a CEA radiation protection expert; it contrasted sharply with what the first generation of French radiation protection experts had argued in the 1950s and 1960s. (see chapter 6) P. Zettwoog, "Cost of Implementing the ICRP Recommandations [sic] in Underground Mines in France, Niger, and Gabon. Analysis of past efforts. Forecasting of the future," CRPM/12.1, typescript prepared for ICRP/CCE meeting in Luxembourg, 5–6 April 1990. CoM-ES archives.

78. P. Zettwoog to D. G. Wymer and J. M. Stewart, 22 November 1992 (including Memorandum by P. Zettwoog and S. Bernhard, 16 November 1992); P. Zettwoog, "Cost of Implementing the ICRP Recommandations [sic] in Underground Mines in France, Niger, and Gabon. Analysis of past efforts. Forecasting of the future," CRPM/12.1, typescript prepared for the ICRP/CCE meeting in Luxembourg, 5–6 April 1990. CoM-ES archives.

79. Technical Committee Meeting on Impact of the New ICRP Occupational Dose Limit on the Operation of Underground Mines, 3–7 June 1991, Final Draft produced after the TC, 7 June 1991. CoM-ES archives.

80. R. M. Fry, "Report of Advisory Group Meeting on the Implementation of the 1990 Recommendations of the International Commission on Radiological Protection (ICRP) in the Specific Area of the Operation of Underground Mines," J1-AG-797, 11 December 1992. CoM-ES archives.

81. D. Wymer, "Report on Two Meetings at the International Atomic Energy Agency (IAEA), Vienna, 7–18 December 1992," 31 December 1992, CoM-ES archives (emphasis added).

82. H. Wagner to T R N Main, Memorandum: IAEA Meeting in Vienna, 12 November 1992. CoM-ES archives.

83. W E Stumpf to T R N Main, IAEA Meeting in Vienna on the Basic Safety Standard for Radiological Protection, 6 November 1992. CoM-ES archives.

84. D. Wymer, "Report on Two Meetings at the International Atomic Energy Agency (IAEA), Vienna, 7–18 December 1992," 31 December 1992; D. Wymer, "Note for the Record: Joint Secretariat for the Basic Safety Standards (Ionizing

Radiation): Working Group on Dose Limitation (Occupational Exposure) in Specific Occupations (Mines, Health Personnel, etc): International Labour Office, Geneva, 29 March–1 April 1993. CoM-ES archives.

85. The first dosimetric reports that I found in the NUFCOR archives dated from 1989, the year after the creation of the CNS. Full compliance, however, seems to have begun in 1993–94. As documented in NUFCOR archives, miscellaneous correspondence in file boxes titled "Dose Reporting" and "CNS and Licensing." Confidential interview.

86. "Report of the Commission of Inquiry into Safety and Health," 63–65.

87. Ibid., 104.

88. Thomas Auf der Heyde, "South African nuclear policy since 1993: Too hot to handle?" (paper presented at National Nuclear Technology Conference, 6–9 September 1998, South Africa). CoM-ES archives.

89. "Draft: South African Energy Policy: Discussion Document: Comment," 2 October 1995, pp. 1, 11. CoM-ES archives.

90. A. H. Munro to M. Golding, 2 March 1995, CoM-ES archives. For the 1990 ICRP recommendations, see *Annals of the ICRP* 21, no. 1–3, especially 25–32.

91. D. G. Wymer, "Note for the record: Meeting between the Chamber and Marcel Golding, Cape Town, 7 June 1995," 14/6/95. CoM-ES archives.

92. AngloGold Ashanti Annual Report 2007—Report to Society (http://www.anglogold.co.za/subwebs/informationforinvestors/reports07/reporttosociety07/radiation-SA.htm).

93. David Fig, "In the dark: Seeking Information about South Africa's nuclear energy programme," in *Paper Wars: Access to Information in South Africa*, ed. K. Allen (Wits University Press, 2009): 56–87. Fig also discusses the struggle of Pelindaba workers to obtain radiation exposure information.

94. McCulloch, "Counting the cost."

95. Elbrecht interview.

CHAPTER 9 AND PROLOGUE

1. The best analysis of anti-uranium activism in Australian politics and social movements can be found in Sigrid McCausland, Leave It in the Ground: The Anti-Uranium Movement in Australia 1975–82, PhD dissertation, University of Technology, Sydney, 1999.

2. Eichstaedt, *If You Poison Us*; Brugge et al., *Navajo People and Uranium Mining*; Masco, *The Nuclear Borderlands*.

3. Four Directions Council, "Discrimination against Indigenous Populations," E/CN.4/Sub.2/1987/NGO/3, 12 August 1987 (ftp.halcyon.com/pub/FWDP/Americas/four_dir.txt).

4. This version is from Kathy Helms, "Experts taking holistic role to uranium health issues," *Gallup Independent*, 5 April 2010 (http://www.gallupindependent.com/).

5. This quotation and those above are from *Poison Fire, Sacred Earth: The World Uranium Hearing* (World Uranium Hearing, 1993). Most of the book's text is viewable at http://www.ratical.com/radiation/WorldUraniumHearing/index.html.
6. CANUC activities are discussed in the wider context of the British Namibia Support Committee in C. Saunders, "Namibian Solidarity: British support for Namibian independence," *Journal of Southern African Studies* 35, no. 2 (2009): 438–454.
7. On migrant labor in Namibia specifically, see R. Moorsom, "Underdevelopment, contract labour and worker consciousness in Namibia, 1915–72," *Journal of Southern African Studies* 4, no. 1 (1977): 52–87; Moorsom, "Underdevelopment and labour migration: The contract labour system in Namibia," working paper, Chr. Michelsen Institute, 1995; R. J. Gordon, *Mines, Masters and Migrants: Life in a Namibian Mine Compound* (Ravan, 1977).
8. For discussion of labor action in the 1970s, see Macfarlane, Labour Control and Bauer, *Labor and Democracy*.
9. This discussion of pre-1985 history is based on the analysis in Macfarlane, Labour Control but is supported by my own archival and oral interview research at Rössing in 2004.
10. RUL/R. Walker, "Social, Manpower Development and Industrial Relations Policy," January 1978, quoted in Macfarlane, Labour Control, 173.
11. This description is based on several dozen interviews and a wide range of archival documents, too numerous to cite individually.
12. Macfarlane, Labour Control, 211; Interview with Bill Birch, 29 Jan. 2004; Interview with Paul Rooi, 30 January 2004; interview with Asser Kapere, 25 February 2004.
13. The unit actually changed names numerous times; for clarity, I have kept the Loss Control designation, since this is how employees referred to it in their discussions with me. This emphasis was, of course, common throughout industry, and not just in southern Africa. See N. Anderson and S. Marks, "Work and health in Namibia: Preliminary notes," *Journal of Southern African Studies* 13 (1987): 274–292.
14. Beginning in 1987, Rössing received the highest possible rating for its performance from NOSA; the following year, it applied for and received a "Sword of Honour' award from the British Safety Council. Rössing annual reports and newsletters, 1980s.
15. Interviews with Rooi, Kapere, Willem van Rooyen, 15 January 2004; confidential interview, February 2004.
16. Van Rooyen interview.
17. Studies on skill and worker initiatives in African mines include Burawoy, *Color of Class*; Moodie and Ndatshe, *Going for Gold*; J. Guy and M. Thabane, "Technology, ethnicity, and ideology: Basotho miners and shaft-sinking on the South African gold mines," *Journal of Southern African Studies* 14, no. 2 (1988): 257–78. On skill and technology in the workplace more broadly, see (among many others) Michael Burawoy, *Manufacturing Consent: Changes in the Labor Process Under Monopoly*

Capitalism (University of Chicago Press, 1979); David Noble, *Forces of Production: A Social History of Industrial Automation* (Knopf, 1984).

18. These reports, and other activities of the Loss Control division, are documented in the Rössing archives; individual documents are too numerous to cite here.

19. Kapere interview.

20. Security and Services Superintendent to Personnel Manager, Confidential Memorandum re: Rössing Uranium Mine Workers Union, 26 February 1986, 2. RAS: Industrial Relations, 1985–87.

21. Bauer, *Labor and Democracy*, 86.

22. W. Groenewald to General Manager, 9 April 1986; M. P. Bates to W. Groenewald, 21 April 1986; Personnel Manager to General Manager, Confidential Memorandum re: Rössing Mine Workers' Union, 24 April 1986. RAS: Industrial Relations, 1985–87.

23. C. V. Kauraisa to B. E. Burgess, P.C. Brown, B. Hochobeb, A. Kapere, and W. Groenewald, Confidential memorandum re: "Rössing Mine Workers' Union," 16 June 1986. W. J. Birch to Senior Security Officers, "Rössing Uranium Mine Workers Union," 26 February 1986. RAS: Industrial Relations, 1985–87.

24. When Ronnie Walker (who served as chairman of the board until 1985) decided to retire, he announced that it would be "appropriate" for a Namibian to replace him and proposed Ngavirue—as an "outstanding Namibian"—for the position. Minutes of the 88th Meeting of the Board of Directors of Rössing Uranium Limited, 23 August 1985. Rössing news release, "Police Action at Arandis," 28 July 1987; Rössing news release, "Rössing Chairman Sees Justice Minister," 30 July 1987. RAS: Board of Directors; Industrial Relations, 1985–87.

25. Kapere was eventually released. Available archives didn't reveal what, if anything, happened to the employees who moonlighted with the special police.

26. Kapere and Groenewald had become members of the MUN's National Executive Committee in 1986, and along with other Rössing labor leaders played a significant role in the formation of the MUN as a national organization encompassing workers in mines across the country.

27. Rössing Mine Workers' Union Statement, 20 April 1988. RAS: Industrial Relations, 1988. "Baaskap" referred to the domination of non-whites by whites under apartheid.

28. The political importance of Namibianisation at Rössing contrasts somewhat with Burawoy's findings on Zambianisation on the Copperbelt (Burawoy 1972); a fuller analysis of these differences must await another venue.

29. Rössing Mine Workers' Union Statement, 20 April 1988. RAS: Industrial Relations, 1988.

30. M.P. Bates to Personnel Manager, Memorandum, 22 Oct. 1987; Namibia Support Committee, "Blockade Southern African Uranium: Solidarity with

Namibian and South African Miners," Oct. 1987 flyer; P. C. Brown, Notes of a Meeting with the Mine Workers Union of Namibia, 17 December 1987, Windhoek. RAS: Industrial Relations, 1988.

31. Interview with Harry Hoabeb, 9 February 2004.

32. "Ben Ulenga," *The Namibian*, 3 June 1988; "Position Paper on Rössing Mine Workers Union (RMU)," 16 February 1987; Personnel Manager to Assistant General Manager, Memorandum re: "Report on Break Away Union Discussion," 28 June 1988. RAS: Industrial Relations, 1988.

33. "Minutes of a Company/Union Meeting Held on Wednesday 16 Aug. 1989 at 15h15 in the Management Conference Room," 22 August 1989. RAMS: Industrial Relations, 1985–89.

34. It is important to see the doctor's work in terms of his own scientific aims; see W. Beinart, K. Brown, and D. Gilfoyle, "Experts and expertise in colonial Africa reconsidered: Science and the interpretation of knowledge," *African Affairs* 108/432: 413–33.

35. Interview with Wotan Swiegers, 18 February 2004.

36. Metcalf interview.

37. B. E. Burgess to Dr. W. R. Swiegers, BEB/bs/Objectives file, 29 December 1980. RAS: Medical Services, 1979–81.

38. Swiegers interview.

39. On corporate and tropical medicine, see, e.g., Packard, *White Plague*.

40. Swiegers interview.

41. Hoabeb interview. This and other testimony supports Lundy Braun's argument about the lasting effects of the deeply racialized history of spirometry in "Spirometry, measurement, and race in the nineteenth century," *Journal of the History of Medicine and Allied Sciences* 60 (2005): 135–69.

42. Interview with Jamie Pretorius, 29 Jan. 2004.

43. Hoabeb interview.

44. Pretorius interview.

45. Hoabeb interview. The company archives document dozens of requests from workers and union leaders for access to medical and personnel files.

46. The company's preparations for independence (along with policy during and after the transition) are documented in the Minutes of the Board of Directors from the 100th meeting (19 August 1988) to the 111th (24 May 1991). By August 1989, SWAPO leaders had formally toured the Rössing site and declared that they would support the normalization of trade as soon as possible. Minutes of the Executive Committee Meeting held on Friday, 18 August 1989. RAS: Independence/MD files.

47. Minutes of the 104th Meeting of the Board of Directors, 21 August 1989. On SWAPO's state-making in the aftermath of independence, see Lauren Dobell, *Swapo's Struggle for Namibia, 1960–1991: War by Other Means* (Schlettwein, 1998 and 2000).

48. Memorandum, Public Affairs Manager to Sean James, Acting General Manager, 1 August 1989 (emphasis added).

49. Minutes of a Meeting of the Board of Directors, 1 June 1990. Hangala had been in exile since the mid 1970s.

50. Minutes of a Meeting of the Board of Directors, 17 August 1990. RAS: Board of Directors. SWAPO had made clear in previous statements (in 1988 and 1989) that it didn't envisage nationalizing industry. Bauer, *Labor and Democracy in Namibia*, 99–100.

51. Minutes of the 104th Meeting of the Board of Directors, 21 August 1989. RAS. Concretely, this approach led to the repeal of the South African Atomic Energy Act and the amendment of the Namibian mining ordinance to include uranium ore and to give the Minister of Mines purview over uranium mining.

52. SENES Consultants Limited, "Proposed Legislation for Uranium Mining in Namibia," November 1990; SENES Consultants Limited, "Review of Occupational Hygiene and Environmental Control Practices at Rössing Uranium Limited," December 1990. This second report did flag potential problems with the eventual decommissioning of the open pit, the waste dumps, and the tailings area, but— unsurprisingly given the nature of the consultancy, and the fact that it was based on discussions with superintendents and managers—recommended only that Rössing "continue its studies of tailings management' and assess "the feasibility of upgrading the acid plant as part of the future phase of operations'); Minutes of the 109th Meeting of the Board of Directors, 23 November 1990. RAS.

53. It took a while—such legislation apparently wasn't a big priority for the new state, and didn't pass until 1994.

54. This parallels Cooper's description of the experience of labor leaders elsewhere in Africa during the decolonizations of the 1950s in *Decolonization and African Society*.

55. Bauer, *Labor and Democracy*. On the transition to independence, including the role of private capital, see also Leys and Saul, *Namibia's Liberation Struggle* (especially L. Dobell, "SWAPO in Office"); D. R. Kempton and R. L. du Preez, "Namibian-De Beers state-firm relations: Cooperation and conflict," *Journal of Southern African Studies* 22, no. 4 (1997): 585–613.

56. PARTiZANS = People Against Rio Tinto Zinc And Its Subsidiaries. In the copyright page of *Past Exposure*, they describe themselves as follows: "Founded in 1978 at the request of Australian Aboriginal communities, PARTiZANS monitors worldwide all the activities and intentions of the world's most powerful mining corporation. It does so by linking together groups from across the globe affected by the company's mines, from Alaska to Zimbabwe."

57. Confidential interviews, 2004. The resulting study, *Past Exposure*, claimed that the union had not provided the documents in question. One set of confidential interviewees, however, stated that certain union members had in fact provided the documents—even if the union hadn't done so officially.

58. G. Dropkin and D. Clark, *Past Exposure: Revealing Health and Environmental Risks of Rössing Uranium* (Namibia Support Committee, 1992).
59. Minutes of a Meeting of the Board of Directors, 20 March 1992. RAS.
60. J. U. Ahmed et al., "Report of the IAEA Technical Co-operation mission to Namibia on the Assessment of Radiation Safety at the Rössing Uranium Mine, 31 Aug.—11 Sept. 1992" (Vienna, 1992), 9.
61. Ibid., 12.
62. Ibid., 9.
63. Hoabeb interview.
64. Leake Hangala, the government representation to Rössing's board, suggested that management "give workers access to health and environmental documents to see for themselves. Such openness would minimize bad publicity and could increase our sales." Minutes of a Meeting of the Board of Directors, 27 November 1992. RAS.
65. Harry Hoabeb interview.
66. Interview with Erich Beukes, 2 February 2004.
67. This detail was mentioned in every conversation or interview in which the Canadian visit was discussed—even those with people who hadn't been on the visit.
68. Hangala reported to the Rössing board in November 1992 that "MUN had been invited to institute an investigation on their allegation of sick people.' Minutes of a Meeting of the Board of Directors, 27 November 1992. RAS.
69. Hoabeb interview. Hangala's reported to the Rössing board in November 1992 that "MUN had been invited to institute an investigation on their allegation of sick people." Minutes of a Meeting of the Board of Directors, 27 November 1992. RAS.
70. Kapere—who by this time had left Rössing for national politics, but who stayed in touch with his friends there and had managed to keep his house in Arandis—had known Zaire when they were both children at the same school.
71. H. Hoabeb to the Company Representative, 6 April 1992. RAS: Zaire file.
72. Kapere interview.
73. C. Algar (Manager Corporate Affairs) to S. James (General Manager), Ref: CAA/2-133/ct, 6 April 1993. RAS: Zaire file.
74. Chief Medical Officer to General Manager, Memorandum, 9 November 1993. RAS: Zaire file.
75. "Dr. R Zaire, Diary of Events," July 1997. Dr. S. Amadhila (Permanent Secretary) to Reinhard Zaire, 15 April 1994; also Dr. E. G. Burger to The Permanent Secretary, 9 March 1994. RAS: Zaire file.
76. Interview with Jamie Pretorius. Environmental Services and Environmental Health Departments, "Work and Exposure Profile of Mr. Edward Connelly, Co No 9679 for the period of 1977-1982. Privileged—Produced for and/or in Contemplation of Litigation," September 1994. RAS: Connelly litigation packet.

77. RAS: Connelly litigation packet. See also the description of the case by one of Connelly's lawyers: R. Meeran, "The unveiling of transnational corporations: A direct approach," in *Human Rights Standards and the Responsibility of Transnational Corporations*, ed. M. Addo (Kluwer, 1999); full text available at http://www.labournet.net.

78. R. Zaire, M. Notter, W. Riedel, and E. Thiel, "Unexpected rates of chromosomal instabilities and hormone level alterations in Namibian uranium miners" (typescript, 1995), 2. RAS: Zaire file.

79. R. Zaire et al., "Unexpected rates of chromosomal instabilities and alterations of hormone levels in Namibian uranium miners," *Radiation Research* 147 (1997): 579–584.

80. R. Proctor, *Cancer War*; Walker, *Permissible Dose*; S. Boudia, "Les problèmes de santé publique de longue durée. Les effets des faibles doses de radioactivité," in *Comment se construisent les problèmes de santé publique*, ed. C. Gilbert and E. Henry (La Découverte, 2009).

81. Chief Medical Oficer to General Manager, Memorandum, 15 March 1996. RAS: Zaire file.

82. A. Hope to J. Leslie, 22 March 1996. RAS: Zaire file.

83. J. S. Kirkpatrick to S. James, 4 April 1996. RAS: Zaire file.

84. R. R. Hoveka, File Note: Presentation by Mr. Zaire at the Arandis Club on 12 July 1996. RAS: Zaire file.

85. Ibid.

86. "Summary of Meeting between Rössing Uranium Limited and the Mineworkers Union of Namibia held on 23 Sept. 1996 at Block F in the Ministry of Health and Social Services," not dated; "Summary of the Meeting between Rössing Uranium Limited and Dr. R. Zaire on the 03 Oct. 1996, at the Rössing Uranium Mine," 3 October 1996. RAS: Zaire file.

87. File Note: "MUN Press Conference-Zaire, Meeting held on 18 Sept. 1996," 19 September 1996. RAS: Zaire file.

88. T. Siepelmeyer and R. Zaire, "High risk of cancer at Rössing" (typescript, not dated), RAS: Zaire file.

89. Ibid.

90. Beukes interview.

91. R. Zaire to A. Muheua, 3 Jan. 1997. Zaire demanded that the union pay his air fare plus a consulting fee of 1,000 DM a day, with a minimum of 6,000 DM. ("Which means that in the event you consult me only for one day during this trip, you will pay me a 6000 DM honorarium.") RAS: Zaire file.

92. File Note: Zaire Meeting held 23 January 1997. RAS: Zaire file.

93. Andrew Hope, File note: "Meeting with Alphaeus Muheua on 27th February 1998," 27.2.1998; File note: "Accord meeting held on 12 June 1998," 15.6.98. RAS: Zaire file.

94. Reinhard Zaire to Charles Kauraisa, 6 October 1996. RAS: Zaire file.

95. Beukes interview.

96. Beukes and Rooi interviews.

97. Interview with Gida Sekandi, 24 February 2004.

98. Interestingly, Rössing's managing director waited until 1997 to raise the subject of Zaire at meetings of the company's board of directors.

99. Beukes and Hoabeb interviews.

100. Hoabeb interview.

101. Sekandi interview.

102. On oncology infrastructure and practice in Africa, see Julie Livingston, *Improvising Medicine: Inside an African Oncology Ward* (book manuscript in progress).

CHAPTER 10

1. See, e.g., *Mining in Africa: Regulation and Development*, ed. B. Campbell (Pluto, 2009).

2. Areva, "Areva in Niger," January 2009. Julie Livingston explores the invisibility of cancer in Africa in *Improvising Medicine: Inside an African Oncology Ward* (book manuscript).

3. Alka Hamidou, quoted in Greenpeace International, *Left in the Dust: Areva's Radioactive Legacy in the Desert Towns of Niger*, April 2010, 48.

4. Ibid., 51.

5. Philippe Brunet, *La nature dans tous ses états: Uranium, nucléaire et radioactivité en Limousin* (Presses Universitaires de Limoges, 2004).

6. Georges Dougueli, "Alain Acker: 'Les choses ont pris plus de temps que prévu,'" *Jeune Afrique* (www.jeuneafrique.com), 15 July 2010 (emphasis added).

7. For a parallel in the world of French nuclear testing, see Yannick Barthe, "Cause politique et 'politique des causes'. La mobilisation des vétérans des essais nucléaires français," *Politix* 23, no. 91 (2010): 77–102.

8. "Radioactivité à Arlit et Akokan: le point de vue d'un spécialiste du CNRP," 20 July 2010 (available at http://www.nigerdiaspora.net).

9. Greenpeace, *Left in the Dust*, 32. The publication of the Greenpeace report prompted the CNRP to hold a public information session in July 2010, during which one of its officials insisted that while activists had indeed "found things," the report had "extrapolated too much."

10. Landry Lebas, *Impacts de l'exploitation minière sure les populations locales et l'environnement dan le Haut-Ogooué* (Brainforest, 2010).

11. U. Schwarz-Schampera, W. Hirdes, and B. Schulte, Mining Potential of Selected Geological Domains of Southern Madagascar, Paper Presented at the International Symposium on "Le Potentiel de Mineralisation de Madagascar," World Bank, Projet de Gouvernance des Ressources Minerales, Antananarivo 17–18 July 2008 (ftp://ftp3.intranet-mdm.gov.mg/). Special thanks to Barry Ferguson for bringing this to my attention.

12. http://namibiauraniuminstitute.com/joomla/index.php?option=com_content&view=article&id=59&Itemid=60

13. "Russia ready to invest in Namibia uranium," 20 May 2010 (http://www
.nuclearpowerdaily.com/reports/Russia_ready_to_invest_in_Namibia_uranium_
999.html).

14. Petronella Sibeene, "Namibia sets up its first Atomic Energy Board," press
release, 20 February 2009, Chamber of Mines of Namibia; Republic of Namibia,
Atomic Energy and Radiation Protection Act, 2005; Stanford Law School and
Legal Assistance Centre of Namibia, "Striking a Better Balance: An Investigation
of Mining Practices in Namibia's Protected Areas," 2009.

15. Hilma Shindondola-Mote, *Uranium Mining in Namibia: The Mystery behind 'low
level radiation'* (LaRRI, 2008).

16. Ibid., 36.

17. Ibid., 42.

18. After Fukushima, Senegal's president canceled plans to build a floating nuclear
power reactor. Ghana, however, seems ready to go full steam ahead, blaming Fukushima only on the earthquake. Justin Gomis, "Crise énergétique: Wade renonce au
projet nucléaire," *Le Quotidien*, 2 April 2011 (http://www.lequotidien.sn); "Ghana
Targets Nuclear Power By 2018," 5 April 2011 (http://www.ghanaweb.com).

19. All numbers in this paragraph are from the World Nuclear Association.

20. Serge Michel and Michel Beuret, *La Chinafrique: Pékin à la conquête du continent
noir* (Bernard Grasset, 2008).

21. For an analysis of possible external support for the Tuareg rebellion, see Jeremy
Keenan, "Uranium goes critical in Niger: Tuareg rebellions threaten Sahelian
conflagration," *Review of African Political Economy* no. 117 (2008): 449–466.

22. Hannah Armstrong, "China mining company causes unrest in Niger," *Christian
Science Monitor*, 29 March 2010.

23. http://www.marketwatch.com/story/uraniums-set-make-waves-futures

24. http://www.u3o8.biz/s/MarketCommentary.asp?ReportID=184497&_Title=
Uranium-stocks-rally-in-advance-of-NYMEX-futures-trading.

25. Linda Sandler, Yuriy Humber, and Christopher Scinta, "Lehman sits on bomb
of uranium cake as prices slump (update1)," 15 April 2009 (Bloomberg.com).

26. For a summary of these measures, see IAEA Safeguards Overview: Comprehensive Safeguards Agreements and Additional Protocols (http://www.iaea.org/
Publications/Factsheets/English/sg_overview.html).

27. As of May 2006, 107 nations had signed Additional Protocols, 75 of which
put them into force.

28. IAEA, INFCIRC/209, 3 September 1974.

29. David Albright, *Peddling Peril: How the Secret Nuclear Trade Arms America's
Enemies* (Free Press, 2010), p. 41.

30. IAEA, INFCIRC/209/Rev. 1/Mod. 4, 26 April 1999.

31. IAEA, INFCIRC/254/Rev.9/Part 1, 7 November 2007.

32. http://www.nuclearsuppliersgroup.org/Leng/02-guide.htm.

33. Arms Control Association, The Nuclear Suppliers Group (NSG) at a Glance
(www.armscontrol.org/factsheets/NSG).

34. These include the Convention on the Physical Protection of Nuclear Material and the Code of Conduct on the Safety and Security of Radioactive Sources, among others.

35. http://www.un.org/sc/1540/index.shtml

36. United Nations Security Council, "Note verbale dated 26 October 2004 from the Permanent Mission of the Republic of Namibia to the United Nations addressed to the Chairman of the Committee," S/AC.44/2004/(02)/36, 5 November 2004.

37. US Embassy Windhoek, Cable ID 10WINDHOEK7, "Namibia's Rössing Uranium—A USG Evaluation," 2010–01–25.

38. The story can be found in the Associated Press archives (nl.newsbank.com) and many other sites; my personal favorite is on Pakistan's *Daily Times* site: http://www.dailytimes.com.pk/default.asp?page=2010%5C04%5C05%5Cstory_5-4-2010_pg4_9.

39. Because "dirty bombs" do not produce a fission reaction, they do not count as "nuclear weapons." The designation begs the question of which bombs qualify as "clean."

40. US Embassy Kinshasa, Cable 07KINSHASA797, "Recent allegations of uranium trafficking in the Democratic Republic of Congo," 2007-07-11.

41. UNSCR 1540 Committee, "Request for Assistance Template (as at 19 November 2007)" (http://www.un.org/sc/1540/assistancetemplate.shtml).

42. In addition to the cited primary sources, this discussion of Katanga relies on Marie Mazalto, "Governance, Human Rights, and Mining in the Democratic Republic of Congo," in *Mining in Africa*, ed. Campbell.

43. Joint UNEP/OCHA Environment Unit, "Assessment mission of the Shinkolobwe Uranium Mine," Switzerland, 2004: 18.

44. Mazalto, "Governance," 202.

45. Cable 07KINSHASA797; US Embassy Bujumbura, Cable 09BUJUMBURA302, "Nuclear Smuggling Incident/Portal Detection at Bujumbura," 2009-07-02.

46. C. L. N. Banza et al., "High human exposure to cobalt and other metals in Katanga, a mining area of the Democratic Republic of Congo," *Environmental Research* (2009), doi: 10.1016/j.envres.2009.04.012.

47. http://www.fnrba.org/ and http://www.uranium-network.org/.

48. For an early history of such tools of visibility, see Joel D. Howell, *Technology in the Hospital: Transforming Patient Care in the Early Twentieth Century* (Johns Hopkins University Press, 1995).

APPENDIX

1. For a nuanced discussion of the place of oral history and testimony in the crafting of African history, see *African Words, African Voices: Critical Practices in Oral History*, ed. L. White, S. Miescher, and D. Cohen (Indiana University Press, 2001).

2. Antoinette Burton, ed., *Archive Stories: Facts, Fictions, and the Writing of History* (Duke University Press, 2005).

3. A 2008 law now requires French archives to reply to *dérogation* requests within a two month delay. In my experience, however, that rarely happens.

4. See, for example, *Paper Wars: Access to Information in South Africa*, ed. K. Allan (Wits University Press, 2009); Helena Pohlandt-McCormick, "In Good Hands: Researching the 1976 Soweto Uprising in the State Archives of South Africa," in Burton, *Archive Stories*.

5. Some of these records did make their way to company archives in France, though a Gabonese doctoral student who requested permission to consult the collection received a familiar stonewalling. See Robert Edgar Ndong, Les multinationales extractives au Gabon: le cas de la compagnie des mines d'uranium de Franceville (COMUF), 1961–2003, PhD dissertation, École Doctorale Sciences Sociales, Université Lumière—Lyon II, 2009.

BIBLIOGRAPHY

ARCHIVES AND OTHER PRIMARY SOURCE COLLECTIONS CONSULTED

ARCHIVES GÉNÉRALES DU ROYAUME, BRUSSELS, BELGIUM The archives of the Union Minière du Haut Katanga, which ran the Shinkolobwe uranium mine in the Congo, are housed here. They offer almost nothing on the subject of workplace conditions and radiation monitoring.

ARCHIVES NATIONALES, ARCHIVES DE LA PRÉSIDENCE DE LA RÉPUBLIQUE (1958–1974), PARIS, FRANCE The *Fonds privé Foccart* contains the papers of Jacques Foccart. Many pertaining specifically to uranium are closed on grounds of national security, but several items were available with a *dérogation*.

ARCHIVES NATIONALES DU GABON. Brief reports on the Haut-Ogooué province (where the Mounana mine was located) informed my thinking, but I found almost nothing specifically on uranium in these archives.

BODLEIAN LIBRARY OF COMMONWEALTH AND AFRICAN STUDIES, OXFORD UNIVERSITY, OXFORD, ENGLAND The papers of the Campaign Against Namibian Uranium Contracts, the British-based anti-apartheid/anti-nuclear activist organization, are held here. They document activists' efforts to reveal illegal Rössing contracts and to stop the flow of uranium from southern Africa to the rest of the world from the mid 1970s through the late 1980s.

CENTRE D'ARCHIVES D'OUTRE-MER, AIX-EN-PROVENCE, FRANCE This center houses archives pertaining to the administration of the French empire. Material found in collections pertaining to Madagascar and to French Equatorial Africa (which included Gabon) informed my thinking but, tellingly, did not include anything specifically on uranium, which was the exclusive province of the Commissariat à l'Énergie Atomique.

CHAMBER OF MINES OF SOUTH AFRICA, JOHANNESBURG, SOUTH AFRICA. The Chamber's official archives for uranium plants and related topics for the period 1952–1970 are available on microfilm. The Chamber's environmental section has a separate documentation room, which includes more recent reports (1990s onward) on environmental and personal exposure levels, minutes of meetings on mine safety, and the like.

COMMISSION DE RECHERCHE ET D'INFORMATION INDÉPENDANTES SUR LA RADIOACTIVITÉ (CRIIRAD), VALENCE, FRANCE The CRIIRAD, founded in the aftermath of the Chernobyl accident, provides "counter-expertise" in radiation and radon monitoring. In 2005 it began partnering with the Sherpa (see below) and other NGOs in France, Niger, and Gabon to investigate human and environmental exposures to radiation in uranium mines in Niger and Gabon. CRIIRAD conducted extensive radiometric surveys in both places, the results of which are held in its offices in Valence, and are now available online.

COMPAGNIE DES MINES D'URANIUM DE FRANCEVILLE (COMUF), MOUNANA, GABON The COMUF had a vast, uncatalogued, unexamined collection documenting more than 40 years of uranium mining in Gabon. Documents included employment and medical records, weekly reports on mining operations and work organization, blueprints and operations manuals for the yellowcake factory, reports on living conditions and social relations by the company's social worker, labor union files, correspondence pertaining to all levels of activity, and much more. I received unfettered access to these documents in 1998, the year before the COMUF closed. Some of these have been transferred to Areva archives in Bessines (see below), but scholars have been denied access to these.

COMPAGNIE GÉNÉRALE DES MATIÈRES NUCLÉAIRES (COGÉMA) (NOW AREVA), BESSINES, FRANCE In the 1970s the Cogéma took over the management of France's nuclear fuel cycle from the Commissariat à l'Energie Atomique. The Cogéma was folded into Areva when the latter was created in 2001. When I consulted the archives in 1998, the Bessines site housed documentation pertaining to the CEA's mining operations in Madagascar from 1946 to 1969 (when CEA mining ceased). These include monthly reports on mining operations, employment records, several in-house sociological studies commissioned to analyze race and ethnic relations in the mines, and more. Bessines now houses what remains of the COMUF papers (see above).

DIGITAL NATIONAL SECURITY ARCHIVE (DNSA) Available online, these consist of a large selection of US government documents collected by the National Security Archive using the Freedom of Information Act. The documents are grouped thematically. The collections I consulted most pertained to South Africa and nuclear non-proliferation.

GOLDFIELDS COLLECTION, CORY LIBRARY FOR HISTORICAL RESEARCH, RHODES UNIVERSITY, GRAHAMSTOWN, SOUTH AFRICA The Goldfields mining company ran uranium shafts out of its gold mines for several decades; its papers are housed here. They include material on uranium sales and safety protocols.

INSTITUT DE RADIOPROTECTION ET DE SÛRETÉ NUCLÉAIRE, FRANCE Archives held at the IRSN's Fontenay-aux-Roses site contain folders refer-ring to radon and gamma radiation readings and data from Gabon, Mada-gascar, and Niger, records of statistical research conducted on radon readings, records pertaining to the development of monitoring instruments, and more. I was denied access to most of these; the staff insisted they couldn't locate the actual folders (despite their appearance in the catalog, access to which was also restricted).

MINISTÈRE DES AFFAIRES ETRANGÈRES, LA COURNEUVE, FRANCE The archives of the Foreign Affairs ministry contain records pertaining to French uranium negotiations with Niger and Gabon. They seem particu-larly rich on Niger. I was denied access to most of these on grounds of national security, but I did find some illuminating items.

NATIONAL ARCHIVES OF SOUTH AFRICA, PRETORIA, SOUTH AFRICA A variety of public (government) and private collections here include records pertaining to the Atomic Energy Board's activities, government policy and legislation, South Africa's participation in the IAEA, the mining, milling, and marketing of uranium, and the early years of the uranium enrichment program. They are freely available.

NATIONAL LIBRARY OF NAMIBIA, WINDHOEK, NAMIBIA Collections include government documents and legislation concerning how the newly independent state dealt with the long-standing presence of the Rössing mine.

NATIONAL LIBRARY OF SOUTH AFRICA, CAPE TOWN, SOUTH AFRICA   Collections include scientific and technical publications on uranium mining and milling in South Africa and Namibia, as well as general scholarship on southern African history.

NUCLEAR ENERGY CORPORATION, PELINDABA, SOUTH AFRICA   This company took over the nuclear fuel cycle from the South African Atomic Energy Board. I received limited access to archives containing material on uranium production and on South Africa's participation in the IAEA, and a few reports and meeting minutes on uranium enrichment and related topics.

NUFCOR (NUCLEAR FUEL CORPORATION), WESTONARIA, SOUTH AFRICA   NUFCOR processed ore for all South African uranium producers and was in charge of selling South African product on the world market. It was also involved in planning a conversion plant for the (then) Atomic Energy Board, and was one of the founders of the infamous uranium cartel in the late 1960s and early 1970s. I received unfettered access to NUFCOR's uncatalogued archives, which included material on production, sales (including contracts that circumvented anti-apartheid sanctions), and working conditions.

RÖSSING URANIUM, LTD., NAMIBIA   This uncatalogued but impeccably organized collection includes material on the mine's daily operations, its contracts with European and Japanese utilities in the 1980s, its dealings with US, British, and French conversion and enrichment plants, labor organization and disputes, and much, much more. I received unfettered access to these archives.

SHAUN GUY PRIVATE PAPERS, JOHANNESBURG, SOUTH AFRICA   In the course of his work for the Atomic Energy Board, Shaun Guy (see chapter 8) collected reports, correspondence, and other documents (produced by the AEB, the Chamber of Mines, the Government Mining Engineer, and others) pertaining to radon exposure in mine shafts from the 1950s to the 1990s. He kindly made copies for me after our interview in 2004.

SHERPA, FRANCE   Founded in 2001, this NGO provides legal assistance to civil society organizations in human rights and global environmental cases. In 2005 it began partnering with the CRIIRAD (see above) and other NGOs in France, Niger, and Gabon to investigate human and environmental exposures to radiation in uranium mines in Niger and Gabon. Sherpa conducted extensive social surveys in both countries. The resulting

report is publicly available, but Sherpa also made available to me the raw
data from these surveys, which complement my own interviews and archival work.

SOUTH AFRICAN HISTORY ARCHIVE, JOHANNESBURG, SOUTH
AFRICA Through its Freedom of Information Programme, this archive
has obtained limited documentation pertaining to South Africa's nuclear
activities.

THE NATIONAL ARCHIVES AT COLLEGE PARK, MARYLAND, USA These
collections include substantial material on US purchases of uranium from
the Congo and South Africa.

THE NATIONAL ARCHIVES, KEW, UNITED KINGDOM These archives house
material from the UK Atomic Energy Authority and the Combined
Development Agency related to uranium exploration and extraction in
South Africa, Namibia, and elsewhere (including Rhodesia). Collections
on Britain's purchase of Namibian uranium while it was proscribed by the
United Nations Council for Namibia have recently been opened. Material
is freely available, subject to 30-year restriction.

US DEPARTMENT OF ENERGY, NEVADA TEST SITE ELECTRONIC
ARCHIVES These collections include material on US AEC missions to
South Africa and Congo.

WORLD NUCLEAR ASSOCIATION (KNOWN AS THE URANIUM INSTITUTE
UNTIL MAY 200 I), LONDON, ENGLAND This association has a substantial
library of "gray" literature on the uranium industry from 1974 to the
present. In addition to free photocopies, the staff gave me a collection of
the Uranium Institute's yearly conference proceedings from the years
1976–1988.

INTERVIEWS

All interviews were conducted by the author. Those including Bruce
Struminger are indicated with an asterisk.

FRANCE
Jean-Claude Andrault: 4 March 1998, Paris and 1 July 1998.
Jean Barreau: 2 March 1998, Paris
Jacques Bernazeaud: 28 January 2001, Pau
Jacques Blanc: 21 May 2007, Paris

Robert Bodu: extended conversations in March 1998; formal interviews 23 and 24 January 2001, Paris
Anne-Marie Chapuis: 1 December 2009, Chatenay-Malabry
Lucien and Lucienne Gabillat: 29 January 2001, St. Nom la Bretèche
André Gangloff: 27 February 1998, Gif-sur-Yvette
Liliane Gruet: 24 January 2001, Paris
Christian Guizol: 26 February 1998, Paris
Yves Legagneux: 24 February 1998, Paris
Geneviève and Jacques Michel: 26 January 2001, Lodève
Jacques Pradel: 22 May 2007, Paris
Jean Randriatoamanana: 2 March 1998, Le Palais sur Vienne
Paul Rasolonjatovo: 25 and 26 January 2001, Lodève

GABON
Zacharye Bondji: 16 July 1998, Mounana
Zacharye Bondji and Ignace Bissikou: 16 July 1998, Mounana
Antoine Boundjengui: 27 July 1998, Mounana
Bernard Keiffer: 22 and 29 July 1998, Mounana
André Kolo: 22 July 1998, Mounana
Mr. Malhaby: 29 July 1998, Mounana
Jean-Marcel Malekou: 16 July 1998, Mounana
Juste Mambangui: 13 and 16 July 1998, Mounana
François Mambangui: 31 July 1998, Libreville
Jean-Blaise Mbazabaka: 22 July 1998, Mounana
Kevin Mouandzuodei: 15 July 1998, Mounana
Mr. Nang-Obé: 29 July 1998, Mounana
Yvon Okanha: 27 July 1998, Mounana
Lambert Ondounda: July 1998, Mounana
Dominique Oyingha: 17 July 1998, Massango
Dominique Oyingha and Marcel Lekonaguia: 17 July 1998, Massango
Marcel Lekonaguia: 21 July 1998, Massango
Mr. Oyo: 29 July 1998, Mounana
Two interviews with people requesting anonymity: July 1998, Mounana

MADAGASCAR
Fanahia: 13 and 14 August 1998, Andolobé (translator: M. Abdoulhamide)

Fanahia and Rebeza: 20 August 1998, Andolobé (translator: M. Abdoulhamide)

Jeremy Fano: 18 August 1998, Tranomaro (translator: M. Abdoulhamide)

Kana: 11 August 1998, Tranomaro region (translator: George Heurtezbize)

Koto: 12 and 13 August 1998, Tranomaro (translator: Georges Heurtebize)

Jean-Louis X: 15 August 1998, Tranomaro (translator: M. Abdoulhamide)

Lakambelo: 13 August 1998, Andranokaolo (translator: M. Abdoulhamide)

Mahata, Rindroy, Fantare and Leboly: 16 August 1998, Tsilamaha (translator: M. Abdoulhamide)

Malakimana and family: 19 August 1998, Andolobé (translator: M. Abdoulhamide)

Joseph Ramiha: 12 August 1998, Tranomaro (translator: Georges Heurtebize)

Refanaliky: 20 August 1998, Tsilamaha 2 (translator: M. Abdoulhamide)

Talingrafy: 8 August 1998, Tranomaro; 16 August 1998, Tsilamaha 1 (translator: M. Abdoulhamide)

Three interviews with people requesting anonymity: August 1998, Tranomaro region

NAMIBIA

Achmet Abrahams: 15 January and 11* February 2004, Rössing

Lucille Abrahams: 6 February 2004, Swakopmund

Wycliffe Amundjeni: 10 February 2004, Rössing*

Johannes Bartlett: 13 February 2004, Rössing*

Erich Beukes: 2 February 2004, Rössing

W. J. Birch: 29 January 2004, Swakopmund

Cbyl Chiron: 5 February 2004, Arandis

Anna-Marie Ellitson: 5 February 2004, Arandis

Veronica Garises: 5 February 2004, Arandis

Kennedy Haimbodi: 27 January 2004, Rössing

Johan van Heerden 13 February 2004, Swakopmund*

Harry Hoabeb: 9 and 11* February 2004, Rössing

C. C. Johnson: 22 January 2004, Rössing

Thomas Kaimbi: 17 February 2004, Arandis

Assser Kapere: 25 February 2004, Windhoek★
Charles Kauraisa: 24 February 2004, Windhoek★
Charles Klaase: 27 January 2004, Rössing
Werner Külbel: 21 January 2004, Rössing
Mike Leech: 30 January 2004, Rössing
Jan Louw: 27 January 2004, Rössing
Alphäus Vehonga Muheua: February 2004, Windhoek★
Zedekia Ngavirue: 26 February 2004, Stanco area★
Ursella Naruses: 5 February 2004, Arandis
Jamie Pretorius: 22 January and 10★ and 11★ February 2004, Rössing
Reiner Przybylski: 29 January 2004, Rössing
Ed Reilly: 3 February 2004, Rössing
Paul Rooi: 22 and 30 January 2004, Rössing
Willem van Rooyen: 15 January 2004, Rössing
David Salisbury: 13 February 2004, Swakopmund★
Gida Sikandi: 24 February 2004, Windhoek★
K. F. Sumaib: 17 February 2004, Rössing
Wotan Swiegers: 18 February 2004, Swakopmund★
Eleven interviews with people requesting confidentiality: January and February 2004; Rössing, Arandis, Swakopmund, Walvis Bay, Windhoek

SOUTH AFRICA
Gert de Beer: 4 April 2004, Pelindaba
Bongani Bekiswa: 15 July 2004, Bizana (translator: Cyprian Stofela★)
Renfrew Christie: 6 October 2003, Cape Town
Piet Engelbrecht: 14 July 2007, Westonaria
Derek Elbrecht: 17 May 2004, Johannesburg
Jozua Ellis: 12 July 2005, Johannesburg★
Stuart Finney: 8 October 2003, Cape Town
Shaun Guy: 12 July 2004, Johannesburg★
Brian Hambleton-Jones: 6 April 2004, Pretoria
Rob Heard: 12 August 2003, Pelindaba
I. D. Kruger: 18 May 2004, Centurion
Mike Lalkhan: 30 April 2004, Johannesburg
Siphiwo Magwaca: 16 July 2004, Bizana (translator: Cyprian Stofela★)
Ntshantsha Marawu: 15 July 2004, Bizana (translator: Cyprian Stofela★)
Gerald Melani: 16 July 2004, Bizana (translator: Cyprian Stofela★)

Dumisani Mcunukelwa: 16 July 2004, Bizana (translator: Cyprian Stofela★)
Philip Mfanekiso: 16 July 2004, Bizana (translator: Cyprian Stofela★)
Michael Moabeng and Edwin Pheeha: 29 April and 14★ July 2004, Westonaria
Kokwana Mpandana: 15 July 2004, Bizana (translator: Cyprian Stofela★)
Goodman Mpinge: 15 July 2004, Bizana (translator: Cyprian Stofela★)
Smangaliso Mteki: 15 July 2004, Bizana (translator: Cyprian Stofela★)
W. E. L. Minter: 9 October 2003, Cape Town
Anthony Jackson: 6 April 2004, Pretoria
A. J. Jansen van Vuuren: Westonaria
James Joubert: 5 July 2004, Stellenbosch★
Donald Sole: 17 October 2003, Cape Town
P. D. Toens: 13 October 2003, Cape Town
Barney de Villiers: 6 July 2004, Cape Town★
Mona Vincilwa: 15 July 2004, Bizana (translator: Cyprian Stofela★)
Seven interviews with people requesting confidentiality: July 2003–August 2004, Durban metro area, Johannesburg metro area, Cape Town metro area

TELEPHONE INTERVIEWS
Charles Scorer, 3 July 2003
Phil Metcalf, 17 June 2004
Barbara Rogers, 20 January 2010
Mike Travis, 3 July 2003

SECONDARY AND PUBLISHED PRIMARY SOURCES

Abraham, Itty. 1998. *The Making of the Indian Atomic Bomb: Science, Secrecy and the Postcolonial State*. Zed Books.

Abraham, Itty. 2006. The ambivalence of nuclear histories. *Osiris* 21: 49–65.

Abraham, Itty. 2009. Contra-proliferation: Interpreting the meanings of India's nuclear tests in 1974 and 1998. In *Inside Nuclear South Asia*, ed. S. Sagan. Stanford University Press.

Abraham, Itty. 2010. "Who's next?" Nuclear ambivalence and the contradictions of non-proliferation policy. *Economic and Political Weekly* XLV (43): 48–56.

Abraham, Itty. 2011. Rare earths: The Cold War in the annals of Travancore. In *Entangled Geographies: Empire and Technopolitics in the Global Cold War*, ed. G. Hecht. MIT Press.

Adas, Michael. 1989. *Machines as the Measure of Men: Science, Technology, and Ideologies of Western Dominance*. Cornell University Press.

Adas, Michael. 2005. *Dominance by Design: Technological Imperatives and America's Civilizing Mission*. Belknap.

African National Congress. 1975. *The Nuclear Conspiracy: FRG collaborates to strengthen Apartheid*. PDW-Verlag.

Akrich, Madeline. 1994. The de-scription of technical objects. In *Shaping Technology/Building Society: Studies in Sociotechnical Change*, ed. W. Bijker and J. Law. MIT Press.

Albright, David. 2003. *Iraq's Aluminum Tubes: Separating Fact from Fiction*. Institute for Science and International Security.

Albright, David. 2006. When could Iran get the Bomb? *Bulletin of the Atomic Scientists,* July/August: 26–33.

Allen, Kate, ed. 2009. *Paper Wars: Access to Information in South Africa*. Wits University Press, 2009.

Allen, V. L. 2005. *The History of Black Mineworkers in South Africa*. Moor; Merlin.

Allman, Jean. 2008. Nuclear imperialism and the pan-African struggle for peace and freedom: Ghana, 1959–1962. *Souls: A Critical Journal of Black Politics* 10 (2) : 83–102.

Amundson, Michael A. 2002. *Yellowcake Towns: Uranium Mining Communities in the American West*. University Press of Colorado.

Associated Scientific and Technical Societies of South Africa. 1957. Uranium in South Africa.

Austin, Brian. 2001. *Schonland: Scientist and Soldier*. Witwatersrand University Press.

Avril, Robert, Charles Berger, Francis Duhamel, and Jacques Pradel. 1958. Measures adopted in French uranium mines to ensure protection of personnel against the hazards of radioactivity. *Proceedings of the Second United Nations International Conference on the Peaceful Uses of Atomic Energy, Held in Geneva, 1 –13 September 1958.* Vol. 21: *Health and Safety: Dosimetry and Standards.* United Nations.

Ball, Howard. 1993. *Cancer Factories: America's Tragic Quest for Uranium Self-Sufficiency.* Greenwood.

Balogh, Brian. 1991. *Chain Reaction: Expert Debate and Public Participation in American Commercial Nuclear Power, 1945–1975.* Cambridge University Press.

Banza, C. L. N., et al. 2009. High human exposure to cobalt and other metals in Katanga, a mining area of the Democratic Republic of Congo. *Environmental Research.* doi: 10.1016/j.envres.2009.04.012.

Bauer, Gretchen. 1998. *Labor and Democracy in Namibia, 1971–1996.* Ohio University Press.

Baulin, Jacques. 1986. *Conseiller du Président Diori.* Éditions Eurafor-Press.

Bayart, J. F. 2000. Africa in the world: A history of extraversion. *African Affairs* 99 (395): 217–267.

Beck, Ulrich. 1992. *Risk Society: Towards a New Modernity.* Sage.

Benally, Timothy, Phil Harrison, and Chenoa Bah Stilwell. 1997. *Memories Come to Us in the Rain and the Wind: Oral Histories and Photographs of Navajo Uranium Miners & Their Families.* Navajo Uranium Miner Oral History and Photography Project.

Bernault, Florence. 1996. *Démocraties ambiguës en Afrique centrale: Congo-Brazzaville, Gabon, 1940–1965.* Karthala.

Bernhard, S., and P. Zettwoog. 1986. Les techniques d'exploitation et de traitement des minerais d'uranium. In *Cycle du combustible nucleaire.* CEA-IPSN-DPT-SPIN-GPMU.

Bess, Michael. 1993. *Realism, Utopia, and the Mushroom Cloud: Four Activist Intellectuals and Their Strategies for Peace, 1945–1989: Louise Weiss, France, Leo Szilard, USA, E. P. Thompson, England, Danilo Dolci, Italy.* University of Chicago Press.

Bodu, Robert. 1994. *Les secrets des cuves d'attaque: 40 ans de traitement des minerais d'uranium.* Cogéma.

Bonner, P. L., Peter Delius, and Deborah Posel. 1993. *Apartheid's Genesis, 1935–1962.* Witwaterstrand University Press.

Borstelmann, Thomas. 1993. *Apartheid's Reluctant Uncle: The United States and Southern Africa in the Early Cold War.* Oxford University Press.

Borstelmann, Thomas. 2001. *The Cold War and the Color Line: American Race Relations in the Global Arena.* Harvard University Press.

Bothwell, Robert. 1984. *Eldorado: Canada's National Uranium Company.* University of Toronto Press.

Boudia, Soraya. 2007. Global regulation: Controlling and accepting radioactivity risks. *History and Technology* 23 (4): 389–406.

Boudia, Soraya. 2008. Sur les dynamiques de constitution des systèmes d'expertise scientifique. La naissance du système d'évaluation et de régulation des risques des rayonnements ionisants. *Genèses* 70: 26–44.

Boyer, Paul. 1985. *By the Bomb's Early Light: American Thought and Culture at the Dawn of the Atomic Age.* Pantheon Books.

Brandt, Allan M. 2007. *The Cigarette Century: The Rise, Fall, and Deadly Persistence of the Product That Defined America.* Basic Books.

Braun, Lundy. 2005. Spirometry, measurement, and race in the nineteenth century. *Journal of the History of Medicine and Allied Sciences* 60 (2): 135–169.

Breckenridge, Keith. 1990. Migrancy, crime and faction fighting: The role of the Isitshozi in the development of ethnic organisations in the compounds. *Journal of Southern African Studies* 16 (1): 55–78.

Breckenridge, Keith. 1995. "Money with dignity": Migrants, minelords and the cultural politics of the South African gold standard crisis, 1920–33. *Journal of African History* 36 (2): 271–304.

Breckenridge, Keith. 1998. The allure of violence: Men, race and masculinity on the South African goldmines, 1900–1950. *Journal of Southern African Studies* 24 (4): 669–693.

Breckenridge, Keith. 1998. "We must speak for ourselves": the rise and fall of a public sphere on the South African gold mines, 1920 to 1931. *Comparative Studies in Society and History* 40 (1): 71–108.

Brion, René, and Jean-Louis Moreau. 2006. *De la mine à Mars: la genèse d'Umicore.* Lannoo.

Brugge, Doug, Timothy Benally, and Esther Yazzie-Lewis. 2006. *The Navajo People and Uranium Mining.* University of New Mexico Press.

Brunet, Philippe. 2004. *La nature dans tous ses états: Uranium, nucléaire et radioactivité en Limousin.* Presses Universitaires de Limoges.

Burawoy, Michael. 1972. *The Colour of Class on the Copper Mines, from African Advancement to Zambianization.* Manchester University Press for Institute for African Studies, University of Zambia.

Burawoy, Michael. 1976. The functions and reproduction of migrant labor: Comparative material from Southern Africa and the United States. *American Journal of Sociology* 81 (5): 1050–1087.

Burawoy, Michael. 1979. *Manufacturing Consent: Changes in the Labor Process under Monopoly Capitalism.* University of Chicago Press.

Burton, Antoinette. 2007. Not even remotely global? Method and scale in world history. *History Workshop Journal* 64: 323–328.

Burton, Antoinette, ed. 2005. *Archive Stories: Facts, Fictions, and the Writing of History.* Duke University Press.

Butler, L. J. 2008. The Central African Federation and Britain's post-war nuclear programme: Reconsidering the connections. *Journal of Imperial and Commonwealth History* 36 (3): 509–525.

Callon, Michel. 1998. *Laws of the Markets.* Wiley-Blackwell.

Callon, Michel, Yuval Millo, and Fabian Muniesa. 2007. *Market Devices.* Wiley-Blackwell.

Campbell, Bonnie, ed. 2009. *Mining in Africa: Regulation and Development.* Pluto.

CANUC. 1986. Namibia: A Contract to Kill. The Story of Stolen Uranium and the British Nuclear Programme. Action on Namibia Publications, Namibia Support Committee.

Cathcart, Brian. 1994. *Test of Greatness: Britain's Struggle for the Atom Bomb*. Murray.

Catholic Institute for International Relations. 1983. Mines and Independence.

Caufield, Catherine. 1989. *Multiple Exposures: Chronicles of the Radiation Age*. Perennial Library.

Cawte, Alice. 1992. *Atomic Australia, 1944–1990*. NSW Press.

Červenka, Zdenek, and Barbara Rogers. 1978. *The Nuclear Axis: Secret Collaboration between West Germany and South Africa*. Julian Freedmann Books.

Chameaud, J., R. Perraud, and J. Lafuma. 1971. Cancers du poumon expérimentaux provoqués chez le rat par des inhalations de radon. *Compte rendu de l'Académie des Sciences* 273: 2388–2389.

Christie, Renfrew. 1984. *Electricity, Industry and Class in South Africa*. SUNY Press.

Clark, N. 1994. *Manufacturing Apartheid: State Corporations in South Africa*. Yale University Press.

Clarke, Roger, and Jack Valentin. 2005. A history of the International Commission on Radiological Protection. *Health Physics* 88 (5): 407–422.

Cohen, Avner. 1998. *Israel and the Bomb*. Columbia University Press.

Comaroff, John L., and Jean Comaroff. 1997. *Of Revelation and Revolution*, volume 2: *The Dialectics of Modernity on a South African Frontier*. University of Chicago Press.

Conklin, Alice L. 1997. *A Mission to Civilize: The Republican Idea of Empire in France and West Africa, 1895–1930*. Stanford University Press.

Cooper, Allan D., ed. 1988. *Allies in Apartheid: Western Capitalism in Occupied Namibia*. St. Martin's Press.

Cooper, Frederick. 1996. *Decolonization and African Society: The Labor Question in French and British Africa*. Cambridge University Press.

Cooper, Frederick. 2001. What is the concept of globalization good for? An African historian's perspective. *African Affairs* 100 (399): 189–213.

Cooper, Frederick. 2002. *Africa since 1940: The Past of the Present.* Cambridge University Press.

Cooper, Frederick. 2005. *Colonialism in Question: Theory, Knowledge, History.* University of California Press.

Cooper, Frederick, and Randall M. Packard, eds. 1997. *International Development and the Social Sciences: Essays on the History and Politics of Knowledge.* University of California Press.

Cooper, Frederick, and Ann Laura Stoler, eds. 1997. *Tensions of Empire: Colonial Cultures in a Bourgeois World.* University of California Press.

Creager, Angela. 2002. Tracing the politics of changing postwar research practices: The export of 'American' radioisotopes to European biologists. *Studies in the History and Philosophy of Biology and Biomedical Sciences* 33: 367–388.

Creager, Angela. 2006. Nuclear energy in the service of biomedicine: The US Atomic Energy Commission's Radioisotope Program, 1946–1950. *Journal of the History of Biology* 39: 649–684.

Crush, Jonathan, Alan Jeeves, and David Yudelman. 1991. *South Africa's Labor Empire: A History of Black Migrancy to the Gold Mines.* Westview.

Daoud, Samira, and Jean-Pierre Getti. 2007. *Areva au Gabon: Rapport d'enquête sur la situation des travailleurs de la COMUF, filiale gabonaise du groupe Areva-Cogéma.* Sherpa.

Debray, R. 1989. *Tous azimuts.* Odile Jacob.

de Grazia, Victoria. 2005. *Irresistible Empire: America's Advance through Twentieth-Century Europe.* Belknap.

Delauney, Maurice. 1986. *Kala-Kala: De la grande à la petite histoire, un ambassadeur raconte.* Robert Laffont.

Dower, John. 1986. *War without Mercy: Race and Power in the Pacific War.* Pantheon.

Dropkin, Greg, and David Clark. 1992. *Past Exposure: Revealing Health and Environmental Risks of Rössing Uranium.* Namibia Support Committee.

Dubow, Saul. 1989. *Racial Segregation and the Origins of Apartheid in South Africa, 1919–36.* St. Martin's Press.

Dubow, Saul. 2006. *A Commonwealth of Knowledge: Science, Sensibility, and White South Africa 1820–2000.* Oxford University Press.

Dumett, Raymond E. 1999. *El Dorado in West Africa: Mining Frontier.* Ohio University Press.

Eden, Lynn. 2004. *Whole World on Fire: Organizations, Knowledge, and Nuclear Weapons Devastation.* Cornell University Press.

Edwards, Paul N. 1996. *The Closed World: Computers and the Politics of Discourse in Cold War America.* MIT Press.

Edwards, Paul N. 2010. *A Vast Machine: Computer Models, Climate Data, and the Politics of Global Warming.* MIT Press.

Edwards, Paul N., and Gabrielle Hecht. 2010. History and the technopolitics of identity: The case of apartheid South Africa. *Journal of Southern African Studies* 36 (3): 619–639.

Eichstaedt, Peter H. 1994. *If You Poison Us: Uranium and Native Americans.* Red Crane Books.

Epstein, Steven. 2007. *Inclusion: The Politics of Difference in Medical Research.* University of Chicago Press.

Escobar, Arturo. 1994. *Encountering Development: The Making and Unmaking of the Third World.* Princeton University Press.

Federal Radiation Council. 1967. *Guidance for the Control of Radiation Hazards in Uranium Mining: Staff Report.*

Ferguson, James. 1990. *The Anti-Politics Machine: "Development," Depoliticization and Bureaucratic Power in Lesotho.* Cambridge University Press.

Ferguson, James. 1999. *Expectations of Modernity: Myths and Meanings of Urban Life on the Zambian Copperbelt.* University of California Press.

Ferguson, James. 2006. *Global Shadows: Africa in the Neoliberal World Order.* Duke University Press.

Fig, David. 2004. *Uranium Road: Questioning South Africa's Nuclear Direction*. Heinrich Böll Stiftung.

Fischer, David. 1997. *History of the International Atomic Energy Agency: The First Forty Years*. IAEA.

Fligstein, Neil. 2001. *The Architecture of Markets: An Economic Sociology of Twenty-First Century Capitalist Societies*. Princeton University Press.

Foccart, Jacques. 1997–2001. *Journal de l'Elysée*. 5 vols. Fayard; Jeune Afrique.

Forland, Astrid. 1997. Negotiating Supranational Rules: The Genesis of the International Atomic Energy Agency Safeguards System. Dr. Art thesis, University of Bergen.

François, Y., J. Pradel, and P. Zettwoog. 1973. Incidences des normes de radioprotections sur le marche de l'uranium. Paper read at Panel on Radon in Uranium Mining, Washington.

François, Y., J. Pradel, P. Zettwoog, and M. Dumas. 1973. La ventilation dans les mines d'uranium. Paper read at Panel on Radon in Uranium Mining, Washington.

Freund, Bill. 2007. *The African City: A History*. Cambridge University Press.

Gaulme, François. 1988. *Le Gabon et son ombre*. Karthala.

Giliomee, Hermann. 2003. *The Afrikaners: Biography of a People*. Tafelberg.

Gilman, Nils. 2004. *Mandarins of the Future: Modernization Theory in Cold War America*. Johns Hopkins University Press.

Giraud, Pierre-Noël. 1983. *Géopolitique des ressources minières*. Economica.

Goldschmidt, Bertrand. 1967. *Les rivalités atomiques, 1939–1966*. Fayard.

Goldschmidt, Bertrand. 1987. *Les pionniers de l'atome*. Stock.

Gomez, Manuel ed. 1981. *Radiation Hazards in Mining: Control, Measurement, and Medical Aspects*, proceedings of an international conference held October 4–9, 1981, Golden, Colorado. American Institute of Mining, Metallurgical, and Petroleum Engineers.

Gordon, Robert J. 1977. *Mines, Masters and Migrants: Life in a Namibian Mine Compound*. Ravan.

Gowing, Margaret. 1974. *Independence and Deterrence: Britain and Atomic Energy, 1945–1952*. Macmillan.

Granovetter, Mark, and Richard Svedberg, eds. 2001. *The Sociology of Economic Life*, second edition. Westview.

Gray, Earle. 1982. *The Great Uranium Cartel*. McClelland and Stewart.

Greenpeace International. 2010. Left in the Dust: Areva's Radioactive Legacy in the Desert Towns of Niger.

Grégoire, Emmanuel. 1999. *Touaregs du Niger: Le destin d'un mythe*. Karthala.

Gusterson, Hugh. 1996. *Nuclear Rites: A Weapons Laboratory at the End of the Cold War*. University of California Press.

Gusterson, Hugh. 1999. Nuclear weapons and the other in the Western imagination. *Cultural Anthropology* 14 (1): 111–143.

Guy, Jeff, and Motlatsi Thabane. 1988. Technology, ethnicity, and ideology: Basotho miners and shaft-sinking on the South African gold mines. *Journal of Southern African Studies* 14 (2): 257–278.

Haasbroek, A. C., and R. S. J. Du Toit. 1973. Radon in uranium mining: Effect of protective controls on uranium resources in South Africa. Paper read at Panel on Radon in Uranium Mining, Washington.

Hacker, Barton C. 1987. *The Dragon's Tail: Radiation Safety in the Manhattan Project, 1942–1946*. University of California Press.

Hacker, Barton C. 1994. *Elements of Controversy: The Atomic Energy Commission and Radiation Safety in Nuclear Weapons Testing, 1947–1974*. University of California Press.

Hacking, Ian. 2002. *Historical Ontology*. Harvard University Press.

Hall, John A. 1965. Atoms for Peace, or war. *Foreign Affairs (Council on Foreign Relations)* 43 (4): 602–615.

Harries, Patrick. 1994. *Work, Culture, and Identity: Migrant Laborers in Mozambique and South Africa c. 1860–1910.* Heinemann.

Headrick, Daniel R. 1981. *The Tools of Empire: Technology and European Imperialism in the Nineteenth Century.* Oxford University Press.

Headrick, Daniel R. 1988. *The Tentacles of Progress: Technology Transfer in the Age of Imperialism, 1850–1940.* Oxford University Press.

Hecht, Gabrielle. 1998. *The Radiance of France: Nuclear Power and National Identity after World War II.* MIT Press.

Hecht, Gabrielle. 2002. Rupture-talk in the nuclear age: Conjugating colonial power in Africa. *Social Studies of Science* 32 (5/6): 691–728.

Hecht, Gabrielle. 2006. Negotiating global nuclearities: Apartheid, decolonization, and the Cold War in the making of the IAEA. *Osiris* 21: 25–48.

Hecht, Gabrielle, ed. 2011. *Entangled Geographies: Empire and Technopolitics in the Global Cold War.* MIT Press.

Helmreich, Jonathan E. 1986. *Gathering Rare Ores: The Diplomacy of Uranium Acquisition, 1943–1954.* Princeton University Press.

Herbert, Eugenia W. 1984. *Red Gold of Africa: Copper in Precolonial History and Culture.* University of Wisconsin Press.

Heurtebize, Georges. 1986. *Quelques aspects de la vie dans l'Androy (extrême-sud de Madagascar).* Musée d'art et d'archéologie de l'Université de Madagascar, Tananarive.

Heurtebize, Georges. 1986. *Histoire des Afomarolahy: clan Tandroy, extrême-sud de Madagascar.* Éditions du Centre national de la recherche scientifique.

Heurtebize, Georges. 1997. *Mariage et deuil dans l'extrême-sud de Madagascar.* Harmattan.

Hewlett, Richard G., and Oscar E. Anderson. 1962. *A History of the United States Atomic Energy Commission.* Pennsylvania State University Press.

Higgott, Richard, and Finn Fuglestad. 1975. The 1974 coup d'état in Niger: Towards an explanation. *Journal of Modern African Studies* 13 (3): 383–398.

Hilgartner, Stephen, Richard C. Bell, and Rory O'Connor. 1983. *Nukespeak: The Selling of Nuclear Technology in America*. Penguin.

Hodge, Joseph Morgan. 2007. *Triumph of the Expert: Agrarian Doctrines of Development and the Legacies of British Colonialism*. Ohio University Press.

Holaday, Duncan A., David E. Rushing, Richard D. Coleman, Paul F. Woolrich, Howard L. Kusnetz, and William F. Bale. 1957. *Control of Radon and Daughters in Uranium Mines and Calculations on Biologic Effects*. US Department of Health, Education, and Welfare, Public Health Service.

Howell, Joel D. 1995. *Technology in the Hospital: Transforming Patient Care in the Early Twentieth Century*. Johns Hopkins University Press.

Hueper, W. C. 1942. *Occupational Tumors and Allied Diseases*. Thomas.

Hunt, Nancy Rose. 1999. *A Colonial Lexicon of Birth Ritual, Medicalization, and Mobility in the Congo*. Duke University Press.

Huntington, Samuel P. 1996. *The Clash of Civilizations and the Remaking of World Order*. Simon & Schuster.

International Atomic Energy Agency. 1975. *Radon in Uranium Mining: Proceedings*.

International Atomic Energy Agency. 1976. *Management of Wastes from the Mining and Milling of Uranium and Thorium Ores: A Code of Practice and Guide to the Code*.

International Atomic Energy Agency. 1976. *Manual on Radiological Safety in Uranium and Thorium Mines and Mills*.

International Atomic Energy Agency. 1989. *Radiation Monitoring in the Mining and Milling of Radioactive Ores*.

International Atomic Energy Agency, International Labour Organisation, and World Health Organization. 1964. *Radiological Health and Safety in Mining and Milling of Nuclear Materials: Proceedings*.

International Atomic Energy Agency and OECD Nuclear Energy Agency. 1982. *Management of Wastes from Uranium Mining and Milling: Proceedings of an International Symposium on Management of Wastes from Uranium Mining and Milling*.

International Atomic Energy Agency, OECD Nuclear Energy Agency, and Inter-American Nuclear Energy Commission. 1980. *Uranium Evaluation and Mining*

Techniques: Proceedings of an International Symposium on Uranium Evaluation and Mining Techniques.

International Commission on Radiological Protection. 1973. *Implications of Commission Recommendations That Doses Be Kept as Low as Readily Achievable. A Report of ICRP Committee 4.*

International Commission on Radiological Protection. 1977. Radiation protection in uranium and other mines. *Annals of the ICRP* 1: 1–28.

International Commission on Radiological Protection. 1986. *Radiation Protection of Workers in Mines: A Report of Committee 4 of the International Commission on Radiological Protection.*

International Institute for Strategic Studies. 2007. *Nuclear Black Markets: Pakistan, A. Q. Khan and the Rise of Proliferation Networks.*

International Labour Organisation. 1974. Radiation Protection in Mining and Milling of Uranium and Thorium, Symposium organised by the International Labour Office and the French Atomic Energy Commission, in cooperation with the World Health Organization and the International Atomic Energy Agency, Bordeaux, 1974.

International Labour Organisation and World Health Organization. 1983. *Radiation Protection of Workers in the Mining and Milling of Radioactive Ores.*

Jammet, Henri, and Marc Dousset. 1980. Les recommandations de la Commission Internationale de Protection Radiologique (CIPR), Application au cas particulier des mines d'uranium. In *Session d'étude sur la protection contre les rayonnements lors de l'exploitation et du traitement des minerais d'uranium,* ed. CEA/IPSN.

Janson, Elana. 1995. The Development of the Uranium and Nuclear Industry in South Africa, 1945–1970: A Historical Study. PhD dissertation, University of Stellenbosch.

Japan Atomic Energy Commission. 1993. White Paper on Nuclear Energy.

Jasanoff, S. 2007. Bhopal's trials of knowledge and ignorance. *Isis* 98 (2): 344–350.

Josephson, Paul R. 2000. *Red Atom: Russia's Nuclear Power Program from Stalin to Today.* Freeman.

Katz, Elaine N. 1994. *The White Death: Silicosis on the Witwatersrand Gold Mines 1886–1910.* Witwatersrand University Press.

Keenan, Jeremy. 2008. Uranium goes critical in Niger: Tuareg rebellions threaten Sahelian conflagration. *Review of African Political Economy* 117: 449–466.

Kempton, D. R., and R. L. du Preez. 1997. Namibian-De Beers state-firm relations: Cooperation and conflict. *Journal of Southern African Studies* 22 (4): 585–613.

Kim, Yung Sam. 1977. Proceedings of the First Conference on Uranium Mining Technology, 1977, University of Nevada.

Kirsch, Scott. 2005. *Proving Grounds: Project Plowshare and the Unrealized Dream of Nuclear Earthmoving.* Rutgers University Press.

Klare, Michael T. 1995. *Rogue States and Nuclear Outlaws: America's Search for a New Foreign Policy.* Hill and Wang.

Klare, Michael T. 2001. *Resource Wars: The New Landscape of Global Conflict.* Metropolitan Books.

Klieman, Kairn A. 2003. *The Pygmies Were Our Compass: Bantu and Batwa in the History of West Central Africa, Early Times to c. 1900 C.E.* Greenwood.

Kodesh, Neil. *Beyond the Royal Gaze: Clanship and Public Healing in Buganda.* University of Virginia Press.

Kreis, Georg. 2007. *Switzerland and South Africa, 1948–1994.* Lang.

Krige, John. 2006. Atoms for Peace, scientific internationalism, and scientific intelligence. *Osiris* 21: 161–181.

Krige, John. 2006. *American Hegemony and the Postwar Reconstruction of Science in Europe.* MIT Press.

Krige, John. 2008. The peaceful atom as political weapon: Euratom and American foreign policy in the late 1950s. *Historical Studies in the Natural Sciences* 38 (1): 5–44.

Kriger, Colleen E. 1999. *Pride of Men: Ironworking in 19th Century West Central Africa.* Heinemann.

Kuisel, Richard F. 1993. *Seducing the French: The Dilemma of Americanization.* University of California Press.

Lagadec, Patrick. 1981. *La civilisation du risque: Catastrophes technologiques et responsabilité sociale.* Editions du Seuil.

Lane, Kenneth. 1988, 1991. *The Economic Definition of Ore: Cut-off Grades in Theory and Practice.* Mining Journal Books.

Larson, Pier M. 1996. Desperately seeking "the Merina" (central Madagascar): Reading ethnonyms and their semantic fields in African identity histories. *Journal of Southern African Studies* 22 (4): 541–560.

Larson, Pier M. 2000. *History and Memory in the Age of Enslavement: Becoming Merina in Highland Madagascar, 1770–1822.* Heinemann.

Latham, Michael. 2000. *Modernization as Ideology: American Social Science and "Nation Building" in the Kennedy Era.* University of North Carolina Press.

Latour, Bruno. 2005. *Reassembling the Social: An Introduction to Actor-Network-Theory.* Oxford University Press.

Lawrance, Benjamin N., Emily Lynn Osborn, and Richard L. Roberts, eds. 2006. *Intermediaries, Interpreters, and Clerks: African Employees in the Making of Colonial Africa.* University of Wisconsin Press.

Lecoq, J. J. 1957. Une perspective minière nouvelle à Madagascar. Les sables à monazite. In *Rapport CEA n. 742.* CEA.

Lee, Christopher J., ed. 2010. *Making a World after Empire: The Bandung Moment and Its Political Afterlives.* Ohio University Press.

Léger, Jean. 1992. "Talking Rocks": An Investigation of the Pit Sense of Rockfall Accidents Amongst Underground Gold Miners. PhD dissertation, University of Witwatersrand.

Lerman, Nina E. 2011. Categories of difference, categories of power: Bringing gender and race to the history of technology. *Technology and Culture* 51 (4): 893–918.

Lerman, Nina E. 1997. "Preparing for the duties and practical business of life": Technological knowledge and social structure in mid-19th-Century Philadelphia. *Technology and Culture* 38 (1): 31–59.

Levin, J. 1957. Concentration tests on the gold-uranium ores of the Witswatersrand for the recovery of uranium. In *Uranium in South Africa*. Hortors.

Levin, Jack. 1985. *The Story of Mintek: 1934–1984*. Mintek.

Leys, Colin, and John S. Saul. 1995. *Namibia's Liberation Struggle: The Two-Edged Sword*. Ohio University Press.

Liberman, Peter. 2001. The rise and fall of the South African bomb. *International Security* 26 (2): 45–86.

Lindee, M. Susan. 1994. *Suffering Made Real: American Science and the Survivors at Hiroshima*. University of Chicago Press.

Lindqvist, Sven. 2001. *A History of Bombing*. New Press.

Lipton, Merle. 1986. *Capitalism and Apartheid: South Africa, 1910–1986*. Wildwood House.

Lipton, Merle, and C. E. W. Simkins. 1993. *State and Market in Post Apartheid South Africa*. Witwatersrand University Press.

Livingston, Julie. 2005. *Debility and the Moral Imagination in Botswana*. Indiana University Press.

Livingston, Julie. In press. *Improvising Medicine in an African Oncology Ward*. Duke University Press.

Lubin, Jay H., John D. Boice, Christer Edling, Richard W. Hornung, Geoffrey Howe, Emil Kunz, Robert A. Kusiak, et al. 1994. *Radon and Lung Cancer Risk: A Joint Analysis of 11 Underground Miners Studies*. National Institutes of Health.

Mabile, J. 1954. Le développement des recherches et exploitations minières du Commissariat à l'Énergie Atomique. *Écho des Mines et de la Métallurgie* 11: 759–762.

Macfarlane, Alastair. 1990. Labour Control: Managerial Strategies in the Namibian Mining Sector. PhD dissertation, Oxford Polytechnic.

MacKenzie, Donald. 2009. *Material Markets: How Economic Agents Are Constructed*. Oxford University Press.

MacKenzie, Donald, Fabian Muniesa, and Lucia Siu. 2007. *Do Economists Make Markets?* Princeton University Press.

Mackenzie, Donald A. 1990. *Inventing Accuracy: A Historical Sociology of Nuclear Missile Guidance*. MIT Press.

Makhijani, Arjun, Howard Hu, and Katherine Yih. 1995. *Nuclear Wastelands: A Global Guide to Nuclear Weapons Production and Its Health and Environmental Effects*. MIT Press.

Mallard, Grégoire. 2010. Crafting the Nuclear Regime Complex (1950–1975): Explaining Dynamics of Fragmentation and Harmonization of Nonproliferation Treaties. Working Paper.

Maloka, Eddy. 2004. *Basotho and the Mines: A Social History of Labour Migrancy in Lesotho and South Africa, c.1890–1941*. Codesria.

Mammadu, Baadikko. 2001. *Françafrique: l'échec: l'Afrique postcoloniale en question*. L'Harmattan.

Markowitz, Gerald, and David Rosner. 2002. *Deceit and Denial: The Deadly Politics of Industrial Pollution*. University of California Press.

Martin, François. 1991. *Le Niger du Président Diori*. L'Harmattan.

Martin, Guy. 1982. Africa and the ideology of Eurafrica: Neo-colonialism or pan-Africanism? *Journal of Modern African Studies* 20 (June): 221–238.

Martin, Guy. 1989. Uranium: A case study in Franco-African relations. *Journal of Modern African Studies* 27 (Dec): 625–640.

Marx, Leo. 2010. Technology: The emergence of a hazardous concept. *Technology and Culture* 51 (3): 561–577.

Masco, Joseph. 2006. *The Nuclear Borderlands: The Manhattan Project in Post-Cold War New Mexico*. Princeton University Press.

Masquelier, Adeline M. 2001. Behind the dispensary's prosperous facade: Imagining the state in rural Niger. *Public Culture* 13 (2): 267–291.

Mavhunga, Clapperton Chakanetsa. 2008. The mobile workshop: Mobility, technology, and human-animal interaction in Gonarezhou (National Park), 1850–present. PhD dissertation, University of Michigan.

Mawhiney, Anne-Marie, and Jane Pitblado. 1999. *Boom Town Blues: Elliot Lake, Collapse and Revival in a Single-Industry Community*. Dundurn.

Mazalto, Marie. 2009. Governance, human rights, and mining in the Democratic Republic of Congo. In *Mining in Africa*, ed. B. Campbell. Pluto.

Mazuzan, George T., and J. Samuel Walker. 1984. *Controlling the Atom: The Beginnings of Nuclear Regulation, 1946–1962*. University of California Press.

McCulloch, Jock. 2002. *Asbestos Blues: Labour, Capital, Physicians and the State in South Africa*. Indiana University Press.

McCulloch, Jock. 2009. Counting the cost: Gold mining and occupational disease in contemporary South Africa. *African Affairs* 108 (431): 561–602.

Medhurst, Martin J. 1997. Atoms for Peace and nuclear hegemony: The rhetorical structure of a Cold War campaign. *Armed Forces and Security* 23: 571–593.

Michel, Serge and Michel Beuret. 2008. *La Chinafrique: Pékin à la conquête du continent noir*. Grasset.

Mitchell, Timothy. 2002. *Rule of Experts: Egypt, Techno-Politics, Modernity*. University of California Press.

Mitman, Gregg. 2007. *Breathing Space: How Allergies Shape Our Lives and Landscapes*. Yale University Press.

Mitman, Gregg, Michelle Murphy, and Christopher Sellers, eds. 2004. *Landscapes of Exposure: Knowledge and Illness in Modern Environments. Osiris*, volume 19.

Mogren, Eric W. 2002. *Warm Sands: Uranium Mill Tailings Policy in the Atomic West*. University of New Mexico Press.

Monsieur X. and Patrick Pesnot. 2008. *Les dessous de la Françafrique, Dossiers secrets de Monsieur X*. Nouveau monde.

Moodie, T. Dunbar, with Vivienne Ndatshe. 1994. *Going for Gold: Men, Mines, and Migration*. University of California Press.

Moody, Roger. 1992. *The Gulliver File: Mines, People, and Land: A Global Battleground*. Minewatch.

Moore, J. D. L. 1987. *South Africa and Nuclear Proliferation: South Africa's Nuclear Capabilities and Intentions in the Context of International Non-Proliferation Policies*. St. Martin's Press.

Moorsom, R. 1977. Underdevelopment, contract labour and worker consciousness in Namibia, 1915–72. *Journal of Southern African Studies* 4 (1): 52–87.

Moorsom, R. 1995. Underdevelopment and Labour Migration: the Contract Labour System in Namibia. Windhoek: History Research Paper No. 1.

Moreau, M. 1960. Cycles uranifères et thorifères à Madagascar. In *Rapport CEA n. 1685*. CEA.

Morikawa, Jun. 1997. *Japan and Africa: Big Business and Diplomacy*. Witswatersrand University Press.

Morrissey, David. 1985. *Coal and Uranium*. Macmillan.

Moss, Norman. 1982. *The Politics of Uranium*. Universe Books.

Mudimbe, V. Y. 1994. *The Idea of Africa*. Indiana University Press.

Mueller, John. 2009. *Atomic Obsession: Nuclear Alarmism from Hiroshima to Al-Qaeda*. Oxford University Press.

Murphy, Michelle. 2006. *Sick Building Syndrome and the Problem of Uncertainty: Environmental Politics, Technoscience, and Women Workers*. Duke University Press.

Nakayama, Shigeru, and Hitoshi Yoshioka. 2006. *A Social History of Science and Technology in Contemporary Japan: Transformation Period 1970–1979*. Trans Pacific.

Nash, Linda. 2007. *Inescapable Ecologies: A History of Environment, Disease, and Knowledge*. University of California Press.

Ndong, Robert Edgar. 2009. Les multinationales extractives au Gabon: le cas de la compagnie des mines d'uranium de Franceville (COMUF), 1961–2003. PhD dissertation, École Doctorale Sciences Sociales, Université Lumière—Lyon II.

Ngolet, François. 2000. Ideological manipulations and political longevity: The power of Omar Bongo in Gabon since 1967. *African Studies Review* 43 (2): 55–71.

Neff, Thomas L. 1984. *The International Uranium Market*. Ballinger.

Nelkin, Dorothy, and Michael Pollak. 1981. *The Atom Besieged: Antinuclear Movements in France and Germany*. MIT Press.

Newby-Fraser, A. R. 1979. Chain Reaction: Twenty Years of Nuclear Research and Development in South Africa. Atomic Energy Board.

Noble, David F. 1984. *Forces of Production: A Social History of Industrial Automation.* Knopf.

Nordstrom, Carolyn. 2007. *Global Outlaws: Crime, Money, and Power in the Contemporary World.* University of California Press.

Nuclear Assurance Corporation. 1973–1990. World nuclear fuel market. Proceedings of the annual International Conferences on Nuclear Energy.

Nuclear Exchange Corporation. 1967–1989. *Monthly Report on the Nuclear Fuel Market.*

Nuclear Exchange Corporation. 1995. *Nuclear Fuel Market Analyses and Price Trend Projections, 1994–95.*

O'Meara, Dan. 1996. *Forty Lost Years: The Apartheid State and the Politics of the National Party, 1948–1994.* Ravan.

Obiang, Jean-François. 2007. *France-Gabon: pratiques clientélaires et logiques d'état dans les relations franco-africaines.*

OECD Nuclear Energy Agency. 1978. Proceedings of the Seminar on Management, Stabilisation, and Environmental Impact of Uranium Mill Tailings.

OECD Nuclear Energy Agency and US Department of Energy. 1982. Uranium mill tailings management.

OECD Nuclear Energy Agency, Radioactive Waste Management Committee, and Committee on Radiation Protection and Public Health. 1984. *Long-Term Radiological Aspects of Management of Wastes from Uranium Mining and Milling: Report of a Group of Experts Jointly Sponsored by the Committee on Radiation Protection and Public Health and the Radioactive Waste Management Committee, September 1984.*

Office of Technology Assessment. 1995. *Nuclear Safeguards and the International Atomic Energy Agency.*

Oldenziel, Ruth. 2004. *Making Technology Masculine: Men, Women, and Modern Machines in America, 1870–1945.* Amsterdam University Press.

Oldenziel, Ruth. 2011. Islands: The United States as a networked empire. In *Entangled Geographies: Empire and Technopolitics in the Global Cold War*, ed. G. Hecht. MIT Press.

Oosthuizen, S. F., W. G. Pyne-Mercier, T. Fichardt, and D. Savage. 1958. Experience in Radiological Protection in South Africa. In *Proceedings of the Second United Nations International Conference on the Peaceful Uses of Atomic Energy.*

Oreskes, Naomi, and Erik M. Conway. 2010. *Merchants of Doubt: How a Handful of Scientists Obscured the Truth on Issues from Tobacco Smoke to Global Warming.* Bloomsbury.

Organisation for Economic Co-operation and Development and European Nuclear Energy Agency. 1965. *World Uranium and Thorium Resources.*

Organisation for Economic Co-operation and Development and IAEA. 1967. *Uranium Resources, Revised Estimates.*

Organisation for Economic Co-operation and Development and IAEA. 1969. *Uranium Production and Short Term Demand.*

Organisation for Economic Co-operation and Development and IAEA. 1970. *Uranium Resources, Production and Demand.*

Organisation for Economic Co-operation and Development and IAEA. 1975. *Uranium: Resources, Production and Demand.*

Organisation for Economic Co-operation and Development and IAEA. 2006. *Forty Years of Uranium Resources, Production and Demand in Perspective: The Red Book Retrospective.*

Organisation for Economic Co-operation and Development, Nuclear Energy Agency, and International Atomic Energy Agency. 1983. *Uranium Extraction Technology.*

Organisation for Economic Co-operation and Development, Nuclear Energy Agency, and International Atomic Energy Agency. 1999. *Environmental Activities in Uranium Mining and Milling: A Joint Report.*

Owen, Anthony David. 1985. *The Economics of Uranium.* Praeger.

Packard, Randall M. 1989. *White Plague, Black Labor: Tuberculosis and the Political Economy of Health and Disease in South Africa.* University of California Press.

Pasternak, Judy. 2010. *Yellow Dirt: An American Story of a Poisoned Land and a People Betrayed*. Free Press.

Paucard, Antoine. 1992. *La mine et les mineurs de l'uranium français*. Editions Thierry Parquet.

Paucard, Antoine, et al. 2007. *La mine et les mineurs de l'uranium français*, Tome IV. Areva.

Péan, Pierre. 1982. *Les deux bombes: Comment la France a "donné" la bombe à Israel et à l'Irak*. Fayard.

Péan, Pierre. 1983. *Affaires africaines*. Marabout.

Péan, Pierre. 1990. *L'homme de l'ombre: Eléments d'enquête autour de Jacques Foccart, l'homme le plus mystérieux et le plus puissant de la Ve République*. Fayard.

Péan, Pierre, and Jean Pierre Séréni. 1982. *Les émirs de la République: l'aventure du pétrole tricolore*. Seuil.

Pearson, Jessica S. 1975. A sociological analysis of the reduction of hazardous radiation in uranium mines. NIOSH technical information. US Department of Health, Education and Welfare.

Perkovich, George. 1999. *India's Nuclear Bomb: The Impact on Global Proliferation*. University of California Press.

Perraud, R., J. Chameaud, J. Lafuma, R. Masse, and J. Chrétien. 1972. Cancer broncho-pulmonaire expérimental du rat par inhalation de radon. Comparaison avec les aspects histologiques des cancers humains. *Journal Francais de Medecine et Chirurgie Thoraciques* XXVI (172): 25–41.

Petryna, Adriana. 2002. *Life Exposed: Biological Citizens after Chernobyl*. Princeton University Press.

Polakow-Suransky, Sasha. 2010. *The Unspoken Alliance: Israel's Secret Relationship With Apartheid South Africa*. Pantheon Books.

Posel, Deborah. 1991. *The Making of Apartheid, 1948–1961: Conflict and Compromise*. Clarendon.

Posel, Deborah. 2000. A mania for measurement: Statistics and statecraft in the transition to apartheid. In *Science and Society in Southern Africa*, ed. S. Dubow. Manchester University Press.

Prakash, Gyan. 1995. *After Colonialism: Imperial histories and Postcolonial Displacements.* Princeton University Press.

Prakash, Gyan. 1999. *Another Reason: Science and the Imagination of Modern India.* Princeton University Press.

Prashad, Vijay. 2008. *The Darker Nations: A People's History of the Third World.* New Press.

Proctor, Robert, and Londa Schiebinger. 2008. *Agnotology: The Making and Unmaking of Ignorance.* Stanford University Press.

Proctor, Robert N. 1995. *Cancer Wars: How Politics Shapes What We Know and Don't Know About Cancer.* Basic Books.

Radetzki, Marian. 1981. *Uranium: A Strategic Source of Energy.* Croom Helm.

Raynaut, C. 1990. Trente ans d'indépendance, repères et tendances. *Politique Africaine (Paris, France)* 38.

Reiss, Mitchell. 1995. *Bridled Ambitions: Why Countries Constrain Their Nuclear Capabilities.* Woodrow Wilson Center Press.

Ringholz, Raye C. 1989. *Uranium Frenzy: Boom and Bust on the Colorado Plateau.* Norton.

Roberts, Alun. 1980. The Rössing File: The Inside Story Of Britain's Secret Contract For Namibian Uranium. Campaign Against the Namibian Uranium Contracts.

Robinson, Pearl T. 1991. Niger: Anatomy of a neotraditional corporatist state. *Comparative Politics* 24 (1): 1–20.

Rodney, Walter. 1981. *How Europe Underdeveloped Africa.* Howard University Press.

Roitman, Janet L. 2005. *Fiscal Disobedience: An Anthropology of Economic Regulation in Central Africa.* Princeton University Press.

Rosner, David, and Gerald Markowitz. 1991. *Deadly Dust: Silicosis and the Politics of Occupational Disease in Twentieth-Century America.* Princeton University Press.

Salifou, André. 2002. *Le Niger.* L'Harmattan.

Samet, Jonathan M., et al. 1984. Uranium mining and lung cancer in Navajo men. *New England Journal of Medicine* 310 (23): 1481–1484.

Sanders, Noel. 1986. The hot rock in the Cold War. In *Australia's First Cold War 1945–1959: Better Dead Than Red*, ed. A. Curthoys and J. Merritt. Allen and Unwin.

Saunders, Chris. 2009. Namibian solidarity: British support for Namibian independence. *Journal of Southern African Studies* 35 (2): 437–454.

Schatzberg, Eric. 2006. Technik comes to America: Changing meanings of technology before 1930. *Technology and Culture* 47 (3): 486–512.

Scheinman, Lawrence. 1987. *The International Atomic Energy Agency and World Nuclear Order*. Resources for the Future.

Schmid, Sonja. 2011. Nuclear colonization? Soviet technopolitics in the Second World. In *Entangled Geographies: Empire and Technopolitics in the Global Cold War*, ed. G. Hecht. MIT Press.

Schmidt, Peter R. 1997. *Iron Technology in East Africa: Symbolism, Science and Archaeology*. Currey.

Schoenbrun, David. 1998. *A Green Place, a Good Place: Agrarian Change and Social Identity in the Great Lakes Region to the 15th Century*. Heinemann.

Schrijver, Nico. 1997. *Sovereignty over Natural Resources: Balancing Rights and Duties*. Cambridge University Press.

Schrijver, Nico J. 1984. *The Status of Namibia and of Its Natural Resources in International Law*. United Nations Council for Namibia.

Schwartz, Stephen I., ed. 1998. *Atomic Audit: The Costs and Consequences of US Nuclear Weapons since 1940*. Brookings Institution Press.

Schweizer, J., and R. Sevin. 1954. l'exploitation des minerais radio-actifs par l'industrie privée dans l'union française. *Écho des Mines et de la Métallurgie* 11: 759.

Sellers, Christopher. 2004. The artificial nature of fluoridated water: Between nations, knowledge, and material flows. In *Landscapes of Exposure: Knowledge and Illness in Modern Environments*. University of Chicago Press Journals.

Sellers, Christopher, and Joseph Melling, eds. 2011. *Dangerous Trade: Histories of Industrial Hazard Across a Globalizing World.* Temple University Press.

Sellers, Christopher C. 1997. *Hazards of the Job: From Industrial Disease to Environmental Health Science.* University of North Carolina Press.

Skeet, Ian. 1988. *OPEC: Twenty-Five Years of Prices and Policies.* Cambridge University Press.

Sparks, Stephen. 2011. Apartheid Modern: SASOL and the Making of a South African Company Town, 1950–2009. PhD dissertation, University of Michigan.

Storey, William Kelleher. 2008. *Guns, Race, and Power in Colonial South Africa.* Cambridge University Press.

SWAPO. 1982. Trade Union Action on Namibian Uranium: Report of a seminar for West European Trade Unions organized by SWAPO of Namibia in co-operation with the Namibia Support Committee, 1981, London.

Taylor, June H., and Michael D. Yokell. 1979. *Yellowcake: The International Uranium Cartel.* Pergamon.

Théréné, Pierre, and Georges Bigotte. 1960. La prospections des substances radioactives en Afrique par le Commissariat. *Bulletin d'Informations Scientifiques et Techniques* 48–53.

Thomas, Lynn M. 2003. *Politics of the Womb: Women, Reproduction, and the State in Kenya.* University of California Press.

Thomas, Lynn M. 2011. Modernity's failings, political claims, and intermediate concepts. *American Historical Review* 116, no. 3: 727–740.

Thompkins, R. W. 1972. Radiation Controls in Uranium Mining: Workshop Course. Australian Mineral Foundation.

Thornton, John. 1990. Precolonial African industry and the Atlantic trade, 1500–1800. *African Economic History* 19: 1–19.

Tilley, Helen. 2011. *Africa as a Living Laboratory: Empire, Development, and the Problem of Scientific Knowledge, 1870–1950.* University Of Chicago Press.

Tirmarche, M., et al. 1993. Mortality of a cohort of French uranium miners exposed to relatively low radon concentrations. *British Journal of Cancer* 67: 1090–1097.

Toens, P. D. 1981. Uranium deposits and their time bound characteristics. *Transactions of the Geological Society of South Africa* 84: 295–297.

Topçu, Sezin. L'agir contestataire à l'épreuve de l'atome: Critique et gouvernement de la critique dans l'histoire de l'énergie nucléaire en France (1968–2008). PhD dissertation, École des Hautes Études en Sciences Sociales.

Tsing, Anna Lowenhaupt. 2005. *Friction: An Ethnography of Global Connection.* Princeton University Press.

Udall, Stewart L. 1994. *The Myths of August: A Personal Exploration of Our Tragic Cold War Affair with the Atom.* Pantheon Books.

United Nations General Assembly. 1980. Report of the panel for hearings on Namibian uranium. Part Two: Verbatim transcripts of the panel held at Headquarters from 7 to 11 July 1980.

United States Congress Joint Committee on Atomic Energy. 1958. *Problems of the uranium mining and milling industry. Hearings before the Joint Committee on Atomic Energy, Congress of the United States, Eighty-fifth Congress, second session.* Government Printing Office.

United States Congress Joint Committee on Atomic Energy. 1969. *Radiation Standards for Uranium Mining. Hearings before the United States Joint Committee on Atomic Energy, Subcommittee on Research, Development and Radiation, Ninety-First Congress, first session.* Government Printing Office.

Uranium Institute. 1976–1990. *International Symposium on Uranium Supply and Demand: Proceedings;* followed by *Uranium and Nuclear Energy: Proceedings (annual).* Uranium Institute.

van Beusekom, Monica. 2001. *Negotiating Development: African Farmers and Colonial Experts at the Office du Niger, 1920–1960.* Heinemann.

Van Onselen, Charles. 1976. *Chibaro: African Mine Labour in Southern Rhodesia, 1900–1933.* Pluto.

Van Onselen, Charles. 1982. *Studies in the Social and Economic History of the Witwatersrand, 1886–1914.* Ravan.

Van Schendel, Willem, and Itty Abraham. 2005. *Illicit Flows and Criminal Things: States, Borders, and the Other Side of Globalization.* Indiana University Press.

Vanderlinden, Jacques. 1991. *A propos de l'uranium congolais.* Académie royale des sciences d'outre-mer.

Verschave, François-Xavier. 1998. *La Françafrique: le plus long scandale de la République.* Stock.

Von Eschen, Penny M. 1997. *Race Against Empire: Black Americans and Anticolonialism, 1937–1957.* Cornell University Press.

Von Schnitzler, Antina. 2008. Citizenship prepaid: Water, calculability, and techno-politics in South Africa. *Journal of Southern African Studies* 34 (4): 899–917.

Waggitt, Peter. 1994. A Review of Worldwide Practices for Disposal of Uranium Mill Tailings. Australian Government Publishing Service.

Walker, J. Samuel. 1992. *Containing the Atom: Nuclear Regulation in a Changing Environment, 1963–1971.* University of California Press.

Walker, J. 2000. *Permissible Dose: A History of Radiation Protection in the Twentieth Century.* University of California Press.

Walsh, Andrew. 2003. "Hot money" and daring consumption in a northern Malagasy sappphire-mining town. *American Ethnologist* 30 (2): 290–305.

Walsh, Andrew. 2004. In the wake of things: Speculating in and about sapphires in Northern Madagascar. *American Anthropologist* 106 (2): 225–237.

Walters, Ronald W. 1987. *South Africa and the Bomb: Responsibility and Deterrence.* Lexington Books.

Weart, Spencer. 1988. *Nuclear Fear: A History of Images.* Harvard University Press.

West, Richard. 1972. *River of Tears: The Rise of the Rio Tinto-Zinc Mining Corporation.* Earth Island.

Westad, Odd Arne. 2005. *The Global Cold War: Third World Interventions and the Making of Our Times.* Cambridge University Press.

White, Luise. 2009. 'Heading for the gun': Skills and sophistication in an African guerrilla war. *Comparative Studies in Society and History* 51 (02): 236–259.

White, Luise, Stephan Miescher, and David William Cohen, eds. 2001. *African Words, African Voices: Critical Practices in Oral History.* Indiana University Press.

Whittemore, Gilbert Franklin, Jr. 1986. The National Committee on Radiation Protection, 1928–1960: From Professional Guidelines to Government Regulation. Dissertation, Harvard University.

Wilson, Francis. 1972. *Labour in the South African Gold Mines 1911–1969.* Cambridge University Press.

Winkler, Allan M. 1993. *Life under a Cloud: American Anxiety about the Atom.* Oxford University Press.

Wittner, Lawrence S. 1993. *One World or None: A History of the World Nuclear Disarmament Movement through 1953.* Stanford University Press.

Wittner, Lawrence S. 2003. *Toward Nuclear Abolition: A History of the World Nuclear Disarmament Movement, 1971 to the Present.* Stanford University Press.

Yates, Douglas. 1996. *The Rentier State in Africa: Oil-Rent Dependency and Neo-Colonialism in the Republic of Gabon.* Africa World Press.

Zeman, Zbynek, and Rainer Karlsch. 2008. *Uranium Matters: Central European Uranium in International Relations, 1900–1960.* Central European University Press.